R. Gary Fulcher
Department of Food Science
250 Ellis Building
University of Manitoba
Winnipeg, MB   R3T 2N2

from *the* farm *to the* table

Publication of this volume was made possible in part by funding from the Experiment in Rural Cooperation, the University of Minnesota Southeast Regional Sustainable Development Partnership.

Culture of the Land

A Series in the New Agrarianism

This series is devoted to the exploration and articulation of a new agrarianism that considers the health of habitats and human communities together. It demonstrates how agrarian insights and responsibilities can be worked out in diverse fields of learning and living: history, science, art, politics, economics, literature, philosophy, religion, urban planning, education, and public policy. Agrarianism is a comprehensive worldview that appreciates the intimate and practical connections that exist between humans and the earth. It stands as our most promising alternative to the unsustainable and destructive ways of current global, industrial, and consumer culture.

Series Editor
Norman Wirzba, Georgetown College, Kentucky

Advisory Board
Wendell Berry, Port Royal, Kentucky
Ellen Davis, Duke University, North Carolina
Patrick Holden, Soil Association, United Kingdom
Wes Jackson, Land Institute, Kansas
Gene Logsdon, Upper Sandusky, Ohio
Bill McKibben, Middlebury College, Vermont
David Orr, Oberlin College, Ohio
Michael Pollan, University of California at Berkeley, California
Jennifer Sahn, *Orion* Magazine, Massachusetts
Vandana Shiva, Research Foundation for Science,
Technology and Ecology, India
William Vitek, Clarkson University, New York

from *the* farm *to the* table

WHAT ALL AMERICANS NEED TO KNOW ABOUT AGRICULTURE

GARY HOLTHAUS

THE UNIVERSITY PRESS OF KENTUCKY

Copyright © 2006 by Regents of the University of Minnesota
Published 2006 by The University Press of Kentucky

Scholarly publisher for the Commonwealth,
serving Bellarmine University, Berea College, Centre
College of Kentucky, Eastern Kentucky University,
The Filson Historical Society, Georgetown College,
Kentucky Historical Society, Kentucky State University,
Morehead State University, Murray State University,
Northern Kentucky University, Transylvania University,
University of Kentucky, University of Louisville,
and Western Kentucky University.
All rights reserved.

Editorial and Sales Offices: The University Press of Kentucky
663 South Limestone Street, Lexington, Kentucky 40508-4008
www.kentuckypress.com

10 09 08 07 06   5 4 3 2 1

All images courtesy of Gary Holthaus.

Library of Congress Cataloging-in-Publication Data
Holthaus, Gary H., 1932-
  From the farm to the table : what all Americans need to know about
agriculture / Gary Holthaus.
      p. cm. — (Culture of the land: a series in the new agrarianism)
  Includes bibliographical references and index.
  ISBN-13: 978-0-8131-2419-3 (hardcover : alk. paper)
  ISBN-10: 0-8131-2419-0 (hardcover : alk. paper)
  1. Agriculture—United States. 2. Farmers—United States—Anecdotes.
  3. Sustainable agriculture. I. Title. II. Series.
  S441.H65 2006
  630.973—dc22
  2006025092

This book is printed on acid-free recycled paper meeting
the requirements of the American National Standard
for Permanence in Paper for Printed Library Materials.

∞ ♲

Manufactured in the United States of America.

 Member of the Association of
American University Presses

For Dick Broeker
(1942–2004)

entrepreneur of ideas,
realizer of dreams

Farmers who do not steward their plants, animals, and nutrients lack the longheadedness, the sense of the future, required to build a republican nation.
—Steven Stoll, *Larding the Lean Earth*

That's How We Came to Have This Place

I came out of that door right there, a Sunday morning in 1974, and I looked across at that field way over . . . and I thought, That tractor looks kinda funny, but I got the car, and the wife came, and we went on in to church. When I got there, the neighbor came right up and he said, "Elton, your son-in-law's tractor's turned over, and it's laying on him." I said, "Is he dead?" and he said, "Yes. We've got to tell the pastor we can't stay." So we talked to the pastor and got the sheriff and the coroner and we went back and got the tractor hoisted up. It broke his neck. Boy . . . I'll tell you . . . That was tough. . . .

Months later, my daughter decided she didn't want to stay, she couldn't handle it all on her own, even with all of us trying to help, and talked to me about selling the place. She felt she ought to offer it to the neighbors first, 'cause that was the neighborly thing, so I said, "OK, you offer it to them and if they aren't interested or won't pay enough, you come back to me." So she did, and the neighbors weren't interested. We made a deal. She needed income every year for a while, so I bought the place on a ten-year note, so much a year till it was paid off. That's how we came to have this place.
—Elton Redalen, Fountain, Minnesota

Contents

Contents

Photo gallery follows page 170

Acknowledgments

"So much depends / on a red wheelbarrow," says William Carlos Williams in a famous poem. In telling these farm stories, practically everything has depended on folks other than me.

Special thanks to the board of directors of the Experiment in Rural Cooperation, who approached me about writing this book and made it happen. Their wise counsel and guidance throughout the project made it—whatever flaws it may still have—a far better book than I could have achieved on my own. Dick Broeker, executive director of the Experiment in Rural Cooperation, not only facilitated the work but enhanced it in every conceivable way and made my job easier than it could have been in any other circumstance. His careful reading, good humor, and perpetually positive outlook strengthened this project daily. His successor, Erin Tegtmeier, has also been supportive and helpful in every way.

Special thanks as well to those farmers and farm families who gave of their time and hospitality to talk with me about their concerns and values, their farms and histories. Their openness and generosity were striking and greatly appreciated.

Gary Snyder started this book toward publication with his suggestions. Jack Shoemaker, of Shoemaker and Hoard, kindly steered us to the University Press of Kentucky, where Stephen Wrinn, director of the press, and his staff have offered their insight, enthusiasm, and expertise. Everyone at the press has been easy and gracious to work with and has my gratitude. Norman Wirzba, series editor of Culture of the Land: A Series in the New Agrarianism, has been helpful and supportive throughout. Special thanks to copyeditor Anna Laura Bennett for her patience, professionalism, and generous spirit.

Gary Nabhan, director of the Center for Sustainable Environments in Flagstaff, Arizona, read the whole text, and his comments strengthened it. Fred Kirschenmann, director of the Leopold Center for Sustainable Agriculture at Iowa State University, read the entire text, made useful suggestions to strengthen it, and took time to visit with me about it when I had questions. Kamyar Enshayan, adjunct professor at the University of Northern Iowa, read portions of the manuscript and offered his encouragement.

Closer to home, Mark Ritchie, director of the Institute for Agriculture and Trade Policy, was enthusiastic about the project from the beginning and read the sections on the global context, saving me from errors. Richard Levins, professor of agricultural economics in the Applied Economics Division of the University of Minnesota Department of Agriculture, met with me several times and led me to other contacts and fresh ideas. Catherine Jordan, assistant professor of pediatrics and neurology at the University of Minnesota, and Nicholas Jordan, professor of agroecology at the University of Minnesota, met with me several times, read the manuscript, and kept me on track. Bruce Vondracek, aquatic biologist in the Minnesota Cooperative Fish and Wildlife Research Unit at the University of Minnesota, called my attention to valuable resources in local libraries around the region. Deon Stuthman, professor of agronomy and plant genetics, talked with me about the state of agriculture as he saw it and also put me on to Internet resources and other articles that proved very helpful. Gary Fulcher, former professor and General Mills Research Chair of Cereal Chemistry and Technology at the University of Minnesota and current head of the Department of Food Science at the University of Manitoba, met with me several times, not only expanding my views of farming but providing good stories and music too. Helene Murray, director of the Minnesota Institute for Sustainable Agriculture, talked with me early on about her perspectives on this project. Deborah Allan and Craig Shaeffer, who had worked with farmers in southeast Minnesota, shared with me their views of the status of agricultural research in the region.

Kathryn Gilje introduced me to Centro Campesino, an organization of migrant workers in Owatonna, Minnesota, and adjacent areas. Victor

Acknowledgments

Contreras, codirector of Centro Campesino with Kathryn, helped me understand the work of Centro Campesino and the circumstances of migrant workers all around farm country. Consuelo Reyes told me about her life and work and showed me migrant housing.

Dean Harrington, at First National Bank of Plainview, read this manuscript—parts of it more than once. His careful readings and comments influenced it greatly. Donna Christison, a hog farmer, read the manuscript and made helpful suggestions. Ed Taylor, a dairyman near Lanesboro, read drafts of the text and met with me several times. His questions and comments always made me think harder, caught me being too opinionated too soon too often, and thus prevented some misjudgments. Barbara Wendland, of Temple, Texas, also read the whole manuscript with an eye for detail that I have learned is characteristic. Her comments and questions made the writing better than it would have been. In the early stages, Prescott Bergh, then with the Sustainable Agriculture Program of the Minnesota Department of Agriculture, offered good counsel.

Reverend Ben Webb, environmentalist, Episcopal priest in Cedar Falls, Iowa, and cofounder of the Regenerative Culture Project, listened to me talk about this project with endless patience and good counsel. Larry Gates, farmer and hydrologist at the Minnesota Department of Natural Resources, spent a whole day showing me around southeast Minnesota and talking about the issues as he sees them. His careful reading raised questions and prevented errors and always triggered long thoughts or raised new questions. He, Jeff Gorfine, then president of the Experiment in Rural Cooperation, and Ralph Lentz, a grass-based beef grower, took me on a drive through the countryside and let me see the land through their eyes, an education hard to equal. It was Ralph's idea to tell the story of farming. He has talked with me all through the process, hanging in there even when we disagreed vigorously, sometimes at the top of our lungs. Larry and Ralph provided a more knowledgeable way to look at the land than I could ever have gotten from research alone. Karen Anderson, of Plainview, Minnesota, graciously transcribed numerous tapes and saved me much of that onerous task.

A number of folks invited me to test these ideas out among public audiences: Metropolitan Community Church in Austin, Texas, was the first, thanks to C. J. and George Taylor and their pastor, Ken Martin.

Acknowledgments

Others included Dorik Mechau and Carolyn Servid of the Island Institute in Sitka, Alaska. Gary Fulcher's class at the University of Minnesota listened carefully and asked helpful questions. Friends in Salado, Texas, engaged me in questions and made comments during an evening at the Salado Public Library organized by librarian Patty Campbell. In Cameron, Texas, the congregation of All Saints Episcopal Church, through their pastor, Don Legge, invited me to join them over a potluck dinner, and their discussion was both stimulating and reassuring. Robert Young at the University of Wyoming invited me to speak with his seminar in American studies, and once again, students and faculty visitors were lively and challenging participants in thinking these issues through. The Land Stewardship Project brought folks together for an evening and conjured up a lively discussion. All of those events helped clarify my thinking and showed me that these farm stories are not confined to their local origins but are meaningful to people across the country. This is not to say that any of these people would agree with or endorse every view expressed here. Any errors remaining, of course, are entirely my own.

Introduction

For the past three years, I have been talking with, and learning from, folks who understand, as best any of us can, how agriculture works. In the process, I've visited with almost forty farm families in southeast Minnesota, northern Iowa, and western Wisconsin. Some of those visits lasted half a day or more and included a firsthand look at the farm. In some cases I've been back several times. I've also spoken with university faculty in our land-grant institutions, talked with county extension educators, attended too many public meetings and farm field days to count, and shopped at local farmers' markets—all for a project initiated by a unique, citizen-led regional organization called the Experiment in Rural Cooperation, most often referred to by locals, and me too, as the Experiment.

The Experiment is one of five regional partnerships created by the Minnesota state legislature and the University of Minnesota. For more than six years now, it has been putting funds into the hands of citizen leaders so they can use the resources of the state's land-grant university in projects that will lead to a sustainable society in this region. The people involved with the Experiment believe that there is a story to tell about farming in southeast Minnesota that is different from the story often told by the media. It is a story of success, at least some success, even on small farms, and of people who are having satisfactory lives, making a living by adapting their farm practices to their particular landscapes and nourishing them to bring health to the land and to the animals and humans who live on it. Together, these farmers, their university, and the

Experiment itself represent an unusual ecosystem, a small habitat for hope amid a tide of issues that threatens to engulf us.

What I have found is myriad stories about farming. Not only does every farmer have a story to tell, but each of the issues that both farmers and food consumers face has a story of its own. I did not set out to make an argument for one point of view or another about agriculture, but as I learned more, I had to develop a point of view, and it shows. Nevertheless, I have chosen to keep a story format, even for thorny issues like genetically modified organisms and the World Trade Organization. I have tried to let them, too, tell their own stories as best I can, while sorting through complexities, controversies, and complications. I hope that I have put together a comprehensible story without ducking any of the issues we need to sort out to create a sustainable culture.

This task has given me an opportunity to see something of the wide diversity of farm practices in our region, for it seems that everyone has a special twist on the conventional approaches. There are commodity producers and vegetable growers. There are rotational grazers, and there are management-intensive rotational grazers. There are confinement operators whose animals never leave the barn and free-range growers whose animals are rarely in one. There are folks who milk the same breed of cows, but some milk twice a day, others three times, and some, so I've been told, even four. Some folks computerize everything and go for the newest equipment. Others drive tractors thirty years old. One farmer I met still has a tractor fifty-seven years old. Sales receipts and catalogs tell us that he regularly has the very best Angus bulls in America. Some folks push for the highest production, whether of crops or of livestock or even of themselves. Others take a more relaxed and easier approach to the earth, their animals, and themselves, working it all together to nourish a healthy life for all three. Some farms seem to encourage erosion and the use of chemicals; others use pasture grasses to naturally hold carbon and nitrogen in the soil.

Working on this project has allowed me to learn something of farming all across America and around the world. We tend to think the issues are how often to milk, whether to milk Holsteins or Ayrshires, what to do with manure, which vegetables will sell best at farmers' markets this summer, how to do this or that or the other, and how to find time in the day to get everything done. In all those daily choices, every farmer differs, and each home place imposes demands that make its inhabitants unique. But from another perspective, one often ignored in global negotiations over free trade, the issues farmers face wherever they live across our globe are essentially the same: learning how to maintain the soil and water, animals and crops; how to take care of the family, keep everyone healthy, and contribute to the community so that one can gain respect and create a meaningful life. These are universal tasks. They could be summed up in a question: How do we, whether farmers or urban folks, sustain ourselves in this place? Looking at this region from that perspective, one can see farming in a universal light.

Along the way, often looking at the land from on foot or from farmers' pickups, always in their company, I have been forced to sort out widely disparate practices, opinions, and information. Among the things I have learned is this: I haven't yet visited long with any farmer I did not come to like a great deal, whether I agreed with his or her views and practices or not. If that sounds a bit too Will Rogers–ish, I'll confess that I have not met all of them yet. Who knows? Someone out there could warp that perception a bit; maybe I'll even run into somebody I don't like at all, though I'm not looking too hard. I also found a high degree of tolerance for divergent views among those who farm in different ways. More than once or twice, I have heard folks offer some version of this comment by dairy farmer and cheese maker Pam Benike: "You've got to remember, Gary, that even those folks who farm totally wrong in my view are still good people."

It is hard to throw rocks at bad practice, for we are all implicated in poor practice, and we are all beneficiaries of good practice. We all burn gas, use wood, waste water, and shop at least once in a while in Wal-Mart or Target. Many of us like ketchup on our hamburger or with our fries, even when we understand that ketchup contains corn syrup—from corn whose cultivation may lead to soil erosion and chemicals, depending on the degree to which its producers follow good practice. We like clean air and so buy ethanol, forgetting that it depends on that same corn that leans toward erosion, depends on chemicals or genetic modifications, and is already grown in surplus quantities. In a peculiar way, then, we are all complicit: participants in a culture that so far remains more a depleter than a regenerator, that takes more of the earth's resources than it gives back. That means our criticisms of the current practice must be tempered by our own involvements—but it does not mean that we must suspend good judgment or common sense. We still can make a stand, cultivate our sensibilities, and try to rein in excessive exploitation that shortens our prospects for a sustainable future.

What follows, then, begins in part 1 by going back—to certain fundamental elements that are as old as human existence. They prod us to remember what's really important as we set out. A brief history then reveals how we got ourselves into the agricultural circumstances that face us. In part 2, "Farmers Talking about Farming," farmers describe for us their practices and, consciously or not, their values. I have chosen these farmers in part because of the range of farming practices they represent, in part because of the range of values they reveal. Individual as they are, each one represents others as well. They show us people who are not provincial but alert to the world, its politics, its violence, and its hopeful possibilities. In part 3, "Farming in America: Who Cares?" we begin to widen our perception, moving out to look at a second circle of issues that surrounds us. That larger circle includes ag scientists and ag economists, migrant workers, genetically modified crops, chemicals, legislators at state and national levels, and issues like hunger and food security. There is an even larger circle of relationships that impacts our farmers and also determines much about the kinds and the quality of life that we all share, whether farmer or urbanite. Part 4, "It All Works Together, or It Doesn't Work at All," looks at the links between our farms and our small towns, suburbs, and cities. It also looks at global factors such as international trade agreements, transnational corporations, the power of poverty and its concomitant hunger, food security, and the diversity of cultures. In part 5, "Alternative Visions, Hopeful Futures," we take a look at how the future might unfold for agriculture in the next decade or so. In part 6, "An Ecology of Hope," we seek the confidence to face the future we all share.

Right up there with the earth, air, fire, and water that make our natural world possible, there is another fundamental element that makes our social life possible: stories. Healthy stories—what many indigenous peoples call "true stories," which "teach us how to be human"—heal the culture and enable it to persist. Unhealthy stories wreak havoc. If you do not believe in the power of stories, consider this: There have been cultures that have persisted for thousands of years without agriculture, industry, banking, and literacy, but there has never been one, as far as we know, without stories, poems, and music. Cultural survival over many millennia appears to lie, in part, in healthy stories about our relationship to one another and to the natural world. Keeping our stories straight and developing balance and harmony rather than discord, our indigenous forebears tell us, are keys to survival. Indeed, we have cultures around us that have survived at least ten thousand years longer than our own Western civilization that prove the point. Our real power in America, largely unrecognized, lies not in our military, nor in our economy, nor even in our agriculture, but in our capacity, limited though it often feels, to tell ourselves healthy rather than unhealthy stories, "true stories" rather than sales, propaganda, or public relations. Because some of those healthy stories for this culture come from the farmers who show up throughout these pages, I've tried to cast each section of this book as a story. I hope that farmers will recognize it as their story, and that everyone will recognize it as our story.

It pleases me, then, to be able to introduce you to some of the folks I've met along the way, and to share their farm stories with you. The stories are filled with hard work, occasional tragedy, hard-learned information, insight, noticeable altruism, and homegrown wisdom, from both harvest and hard times. As always with good stories, in these lives we recognize something of our own, whether we fancy ourselves urban or rural. And we may come to understand why we cannot create a sustainable community or a lasting culture unless all of us, no matter how far we live from the nearest farm, support a sustainable agriculture.

Part I

In the Beginning

CHAPTER 1

Fundamentals

It seems right to begin with the oldest elements. From the beginning, the Sumerians were right, the ancient Greeks were right, the American Indians were right, the Chinese were right: in the beginning, there were earth, air, fire, and water. We may all know these, but some in our cities and urban bureaucracies—and even some farmers—may have forgotten them. It is no disservice to either language or thought to speak of soil as earth, and light as fire, for soil provides the earth a skin of healthy nourishment that enables life, and light takes its origin in distant fire, is but fire spent by distance. For farmers, soil, air, sunlight, and water are perhaps the more pertinent names for the ancient elements, for any farmer knows profoundly that everything depends on them.

Lao Tzu thought water offered a good model for human behavior because "it does not contend," as one translator ends chapter 8 of *Tao Te Ching*. "The best way to live / is to be like water / For water benefits all things / and goes against none of them," Jonathon Star begins his translation of that chapter. "No fight, no blame," concludes another, by Gia-fu Feng and Jane English.[1] I love that chapter, especially its central description of a way to live, and have turned to it often over the years, grasping, as always, for straws that may help me create a life worth living. Farming is clearly a life worth living, but most farmers I know are on a constant quest to find "the way" to live it. Nevertheless, Minnesotans—and practically all others—know that Lao Tzu was wrong. Water does contend. It contends with earth as spring's snow melt floods our rivers and summer rains move tons of topsoil off our fields. Water not only contends but often wins, rearranging earth with a power that increases geometrically as its volume increases arithmetically, moving all but the most basic geologic forms before it in that eternal war that Heraclitus said was the beginning of everything. In northern climes, even those rocks we think eternal surrender to water, which seeps into cracks, freezes, and breaks them apart with its sheer expansive power. Water and soil are in balance when we are fortunate and do not abuse either, in contention when we are careless or uncaring, stripping the cover off soil, exposing it to the power of rain's erosion.

As contentious as water has been and increasingly will become, "there is a lot of nonsense about water being our most important resource," says law professor Charles Wilkinson, who specializes in western water issues and American Indians' water rights. He convinced me with one question: "Which would you rather be without for the next half hour, water or air?" How we love air. To fill one's lungs with air after exertion, to come up from under water not sure if our lungs will burst before we reach the surface, to inhale deeply after an asthma attack: such experiences remind us of the sacredness of air. Without air there can be no water, and fire suffocates and dies. Air is the great respirator, for fire and for plants and animals and soil. When air combines with bacteria that cling to the roots of plants in healthy soil, nitrogen is formed, and all life becomes possible. So air is another of the great elements essential to life, and it too has its own power. Air in motion can wear down rock, pick up earth enough to hide the sun, carry strontium 90 to warp the genes of our children, or bear the pollen dust from genetically modified crops to corrupt our vegetation. And when it really kicks off its shoes and starts to dance, it can knock houses off their foundations, throw cows over phone lines, and level towns. But do without air? Not a chance.

Divine fire, nurtured by air, was wrested from the light of the gods by Prometheus, according to one useful early story. Prometheus pays dearly for that service to us, spending eternity chained to a rock while birds pluck out his liver every day, only to have it grow back every night, repeating the agony over and over. Imagine those birds clawing right now, as you read this. Think of the pained body, chained to rock, knitting through the long night, trying to heal itself before the next day's tearing of flesh. The interpretations of the Prometheus story I have read stress his arrogance and his desire to usurp the power of the gods. But the story also makes clear that the theft of natural resources, our modus operandi today, is not to be taken lightly. We will pay for our thoughtless exploitation. Call it fire or light, sun's gift to us, via Prometheus, is the bringer of warmth, creator and transformer of everything green. Without fire, nothing works, and we are chilled to the bone. Better believe it. Norman, an Eskimo friend, once tried to walk from Nome to Teller at minus forty degrees and the wind blowing. He strayed from the trail and disappeared into a long, fireless night, till his body was eventually found. Too much fire and we crisp; only a shadow remains, burned into a wall in Hiroshima. Too little and we freeze. Fire to see by, fire to contemplate and learn from; who can resist looking into its bright leaping and tossing? Fire in the belly to ignite the heart and balance the light of the mind. The power inherent in fire strikes fear, or awe, into us, yet there is the renewal of fire: "From the ashes, a fire shall be kindled; A light from the shadows shall spring," says J. R. R. Tolkien, sounding cadences akin to the prophet Isaiah's.[2] Destroyer and builder, fire cleanses the earth in a flash, provides heat needed to germinate seeds, clears away leaves and branches that block the sun, and frees the earth to flourish again, its ash recreating and renourishing soil.

Soil. Dirt. The earth, from which we come and to which we return. Source of all we raise, and of myriad healing plants we neither sow nor tend, many of which we do not yet know. Soil is the other essential, always primary; seed is always secondary —purebred, hybrid, or mongrel GMO—no soil, no crop. Healthy soil is one element that is but a combination of all the others: water, air, and fire; plants, animals, minerals, and that warm light from the sun that speeds decomposition. Soil's power, too, has a name; call it germination. Immanent in healthy soil lies the source, perhaps, of all creativity, a source of food for all: microorganisms—one farmer I know insists we use their "scientific" name, "critters" —and all the myriad species of vegetation, and all the creatures that depend on plants for food or a home. All species have this dependence in common. Both praying mantis and human live within plants, the mantis poised on its green stem, sheltered and shadowed by the leaves and branches above, just as we live within the trees and grass that frame our homes and thatch our roofs. Each has its uses in this great, laughing, complex, lively scheme of existence. The soil that supports us all, soon or late, consumes us all, and we are all one in it. Like rain, soil cares not if we are just. Ultimately, it dissolves us all, and absolves us as well. Soil, when healthy, is the ultimate giver, giving its all to generating and nourishing all. I mean no disrespect when I suggest that soil, Lao Tzu, not water, should be our model.

And the interdependence of these four ancient elements—not one can exist without the others—offers a clue to our own interdependence, regardless of culture or language, religion or color, and to our absolute dependence on the earth's own elements, however many there may be. Whatever we need to know to survive and flourish we may learn from the earth itself and all its interdependent species, including humans. Indigenous peoples have understood those relationships for thousands of years. Some of us are unwilling to acknowledge the connections yet. The day after Prometheus stole fire, he tried to hide it from the searching gods. In the process, he smothered it, and it went out. He stole it again next day, separated it into several fires, and ran his first scientific experiment. He allowed it air ever after. . . . No, you're right, that's not in the story, but since we humans often seem to learn best the hard way, my story is just as likely to have been the case as the older one.

Nothing much has really changed, even after nine thousand years of applying the scientific method in agricultural experiments. Yes, we've added a few elements to the periodic table, but the old rules still apply: we must end our war against the elements, our best knowledge still insists, and generate balance and harmony—the great ritual linchpins, from the teachings of old Confucius to those of modern Navajos—or we die. Our choice. Work with life's fundamentals: nourish the soil, maintain the water, protect the air, and either block the sun or open oneself to it, as appropriate. That's all that is required of us, but these are not elements to mess with, and we dare not shirk our responsibility to them. Ask Prometheus. Ask anyone whose aircraft, for whatever reason, has lost its lift. Ask my friend Norman. Ask Napoleon and Hitler and their invading armies. Ask any farmer who plants too early, or too late, or who watches the rain wash his topsoil into the creek, to be swept into the great river, and sometimes into the water supply. Soil, water, air, and light. These are still the things without which neither agriculture nor a society of any kind can begin or continue. Talk about self-interest! Our care for them is the outward and visible sign of our care for ourselves, and one indicator that we have a future on this planet.

Yet agriculture is innately destructive, an earth-depleting activity. Whatever plants we raise, for food or beauty or healing, take chemicals out of the soil in that great reciprocal exchange that marks every natural process, whether cosmic, atmospheric, geologic, or human. The brighter the light, the darker the shadow, Tolkien might have said. We have understood for millennia that we have to put back into the soil what we take out. But if we do not exercise care, our replacing of those chemicals pollutes our streams. And now we are discovering that, without great care, our confinement of large numbers of animals exchanges life-giving air for toxic methane. There go soil, water, and air—three strikes and we're out.

However our agriculture works, when it works, chances are good that it works because people thought very carefully about what they wanted to accomplish and tried a variety of things before they hit on practices that brought the desired ends. What farmers know is that whatever practice works now may not work next season, because the frost will come late or the rains early, and heavier than expected, or not at all. The most carefully considered plan will have to be revised. And they all also seem to know that just the fact that it works on this place is no sign it will work on yours, or on that other one a drainage over—you know, that 360 down on the county line.

While I may have made working with the basics sound simple, there is nothing simple about it. In the nature of this cosmos lies a conundrum: everything is related, so what appears simple inevitably has a context that makes it incredibly complex. Every connection, benign or malignant, metastasizes: earth's soil related to sun's light; sun's light related to bacteria; bacteria and light related to nitrogen; nitrogen related to plants; plants related to the respiration of everything living, permitting and sustaining our human lives. The cosmic and the most microorganic are thus related: the long-lived light of stars related to the brief life of the tiniest bacteria, some so far underground they never see the light, yet they absorb its presence. There is no escape, and there are no exemptions from this system. When the system goes down, we all go down, from the very top of the food chain to the very bottom. If we do not nourish the smallest creatures, we dismantle the mantle of earth, knock the props right out from under ourselves. Perhaps that is why Tlingit Indian elder Austin Hammond says, "You have to remember that Ground Squirrel is Grandfather to Bear." And our human lives are all related, not only to those fundamental elements of nature that tend toward balance but to all other humans. We are all caught up in the same natural processes, and we are all equally caught up in those social processes that yearn for harmony at the same time that we thwart it, acting too often out of blind self-interest that refuses to see where harmony lies. So we kill each other, dominate each other, exploit each other, refuse to cooperate, and seek our own advantage, or try to, knowing all the while that societies, like the more fundamental elements, are created for a gentler balance and harmony.

Farming today is a matter of dealing not only with the complexities of earth, air, fire, and water, but also with the complexities of nation-states, transnational corporations, trade policy and so-called trade barriers, markets or lack of them, supply and demand, our human greed and our human compassion, our excitement at competition and our pleasure in cooperation. Given the complexities, farming in our time is never about farming only. Whatever else it might be, the story of farming is a story of connections. Those connections are not only biological or geographical but historical, taking us far back in time. "All narratives require a scale," says Richard Fortey in *Life: A Natural History of the First Four Billion Years of Life on Earth*.[3] Indeed, historical time is too short to measure the present. But even Fortey's geologic time gives us a scale too brief for understanding the present moment. The photosynthesis taking place in your garden, pasture, alfalfa, or oaks this instant began not eras or eons ago, as geology measures time, but light-years of time and distance away, connecting us to a past—and a cosmos—all but unfathomable. Since those four elements with which we began are involved in the growth of plants and the health of animals, including humans, farming's connections extend to the farthest stars, to light and times far older than our oldest stories and our deepest geologic strata.

The connections we need to acknowledge are also economic and political, for farmers around the world today are connected by a single marketing system. There is no farmer anywhere who farms in a vacuum, or in the old dream of independence. Though farm practices may differ enormously from one region to another, from one country or continent to another, one thing the North American Free Trade Agreement (NAFTA) surely illustrates is that Mexican subsistence farmers, American farmers—whether organic or agribusiness—and Canadian cattlemen and grain farmers are all involved in the same economic and political process. In that process, no individual American farmer's combine or high cab tractor gives her an advantage over a subsistence farmer's hoe, and NAFTA unites American, Mexican, and Canadian farmers more powerfully than our Constitution ties us to the United States. What we mischievously call "free trade" is the great leveler and oppressor, and its globalization unites the people of the land in a single system that turns Jefferson's independent yeoman farmers, wherever they live on this earth, into serfs for transnational corporate profit. This in a cosmos that has no use for single systems and dooms them to a short life.

All these issues become essential elements of the farm story all across America and wherever farming happens in the world. There can be no separation because we are all related, all of us ingredients in the great mystery that somehow turns light and leaf into food that creates and sustains us and allows us to breathe. We must learn to cooperate with that mystery; if we do not, Steven Stoll reminds us, we lack "the longheadedness, the sense of future, required to build a republican nation."[4] In our own time, we could add "or even to survive." It's our choice. Every day.

CHAPTER 2

Histories

Immigrants in 1846 followed wagon ruts all the way from Chicago to Red Wing, Minnesota. The army had worn the ruts into the rolling hills and prairie during the Black Hawk War. At Grand Detour, a common stop along the way, the Anderson family halted for a rest. They saw a plow leaning against the blacksmith shop, gleaming silver in the sun. They were struck by it and inquired after it. Their respondent gestured toward the smithy and said, "He is the only man in the world who can make a self-polishing ploughshare. Out here, in this new soil, a farmer has to spend half his time pushing the dirt off his ploughshare."

"You mean that blacksmith can really make a plough that scours?" the Andersons asked.

"Yes siree! Wouldn't be surprised if someday you hear more about him. His name is John Deere."[1]

Early Swedish settlers thought the prospects on the other side of the river looked promising, according to James Banks in *Wing of Scarlet*, an early history of Goodhue County. They saw "a territory that was rolling, covered with heavy timber, and had rich soil. The rolling terrain provided self-drainage. The heavy growth of timber provided building material and fuel and cover for game." Banks notes that wild nuts, berries, and fruit were abundant, and he also identifies a wide variety of wild animals, some of them more appreciated than others. He mentions beaver, mink, bobcats, timber wolves, gray squirrels, rattlesnakes, skunks, rabbits, and deer. He also calls attention to a universal male settler characteristic. "In those days one of the first ambitions of a young man was to grow a beard. It was a custom for all men to display the heaviest growth of whiskers possible."[2] We could note that what attracted people, whether American Indian or European, was the natural biodiversity inherent in this natural landscape—at least till we get to what appears to be a monocrop of whiskers. Those early, bearded homesteaders soon cut the timber, stacked the logs, cleaned out the stumps, and began to farm. But they were far from the first to farm the region.

The first settlers arrived perhaps 12,000 years ago. They were hunters of mastodons, caribou, and bison and harvesters of green plants that grew in wild profusion. Archaeologists discern a change in the physical evidence around 2,500 years ago that suggests a change in culture, one so great that it perhaps represents a whole new set of invaders rather than a cultural evolution. Within a few hundred years, the hunters of ancient animals now long disappeared and the foragers of wild plants in abundance also became harvesters of domestic crops, beginning an agricultural adventure that has never ended, though it has surely waxed and waned.

That culture, called the woodland stage of the immense Mississippian culture that extended from central Minnesota to southern Florida, flourished for 250 years during the twelfth and thirteenth centuries, extending back to a time when buffalo and grizzly bears were still abundant and elk were still a swift plains animal rather than a majestic mountain creature. Villages were numerous along the bluffs above the Mississippi, and the farmlands extended miles westward along small creeks. A couple of years ago, an archaeologist, standing just across U.S. Route 61 from the Anderson Center on the edge of Red Wing, described five thousand teepees covering a small plain and stretching along creek bottoms, all within view of where we were standing. These early settlers harvested more than forty species of mussels for food and used their crushed shells to temper pots. The Mississippian Indians honored their dead by building mounds, many of them effigies of animals and birds, that reveal a rich ritual life. They also developed the bow and arrow, built log fortifications, and laid out large cities as carefully as any contemporary urban planner. But they gathered in this place to farm. Several varieties of corn were among the primary crops, along with squash and beans. Though game was abundant, farming then represented the core of the economy—as it does now, despite our technology.[3] We forget that core of agricultural economy to the peril of our own cultural enterprise, including warfare and computers.

The earliest European settlers followed a pattern not so different from the Indians'. They not only cleared some land for farming, but they also lived a subsistence life that supplemented the farm products with game and fish and birds still available in the area. But it was not many years before the expansive nature of farming altered the landscape and the sociology of Goodhue and all the other counties of the upper Midwest. The Indians were pushed aside—slowly, it must have seemed at the time, but looking back on the first fifty years of the Europeanization of Minnesota, it now seems that it happened with eye-blink speed.

In his book *Geographical and Statistical Sketch of the Past and Present of Goodhue County*, published in 1869, W. H. Mitchell, a chauvinist for sure about Minnesota, declares forthrightly, "The Agricultural Capacities and advantages of Minnesota can hardly be over-stated." One can imagine him bending down to sift the dark prairie soil, rich in organic matter, then rubbing it between his hands, letting it drift away on the breeze. He knew something about soils. "Long ages of growth and decay of vegetable matter on the wide-spread prairies of Minnesota, make up the organic ingredients of a soil abounding in all the most productive elements, the prevailing feature of which is a dark calcareous, sandy loam with a strong admixture of clay." Mitchell provided comparative statistics to show just how quickly agriculture was growing and the important role it played in the state's economy.

	1860	*1862*
Rye	21.56 bushels per acre	24.00
Barley	33.23	34.00
Buckwheat	15.73	26.00

"In 1865," he reported, "the wheat crop of Minnesota exceeded 12,000,000 bushels, somewhat more than 46 bushels to each man, woman, and child in the state." Other crops prospered as well. "Potatoes, in this climate, attain their highest excellence, and in flavor and rich farinacious qualities are superior to those of any other section." Not all of this was hyperbole. The climate, the pastures, the hardy prairie grasses—not to mention the burgeoning number of immigrants—allowed crops to flourish and livestock to increase at a remarkable rate. It must have seemed that all nature was on the farmer's side. "In 1860 the whole number of sheep in Minnesota was only 5,941; in 1864 there were 92,612, while in 1868 it was estimated that there are not less than 200,000."[4] Surely none of the growth reflected in the statistics Mitchell compiled came as easily as his enthusiasm for Minnesota's agriculture.

Reverend J. W. Hancock, who lived near Vasa, Minnesota, was both a clergyman and a farmer. He described farm life as a bit less glorious. In his diary for 1869, one can find the following:

April 16, 1869
Plowed about 1 acre this P.M. Frost in some places. The mud also not quite dried up enough to make it easy plowing.

April 20
Snowing a part of the day and very cold for the season.

Nevertheless, there were also some pleasures along the way.

April 21
Plowing in the morning. Went to the sociable in the evening at Mr. Brown's. Had a good time.

It wasn't till May 18 that Hancock began to plant his corn. And fall came early, wet, and cold.

Sept. 11
Do some plowing. Found the ground almost too wet.

From September 20 into October he was putting up corn in shocks. But October was that year's cruelest month.

October 30
Our cold weather comes early. Many have been caught with vegetables frozen into the ground. The plowing is not one third completed and now the ground's frozen several inches in depth. My turnips are many of them in the ground.[5]

It does not seem to matter, when you turn the leaves of history, which date shows up. Farming every season, every year, is a blend of many long hours of work, a few hours of pleasure, and a gamble every minute.

TECHNOLOGY CHANGES HISTORY

One way to chronicle the history of agriculture is to follow the development of technology. Steven R. Hoffbeck uses this system effectively in *The Haymakers: A Chronicle of Five Farm Families.* Hoffbeck traces the shifts in equipment used in haying in five Minnesota counties.

His story begins in 1862, with horses and scythes. Andrew Peterson, forty-two, needed hay for three cows, "each with a calf, and one yearling heifer." Peterson also had "two adult oxen, two young bulls, two ewes with lambs, and five pigs." He needed at least fourteen tons of hay to get his animals through the winter, all of it cut by hand and stacked or hauled with oxen.[6]

Farmer Oliver Perry Kysor came to Otter Tail County as a three-year-old in 1832. In 1883, at the age of fifty-three, he had his own place and the new equipment to operate it, including a mowing machine for cutting hay and a reaper for wheat. He sowed the wheat and some oats on April 12 and planted potatoes, corn, sweet corn, cabbage, and beans from then till the end of May. He used horses rather than oxen to pull the mowing machine and the reaper. "These devices," Hoffbeck says, "invented in 1831, and in widespread use by the 1860's allowed a farmer to cut ten times as much hay in one day as a man using a scythe."[7]

Gilbert Marthaler, a German American farmer, began haying with his family when he was seven years old. In 1924, the family acquired "a full line of machines powered by horses: a mower, a side-delivery rake, and a hay loader." There were six horses to do the work. The family also "had purchased a small Fordson tractor in 1921," but none of the haying equipment was rigged to work with a tractor. The mower hung up frequently because the hay was too wet, or it jammed with soil jutting above gopher mounds. The side-delivery rake worked more effectively, and the hay loader saved everyone from making or hoisting haycocks by hand. The farm put up more than fifty tons of alfalfa and clover hay that summer, enough to winter over the six horses, twenty-five Holsteins, and assorted other animals.[8]

By 1959, the Rongens of Polk County were using an International Harvester H tractor and a new International Harvester baler, "tangible evidence of an extraordinary change for both the Rongens and for American Agriculture," writes Hoffbeck. The new equipment represented "the replacement of muscle power in farming by engine power." There would be no turning back; the Rongens had sold their horses, and they owed on the machinery. "The machine method and the chemical method were supplanting the old 'armstrong' method of raising crops and livestock," according to Hoffbeck. This period also saw increasing uses of pesticides, herbicides, and fertilizers, many of them the products of technologies developed in World War II.[9]

Larry Hoffbeck, the author's brother, farmed in 1984, making hay with a swather, tractor, and chopper. Blue silos went up, along with an increasing debt load, and longer hours were needed to get the work done. There was also greater danger from the more sophisticated heavy equipment and the stress of impossible workloads. In a careless moment, as Larry lay under the swather without blocks to hold it, the head came down on his chest.[10]

From the 1820s, when the first stumps were pulled, to the present, a pattern of ever-increasing speed and desire for higher production is clear. Whether the shifts in mechanical devices to ease the human burden of work actually made it easier was another matter. As the mechanical equipment increased in size, complexity, and speed, the source of danger shifted from large animals to belts and gears and finally to sheer stress, which contributed to accidents such as Larry Hoffbeck's, who had had to work more and more hours per day to make the new equipment pay.

HISTORY UNFOLDS IN A FARMER'S LIFE

Another way to look at the history of agriculture is through the eyes of a single farmer who witnessed most of the major transitions. H. C. Hinrichs was a farmer who lived most of the twentieth century in Goodhue County, on a farm that went from producing "10 hogs a year" in 1910 "to the farm that produced 2,000 hogs a year" in the 1970s. All four of Hinrichs's grandparents were German immigrants who found their way to a community so new that it was "without boundaries," Hinrichs writes in "As I Remember: A Treatise on Early Rural Life in Goodhue County, Minnesota."[11]

In 1910, Hinrichs's father bought a farm in Featherstone Township. The family moved from a German community that stressed work to an English community "that stressed reading, writing, arithmetic, and spelling," where no one spoke German. There was not a lot of visiting. "Often we would see no one outside of the family members for a couple of weeks. The mailman drove by six days a week, but we met him only when we needed a money order or stamps." Everything used on the farm was grown on the farm: cornhusks filled the mattresses and were changed every year at harvest time; breakfast was "usually fried potatoes covered with eggs" and homemade bread with hand-churned butter and jam. After a hog was slaughtered, "gritwurst" was added to the menu. It was "made from the head meats of a hog, steel cut oats, and seasoning. To make it extra good mother put in a lot of raisins." Potatoes were the staple. "Potatoes three times a day was the rule except when there were not enough holdovers." Cows were milked by hand twice a day, and when chores were finished in the evening, a special treat was to put a pail of butternuts on the kitchen range until they split open. "Then we would all sit around the table and eat butternuts. Nine o'clock was bedtime." The range provided both cooking and heating; the rest of the house stayed cold. There was no refrigeration, but a "basement room under the front parlor stored 50–75 bushels of potatoes, cabbage, carrots, preserved meats and canned goods."

Changes on the farm marked the historical changes in agriculture generally. The Hinrichs family liked to keep up when they could afford to, so as technology changed, they changed; as knowledge grew, their practices shifted in its light. By 1918, "The principle crops grown on our farm were wheat, barley, flax, and occasionally winter wheat and rye. These were grown for market and to pay off the mortgage. Oats and corn and hay were grown as livestock feed." There was a silo 14 feet in diameter and 40 feet tall, "supplying adequate room for corn silage throughout the winter season. . . . To fill the silo, the corn was cut with a corn binder that tied it into bundles. The bundles were loaded on hayracks by hand and hauled to the silo, there to be run through a belt-driven silo-filler, cut up and blown into the silo. This was always the hardest work of the year for the men. It also meant better feed for cattle and more convenient feeding in the winter."

Moving with the times, Hinrichs's father bought his first tractor in 1918, a Happy Farmer model built by the La Crosse Manufacturing Company. "It had one gear forward and one gear reverse, and the heavier the load the slower the motor would run." That Happy Farmer tractor was pretty primitive, but the difference between the tractor and the team was quickly apparent. "With a tractor we could plow as much as fifteen acres in a day, but with horses we could only plow about an acre per horse. . . . Belt work was coming into its own, and we needed a tractor for feed grinding, silo filling, and threshing with the small threshing machines." Stack threshing came to an end as the new twenty-inch Racine thresher "proved practical, and soon all the farmers banded together in groups of two to five to own their own threshing rig." Though the Happy Farmer tractor was mechanically unreliable, they kept it running until 1926, when they bought a McDeering 15-30, "the first of the better-built tractors." Hinrichs's father continued to keep up with new developments, moving in 1924 to certified seed grains and purebred milking shorthorns. When hybrid seed corn came along, Hinrichs and his brother became producers of Minhybrid Seed Corn. In the 1980s, Hinrichs wrote proudly that his "sons were still growing certified seed."

Hinrichs's father was progressive in his politics as well as his farming, and he played an active role in the Nonpartisan League. "Everybody in Featherstone Township joined," Hinrichs notes, and, "as time went on, the officials elected by the League became known as the farm block in Washington and, as such, held the balance of power. Neither Republicans nor Democrats could pass legislation without the help of the farm block."

Hinrichs married in 1929 and began to farm on his own. When the Great Depression hit shortly thereafter, he was soon insolvent. One year, their "total *gross* farm income was $954" (my emphasis). One-third of Hinrichs's crop went to the "landlord's share." Rent that year was $250, the tab for threshing was $273, and cost of fuel for the tractor was $276. Those were just the big tickets, not including other essentials: feed, seed, repairs, insurance, and groceries. He paid for the threshing and paid the rent, but there was no money for the fuel bill, and none that he could borrow. Perhaps it was a measure of the times, the equivalent of one of the social indicators of a sustainable community that we use today, that Hinrichs was able to strike an arrangement with the fuel man—after Hinrichs bore his initial wrath—that allowed them both to stay in business through such hard times. That transaction, despite its acerbic beginning, worked. Hinrichs got the fuel he needed, and the fuel man eventually got paid. Hinrichs never forgot his benefactor's generosity, and the fuel man became aware of Hinrichs's hard work and honesty.

Ultimately, the agreement was a good investment for both, and they became lifelong friends.

In the early 1930s, Bang's disease hit the cows and spread from farm to farm. Prior to the Depression, Hinrichs could have sold the cows for $150 each, but in those hard times he sold the cows for meat (which was unaffected by the disease) for no more than $20 each. A federal program finally helped bring the disease under control. But then encephalitis struck the horses, which were still a primary source of power. Once again, the federal government stepped in to help with research, and Franklin D. Roosevelt's election began an era of farm programs, starting with corn, hogs, and wheat. It was then that the Triple A, as folks called the Agricultural Adjustment Administration, began.

We learned some other things during the 1930s as well. Banking, debt, and soil conservation were linked in those years, in part because of the Depression, and in part because American farming till then had often been cavalier. When the land's fertility ran out, the farmer simply moved west, leaving unproductive soil behind. Soil depletion is not an asset but a loss in the economic and environmental columns. The issues were clear not only for young Hinrichs but for everyone, including both banker and loan applicant. The 1938 *Soils and Men*, an annual report of the U. S. Department of Agriculture, ties credit directly to the health of the soil: "Credit has a definite place in soil maintenance. . . . Soil maintenance is a secondary lien on farm income." A lender needs to keep loans from being too onerous: "Because soil maintenance is an expense that can thus be deferred, it is one of the first currently eliminated whenever other expenses begin to absorb farm income." To avoid that outcome, "the average total annual payment required to amortize a mortgage should not exceed the average of the farm income that can be devoted to payment of interest and principal and still leave enough for farm operating expenses and an adequate amount for family living."[12]

The rationale behind these caveats has three clear principles: (1) Careful arrangements of credit are necessary because everyone's future depends on healthy soil. . . . Good soil allows the farmer to produce enough to cover a loan. If the soil is depleted, the lender may have to take back a farm that has been devalued. (2) "Since the loan is always based on the farm, the care and management of the farm is of first importance." Loans should build in provisions that assure "the loan contract

18

may be a further aid to conservation." (3) The identity of interest between the lender and the recipient means that both should include "certain provisions giving definite assurance that all necessary measures will be taken to conserve the fertility of the soil." Such provisions might include "the kind of crops and their rotation, the use of livestock on the farm, and possibly other practices." The farmer thus assures the lender of his ability to repay the loan and "of his intention and capacity" to farm "according to the best practices," so that the farm they both own will not decrease in value because of soil depletion. The borrower and the lender are mutual investors, so whether the farmer can repay the loan is not the only question to be asked by the bank. The investment for both is not in the product, whether that be corn or beans, beef, pork, poultry, or flowers. The investment is in the farm—and not only in the dollar profit on the farm, but in the farm and the farmer that create the product and the dollars. Borrower and lender both lose if the farm loses value as the soil loses productivity.[13]

Hinrichs held that, despite the problems caused by the Depression, "There was also a lot of good that came from it. It made friends and neighbors. It made people considerate of each other and they shared their joys and sorrows. It made good citizens out of everybody and created a vital interest in the affairs of our community, our state, and our nation." An assessment like this could lead one to speculate that, if the government wanted to create a really good social program, it should simply let Wall Street go down the tube.

In the midst of the Depression, hybrid seed corn entered the market. Hinrichs and his brother began to produce for Minhybrid and then developed their own "Hinrichs Minhybrid." The new hybrids were "increasing the county average yields by leaps and bounds, from 39.6 bushels to 60 and then to 75 bushels per acre. Hybrid seed was here to stay," Hinrichs realized. Accompanying the new corn was commercial fertilizer. By the mid-fifties, Hinrichs began using phosphate "at a rate of about 70 pounds of 0–54–0 per acre put on with a fertilizer attachment on the corn planter." Hinrichs was one of those, described by Ron Scherbring in chapter 5, who embraced the new technology and were quick to adopt each new advance. "Soon it was discovered that a mixed fertilizer like 4–24–12 was better and we used it in ever increasing amounts." Yields of 125 to 150 bushels quickly became common.

Cropping systems changed in these years as well. Oats came to the fore for a time, taking more and more acreage. Barley and wheat had become problematic because of various diseases, and flax, deemed hard to handle, fell from popularity. But oats suddenly ran into its own problems: smut, rust, and helminthosporium. A system was now in place, however, that sounds like our own times: "Plant breeders got busy and developed new varieties resistant to the diseases." The new varieties generated talk that foreshadowed current conversations about Bt corn and Roundup Ready soybeans: "Resistance was good only for a few years, however, and then a new variety had to be introduced" as the bugs found ways to survive the new threats to their existence. Finally, the system really did become contemporary: "As corn yields increased and soybeans became popular as a crop, the cereal crops went out of the picture." Hinrichs, writing in 1982, concludes, "Now oats are grown only as a nurse crop for alfalfa." His memoir immediately shifts to describe people giving up farming and moving off the land and the farm auctions that inevitably follow.

Indeed, it almost seems as if Hinrichs's attitude from the beginning was that of a contemporary farmer: he grew and changed in his views of farming, trying always to think ahead, seeking to increase production, expand holdings, utilize the newest technology, increase speed, and—at least in theory—decrease workload. After climbing out of debt following the Depression, his first thought was that "it was time to expand," and he did: "So we rented the old Bennett place." Other improvements quickly followed. Before rural electrification, Hinrichs already had a thirty-two-volt light plant, "but it was dim and quite a chore to keep working and insufficient to provide power of any kind." When the Rural Electrification Administration came to Goodhue County in 1937–1938, he immediately wired the house and acquired "a refrigerator, electric iron, and a washing machine, and power to turn the cream separator and the fanning mill." In 1939, Hinrichs's old Farmall gave out, and he bought the first model M McCormick tractor in the county. In 1941, he bought a combine, "a Case A6 pull-type powered by a four-cylinder Wisconsin engine mounted on top. . . . We had graduated from a two-plow tractor to a three-plow tractor and more speed and comfort." Naturally enough, "this marked a big change in our harvesting system. . . . No more cutting grain with a binder and shocking and threshing. We windrowed the grain to dry and then came along with the combine and threshed it, leaving the straw behind in a windrow to be picked up by the baler." Hinrichs's brother bought the baler, which changed their haying as well. The early mechanical corn pickers could "pick as much corn in a day as six men could husk by hand." But they were "a cumbersome machine with many breakdowns," so husking by hand continued until 1942, when a "reasonably satisfactory corn picker became popular." Also in the early 1940s, the family bought a milking machine, built their first seed corn drying and processing plant ("Now we could increase our production"), and the whole family, including father and brother, showed cattle, sheep, and seed corn at the county fair. Their milking shorthorns were sold for breeding stock and for beef. "We were providing farmers in a fifty-mile radius with their bulls," says Hinrichs.

However, Hinrichs's foresight was not always so reliable. As late as 1942, he went into horse raising, "not riding horses but draft horses." Though the number of draft horses was diminishing, Hinrichs reports that he could not understand "how a farm could survive the average winter without a team of horses to get around in the snow." He had two brood mares from which he got five colts. His report makes one wonder why anyone *ever* raised horses: "One developed into a good horse; another broke well and drove well but could never be trusted for biting or kicking. The third one developed into a heinous outlaw that sent me to the hospital with a broken leg." That might not have been enough to put a stop to Henry's nonsense about horses, but "the prices kept going down" until, he confesses, "I finally sold them all." Hinrichs also recalls, somewhat ruefully, another poor venture: growing Christmas trees. "I reasoned that the poorest soil on the farm could produce a thousand trees in ten years. At the prevailing price this would be very profitable." He even dreamed of an advertising scheme: "Bring your children out for a sleigh ride to the Christmas Tree Forest and cut your own tree!" Everything seemed as though it had to work. "It sounded and looked good and the trees grew." But when the harvest time came, "we no longer had horses— nor buyers for the trees," and one suspects the sleigh was gone as well.

In 1950, the scene began to take another discernible turn, one reflected in changes in our small towns. This is the real beginning of the loss of small farms, and it reveals the impact that changes in agriculture had on our communities. Till then, Red Wing had been thought of as an agricultural town. "It had two flour mills that bought wheat from farmers, a flax processing plant that bought flax, and two malting companies that bought barley." The town still had two produce stores, three feed stores, and two implement dealers. But the creamery had gone, and soon the rest of the farm-oriented businesses disappeared as well. From then on, Red Wing, surrounded by a thousand square miles of farmland, turned its back on agriculture and focused on industry and commerce.

The year 1950 also saw the introduction of the field chopper and of the portable elevator, which made loading silos much easier and reduced the need for hired labor. The two-row corn picker came into common use, harvesting two rows at a time and pulling the corn wagon behind. The changes again showed up in the cropping patterns and in the whole culture of farming. "Corn and soybeans were commanding increasing acreage and small grain was growing less profitable. The change in cropping required new equipment." The new self-propelled combines had larger capacity and could harvest corn as well as beans and small grains. Hinrichs saw a cause-and-effect relationship between the increase in corn and beans, the use of new equipment, and the loss of small farms. "Farmers began expanding their operations. When a farmer quit farming and had an auction, the farm was no longer taken over by a new farmer. More than likely the farm was purchased and absorbed by a neighboring farmer, making his farm that much larger." There was a whole new agriculture coming to life. Hinrichs summarizes it this way:

> Diversified farming was giving way to specialized farming. Sheep and poultry gave way to dairy, beef, hogs or crops. Most farmers discontinued the lesser of these productions and increased one of the others so it became their specialty. The dairy farmer milked 50 to 100 cows, the hog farmer raised 1,000 to 2,000 hogs annually, and the beef farmer fed 500 steers or more. Except for the crop farmers, the average farmer utilized his entire crop production for feed. The crop farmer marketed corn and soybeans. Sheep were generally unprofitable and the few farmers that kept them had small flocks to keep the weeds under control and scavenge around the farmstead. They were not profitable until the late 70's. . . . The four-bottom plow gave way to five-bottom equipment. The two-row combine was replaced with a four-row. The hatch corn dryer was replaced with a continuous flow dryer and more storage had to be built.

The evolutionary changes in technology and scale became built into the system. One would only have to change the number of dairy cows from one hundred to three hundred or a thousand, for example, to fit today's similar pattern.

The sociology of the farmer's life changed along with the cropping and the farm expansion. Hinrichs explains, "Farmers began to take off a weekend or two during the year or even perhaps a week. . . . Prepared foods and household conveniences relieved the farmer's wife of many of her household duties. Many . . . found themselves with time on their hands and sought employment in the city. . . . The farm was supporting two family automobiles. When the children grew old enough for a driver's license, they, too, bought an automobile. Their growing up made further expansion of the farm operation possible." As one would expect, the economics changed too. Henry Hinrichs's brother Erwin decided to sell his farm and retire. "To this time, the effect of farm prosperity and inflation can best be expressed by the fact that he bought the farm in 1941 for $18,500 and sold it again in 1971 for $100,000." Henry marvels, "Eighty-one thousand five hundred dollars appreciation!"

One can see throughout the story of this farm family the roots of trends in present-day agriculture. Hinrichs's farming life began at the dawn of the twentieth century and ended when it was more than 80 percent over. Nearly all the forces now at work came into play over the course of this farmer's lifetime, and his attitudes as well as his actions mark the century's changes. Hinrichs's easy-to-understand delight at the speed of new technologies is obvious, as is his amazement at the amount of work accomplished and the increase in crop production. Yet speed and efficiency are not all benefit. They have a downside as well, as Hoffbeck's story of farm machinery development reveals and as Eric T. Freyfogle, professor of natural resources, property, and land-use law at the University of Illinois, Urbana-Champaign, explains in *The New Agrarianism: Land, Culture, and the Community of Life.* "Good work, agrarians recognize, often takes time, and some jobs cannot safely or wisely be speeded up. Bad work, on the other hand—bad in terms of adverse effects on the land community and the social order—can happen quickly and leave enduring scars in its wake."[14] The untested effects of new seed varieties that came along in Hinrichs's early years, and the untested effect of chemicals over the long term, would gradually reveal themselves after Hinrichs's day. Even an early version of globalization expressed itself as part of the "green revolution," which Hinrichs witnessed. But Hinrichs could not yet see the effects of these and other changes. Some would occur soon after he wrote his memoir; the increased debt for combines and high cab tractors and the drop in farm prices, especially the collapse that ruined so many in the late eighties, were just beyond the range of his otherwise ample foresight.

The complexity of the farm story has to do with cultural changes in the distant past as well as some changes that are more recent. Don Worster, historian at the University of Kansas, believes that "the most important roots of the modern environmental crisis lie not in any particular technology of production or health care—the advent of medical inoculations, for example, or better plows and crops, or the steam engine, or the coal industry, all of which were outcomes more than causes—but rather in modern culture itself, in its world-view that has swept aside much of the older religious outlook." Worster names that worldview materialism and holds that it is accompanied by another component, secularism. The former had a powerful impact on the biological world because the secular goal of its driving economy was "achieving more comfort, more bodily pleasure, and especially a higher level of affluence." This became the great good in life, "greater than securing the salvation of one's soul, greater than learning reverence for nature or God." The secularism that accompanies materialism in contemporary culture, Worster says, "undertook to free people from a fear of the supernatural and tried to direct attention away from the after-life to this-life and to elevate the profane over the sacred." That secularism "even invaded the very core of religious expression," subverting the church and distorting its expression, so that "today we can find unembarrassed Hindu gurus buying fleets of Rolls Royces or Protestant television evangelists selling glitzy condos in a religious theme park." According to Worster, the idea of "progress" is also implicated in our modern view that defines progress "mainly as an endless economic or technological improvement on the present."[15]

Those shifts in our culture's worldview and mind-set have altered our view of our farms and what we hope to get out of a life in farming. A materialist culture rewards material progress and high production that, at least for the short term, brings material prosperity. Sometimes that comes at a price, the price being land and animal health, both of which get sacrificed so we humans can have not health but wealth, so we can be more comfortable and own more things. Freyfogle holds that the agrarian view strives to move in a different direction from the general culture of materialism: "At its best—and its best, to be sure, is often not fully attainable—the agrarian life is an integrated whole, with work and leisure mixed together, undertaken under healthful conditions, and surrounded by family. As best they can agrarians spurn the grasping materialism of modern culture; they define themselves by who they are and where they live rather than by what they earn and own."[16] The farm story of this region, perhaps of any region, is also more complex than technological developments might indicate, for as Worster and Freyfogle point out, we are all caught up in historical forces that we may not be aware of, participants in a changing culture the implications of which are especially difficult to discern.

FARMING RESPONDS TO SUPPLY AND DEMAND

Another way to look at our farm history is to trace its economic development. Willard W. Cochrane, professor emeritus at the University of Minnesota, sums up the previous century's agricultural economic history in one long, Faulknerian sentence: "Farmers experienced a wonderful economic high during the first two decades of the 20th century; fell into a depression in the 1920s; fell into a deeper depression in the 1930s; once again enjoyed economic prosperity in the 1940s; experienced falling prices and economic hard times in the 1950s; experienced moderate prosperity in the 1960s and early 1970s; then farm prices fell and hard times returned in the late 1970s; hard times returned in the middle 1980s; and farmers are once again experiencing economic depression in the late 1990s." Cochrane's analysis of that history is interesting:

> Each downswing is attributed to some specific cause. But why do these specific causes induce a "feast or famine" type of behavior in the food producing industry—in farming?
>
> They do so because the food producing industry is an inherently unstable industry resulting from the highly inelastic aggregate demand for food on the one hand and, in the short run at least, a highly inelastic aggregate supply of basic food products on the other. Thus, with any *small* shift in either the aggregate demand for food or the ag-

gregate supply of food products, we get a *large* price response up or down depending upon the nature of the shift in either demand or supply, with a consequent change in farm incomes. These specific causes (e.g., war, or peace, or great drought, or great technological breakthrough) are always at work shifting either aggregate demand or aggregate supply. Unfortunately, the reality that the food producing industry is basically an unstable industry as the result of the economic forces noted above, is something that most political leaders, farm leaders, farmers themselves and agricultural economists either don't understand or don't want to understand. [Cochrane's emphasis]

The lesson to be learned from that history, according to Cochrane, is simply that "the food producing industry cannot and will not level off at some desirable economic level and stay there. The economic forces, inelastic aggregate demand and inelastic aggregate supply, won't let it."[17]

In the following chapters, we will look through the eyes of farmers at the political history of farming and some of the complications that have resulted from changes in farm policy. These farmers' stories are filled with another kind of history, personal or vernacular history, that reveals itself in the "that's how we came to have this place" stories that I heard often. One thing all these folks, of different scales and widely diverse practices, have in common is this: They all believe they are following practices that are sustainable. They all believe they are contributing to a more secure future for themselves, for farming, and for the environment.

Part II

Farmers Talking about Farming

CHAPTER 3

Two Views, One Farm

VANCE AND BONNIE HAUGEN

Driving from Red Wing, Minnesota, down to Mabel, Minnesota, I take an inland route, a big semicircle cutting southwest across the bluff country along the Mississippi, heading up into the borderland between those steep, tree-covered hillsides and the beginning of the rolling prairie that will soon taper into the Great Plains. In the hills leading away from the bluff lands along the river, the soil lies lightly on the ridges, a porous sandy silt called loess, easily eroded, and not very deep to start with. Driven by ancient winds following the most recent ice age, about ten thousand years ago, it covered the higher ridges, twenty feet deep in places, but lay more thinly, sometimes less than a foot deep, where wind scoured the side hills. Such soil covered all of Wabasha, Winona, and Houston counties, and a little over half of the southeast Minnesota region. This soil's name sounds as if it came from *Star Wars* or from the magic kingdom of Narnia: Udalfs—a suborder of Alfisols, which are characteristically forested. It's this particular soil that supports this western edge of our American deciduous hardwood forests.

This green landscape now wears a bucolic mantle of farmed domesticity that belies its dramatic wilderness history. Though 80 or 90 percent of it is now under cultivation, the land still allows us glimpses of its wild past. The wind that drove that soil across steep hills to drift over high ridges bespeaks a turbulent, tumultuous time when everything loose was up for grabs. Wind's descendent, the tornado of today, reminds us of that earlier time. Today I am driving along an ecological border country between grasslands and deciduous hardwoods, moving in and out of vestigial remnants that hang on despite our agriculture. Experts would probably call it a transition zone. Along Minnesota 58, we move in and out of noncontiguous bits of tallgrass prairie and maple, oak, basswood timber, both zones mostly broken up by agriculture: corn and beans in places, dairy cattle on some hills, pasturelands and contoured strips of grain and alfalfa. The bluff lands along the eastern half of Goodhue County are covered with this soil; the western half is rapidly becoming prairie. This landscape, like much of the country around here, is called the "driftless" area because it escaped the last wave of glaciers that drifted over the area, leaving the land free of the glacial till and debris that inevitably mark a retreating glacier's path. In the fifteen miles between Red Wing and the town of Goodhue, the shift from river bluffs and timber to tallgrass prairie—big bluestem and Indian grass—is so abrupt that it is apparent even to my untrained eye, even if the bluestem is now gone along this stretch, replaced by corn and soybeans.

Beyond Zumbrota, the arc I'm driving swings to the south, down U.S. 52, through Rochester, held up by an ancient bedrock called the Rochester Plateau, and continues on down through the hills and sinkholes at Fountain and Preston, where the karst limestone ledges rise occasionally above the road. In this area, some of the rotting limestone beneath has collapsed on itself, forming caves and underground streams and lakes, groundwater we depend on for drinking, agriculture, and industry. On some of those exposed limestone layers, ancient striations and grooves, clear as road signs, lay out the direction of glaciers earlier than our last wave, apparently headed south for the winter. Inside the rock, fossils of marine plants and animals tell us cool northerners that our land once endured an equatorial climate in that geologic time called Silurian. How our dour, stoic pleasure in our ability to stand the cold would have suffered in such heat! Our summers now are short, and we forget stoicism to complain bitterly about the humidity. Equatorial heat year-round would surely melt our taciturn endurance like ice cream in August, raising questions about our very nature. Who would we be in such a place?

Because the most recent Wisconsin glaciation bypassed the region, the glacial evidence here speaks of older events, even more ancient movements of rock and ice. Here fossils reveal earlier plants and animals that spent their lives in the shelter of glaciers far older than those that swung down through Minnesota just a bit west of here and south as far as central Iowa.

South of Preston, the twisty road finally rolls out into broader, gentler curves and prairie and into Amish country. The Alfisol soil here shares the prairie with Mollisol, that dark, loamy soil rich in organic matter that my German immigrant grandfather loved to crumble in his hand a few miles south of here, in Delaware County, Iowa. This ecosystem is savannah, and this particular portion of it is known as Southern Oak Plains. One county to the west, the Wisconsin glaciers left still another soil. True tallgrass prairie rose there, miles of it, stretching west to the one hundredth meridian, where the country gets more droughty and the grass gets shorter. It is easy to imagine the soil deep on this rolling land and to imagine the oak savannah that once reigned over much of southeast Minnesota, thanks to systematic and knowledgeable burning by local tribesmen. "Native Americans, they did a lot of it," says soil scientist George Polk, during a conversation in his kitchen. "If they hadn't burned it, the woods would have taken over everything.... Only the bur oaks—they are really tough, with a thick bark—survived." How do we know there was once oak savannah here? "You can tell just by looking," says Polk. "Forest biomass is above ground. It leaves behind a very light, sandy soil." With prairie soil, "the biomass is two-thirds underground. It's dense, and the prairie soils are much less erodable. So all you have to do to see what was where is look at the soils." Still, it would help to have a more educated eye than mine.

There are Amish families scattered throughout this region, but here enough Amish are concentrated that the highway department has added special lanes on the side of the highway for their horse-drawn buggies. As they trot in for parts or groceries or head home in the dusk, the black buggies and their sorrel or dark bay horses would be especially vulnerable to accidents on the regular pavement, despite the bright, incongruous triangular yellow highway department caution signs fashioned to the back of some. Even that safety feature's concession to modernity is too much for others, who eschew both signs and safety in trade for a secure tradition. Stopping for gas along the way, I think the town name, Harmony, reflects something of that Amish influence as well. Or is it only settlers' hope?

Where 52 turns ninety degrees south and heads into Iowa, I turn north on the gravel, go half a mile down into a creek bottom, cross a narrow bridge, and wind up into hills rapidly becoming steep again, back toward that old Paleozoic Rochester Plateau, the creek bottoms and slopes again covered with the loess that spawns the deciduous oaks and maples that crowd the road. Coming up out of the creek, past an abandoned, white-frame building that locals refer to as the Amish schoolhouse, I turn again at a tiny, weathered outbuilding where children once waited for the school bus, into Vance and Bonnie Haugen's farm atop the ridge. The farm is not quite on the Iowa border, but close enough that a young pitcher with a long arm and good windup, a coming Bobby Feller, say, might dream that he could throw a rock from here and hit Iowa right in the strike zone.

I'm here to go on down to Iowa, near Decorah, for a field day on the farm of Dan and Bonnie Beard, longtime friends of the Haugens. While waiting for the Haugen Bonnie (as opposed to the Beard Bonnie), Vance and I stand outside where we have a pretty good view of the farm. Pointing and gesturing, Vance tells me his view of it, showing me what I'm looking at, laying it all out from his unique perspective and our mutual vantage point.

I never thought we'd be farming here.... My father-in-law had an Amish hired man, and when I'd come to visit, he'd say, "You know, there's this farm for sale." So we finally came over and looked at it. The buildings were in bad shape: there was two feet of water in the basement of the house; there was trash everywhere. I thought, Oh God, this will never work, but the price was right, so we took it over. I told Jake, Jake Yoder, a friend of mine, what we were going to do—how we were going to turn it into a grazing farm and make it work that way. He said, "Well, you can't do any worse than the guy who was there." By that he meant the guy had gone bankrupt, so we couldn't do any worse than that.

We do dairy. I want to show you—it looks pretty good. We're conserving the soil. I feel pretty good about that. Twenty years ago that hillside there (*pointing to a sloping field north of us*), it was a mess. There was an awful lot of erosion here before. Now there's almost zero. There's 230 acres—I'd say there's about 190-some acres grazeable, with a little woodland and a building site. On the one corner down over here (*pointing again, to our right*), there's a little creek that goes through. And then we have a two-acre pond on the other land. It's kind of nice.

And you'll notice, you don't see any paddocks out there. What we did, this is something that I've developed, we fence on the contour, so I've got high-tensile fence. If you look really close, you can maybe see, up where that water tank is, there are actually wood posts. We fence there, and then we can make the paddock any size we want. The other thing, then, I take a board or a post and we lift the wire up and the cows walk under it. That way, you can have the cows enter or exit any place. And you don't have to have lanes—we still do have some lanes—but you don't have to have permanent lanes; you don't have permanent places because oftentimes, that's where you have erosion. Some people really object to the "cow path" look. I can understand that. There's a downside to it. Occasionally, a cow will knock the post down, and I know Art [a dairyman not far away], he teases me terrible about this. "Oh," he says, "you've got to have gates. When you open a gate, you know that you've got it right." True! That's very true. But in most of these, you know the cow has to go through that area. Sometimes you get mud there. . . . I like it this way, but it does take a little bit more work. I have to admit that. But it's less expensive.

This used to be a two-family operation, and then the other family decided they didn't wish to be here, so we bought them out. At the height, we're at 150 milk cows. And now we're down to about 64 this month. We could easily support—I say easily—but we were supporting 150 OK. But then we didn't grow any grain or corn silage on this farm, we had just grass, and now we do grow corn silage. I have twenty acres of corn silage on here. We still buy all the grain because it's just not worth it. So it's kind of nice. I like the corn silage for a lot of different reasons, and I also like growing it in that we can renovate—so that if we have a field that we feel isn't doing well, we tear it up and seed it down again. So for us, it works out OK.

We're here, this is our tenth or eleventh—tenth—year, and we're still making all the payments. I still work off the farm, but I wouldn't really have to, economically. We could make it. It just makes things easier. It's one of those things. . . . But it's frustrating, you know, because I see my neighbors over there. That's corn and beans. Nice guy. I don't think he's making any money. If he is, it's all from the government. And there's erosion.

I'd like to see a hundred farms like this. You look up and down here and we're losing farms right and left. I think this is a viable option for people—if they want to be dairy. Now, I have a good friend who farms just over there. He's a real nice guy. I always tell him, "Friend, you're not making any money with corn and beans. You should have cows. We could really set up a nice grazing farm and—" "Nah," he says, "then I'd have to milk!" (*Vance laughs.*) I tell him, "Yes, I guess that's a problem." It's not for everybody, that's for sure. But we do OK.

We started out with Holsteins. Now we've got a mixture. We bought a little over a hundred Holsteins and about thirty Jerseys. What we've done is, we've gone to crossbreds. So we have predominantly Jersey-Holsteins. Probably fifty of the cows out there are Jersey-Holstein crosses. They're doing just really great for us—they do a very nice job. We like the milk better. The components are higher, and it's a smaller animal. It's about an eleven hundred–pound animal with those cross-breds. They just do really well. They seem to have a little better fertility. We've only been doing it for eight, nine years. So I don't want to say it's "the thing," but it's been working well for us.

Vance sounds like others I've talked to who have tried this or that for a number of years and are happy with the results of their experiments. Yet I've never heard one say "It's the best way" or "It's the only way." The biggest claim is "It seems to be working on this place."

I was looking at some bull stud information, and they were talking about this cow that had a 64,000-pound average, and they were touting her son and all, you know. I like looking at machinery, especially cars, so looking at a Ferrari, that's kind of fun, but I sure wouldn't want to be driving one of those down these gravel roads! So I'm thinking, they're talking about that, and you know, we've got different cars and different vehicles for different things, and I kind of think, perhaps they've lost track of it. I mean, they're trying to push so hard, you've got to have that top production, and I was talking to a friend of mine, and he said, "You know, you have to have that top production."

And I said, "Why?"

"So you can be profitable."

"Oh, so what it costs you to get that top production doesn't matter then?"

"Well, yeah, it does, but if you get top production, you always make money."

And I said, "Baloney sausage." It isn't going to happen. You have to be careful on what you pick and choose. But I take a look at—if you can have low-cost forage, you know, and let the animal harvest it for as many months as possible. . . . It depends on how you do the figures, but we can get break-even costs, just cash costs, we can get them down in that $6.00–$7.00 area. You add in depreciation and the rest of it, I think it gets in at around $10.00–$10.50. We can stay fairly competitive. I generally look at some of the larger confinement operations—they're talking $11.50–$12.00. Well, if that's what they need to break even, and if Land O'Lakes and the rest of them are going to keep giving them that price, well then, if I can keep underneath it, I can stay in business. We're doing OK.

Bonnie's ready.

Shall we go over and take a look at that other farm?

Vance and I get into my car so we can continue the conversation, and Bonnie drives their vehicle. Vance and she will stay with the Beards longer than I can, and he will drive back with her. Along the way, he gives occasional directions since I've never been to the Beards' farm.

Dan's got an interesting operation. You know, the folks that are still in this business are guys that are interested in trying something different, generally. If you've always done what you always did, you always get what you always got.

This is a comment I hear often, in one version or another. I've heard it from both agribusiness producers and farmers taking a sustainable approach. One of the ties between them, beyond a region they all love, is this way of talking about farming. The practices they follow may differ widely, but the way they talk about their work is often the same. Adaptability and flexibility are requisite characteristics for staying in business on the land today. Any day.

I get to see an awful lot because I work off the farm as a county agent in Wisconsin. So I see an awful lot of differences, and of course similarities, between the two states, and also traveling through Iowa, I get to know a lot of the folks here.

Vance and Bonnie both understand the nature of this landscape. They have taken its vagaries and peculiarities into account as they seek the best way to farm it.

This a real interesting area, this driftless area. It's neat to see people trying different things. Of course, my big thing has been the grazing and dairy. Not only did we do it, but that's one of the things that I've worked on or teach. We can't be proponents of it, working for the county, but I think it's a great system—it probably will work for more people than our large confinement operations will. There are some days when you have mud—when you have real bad weather, you kind of wonder. For the most part, to have an animal on concrete during its entire life—never getting off—doesn't make a whole lot of sense to me. And the pork, too; I mean, we've got some pasture pork. My son is raising some. There's a lot of really good—we've got Tom Frantzen over here, for instance—he's a pork producer down here in Iowa, and he does it out on pasture. He claims he's making money and he's still in business, so I kind of believe him.

But things change; the technology changes or the economy changes, and farming has to change too. The thing is, the technology works; it works, there's no doubt about that. On the other hand, can you pay for it? If you set things up for a ten- or fifteen-year note and all of a sudden things change dramatically in three or four years, and then you can't pay for it—that's when stress becomes almost unbearable. And some of these guys have killed themselves over that, or killed themselves working.

Some people say, "Strive for the maximum production." Well, yes, you can do that, and you can buy the technology for it, but does it really pay? And the other thing is, if things go against you, if you have a bad year.... If everything goes perfect and you've got it all worked out, yeah, fine. But what if you have a drought year? What if an '88 shows its ugly head? We're not going to get an '88 this year [2002], but it could be. We could have reduction in yields because of drought. For some of these guys, it can be devastating.

I worked at Fargo for ten years, and before that I lived up near Thief River Falls. So I saw some of those rich beet farmers, and I saw some of those problems. They went from wealth to nothing—worse than nothing.

Getting back to chickens, hogs, dairy: As far as having more of a natural system, one thing that bothers me is when I go to these really big confinement systems and look at them. They're imposing an industrial system on a biological one. You can see it is working, but it's not a good way to do things. Farming's not natural, to begin with. People say they want to go back to nature, well, farming is totally against nature, but you can do it within degrees. You can be totally radical or not quite so radical. I think those guys have gone completely off the deep end in some of their stuff.

In order to make farming work, you have to change it. But you have to change it within reason. And you also have to take a look at what's the sustainability on this? That's where I get concerned about the corn and bean rotations that we've got up here—is it sustainable? They say, "Well yes, we'll make sure it's sustainable." But how long? Ten years? Twenty years? One hundred years? And are you talking about sustainable from the soil point of view, or [are] you talking about it from a community point of view?

Like ethanol—that bugs me, how ethanol is touted as a nice, renewable fuel. But it takes corn. So I'm looking at it and thinking, well, if we can raise enough corn so that we're not hurting the environment, well then, maybe that's a good thing. But then, what about the energy it takes to get the chemicals and the tillage and herbicides and all the rest of the stuff? And I'm thinking, so it takes more energy to make ethanol than the ethanol saves; you know, maybe this isn't the greatest idea. It's totally wacko! And look, on top of it, here's a person who's putting a badge of honor on their chest, saying, "Look—I'm using ethanol," and they're driving a suburban assault vehicle. It's ridiculous. And then we have our president saying, "You really don't need to conserve, we'll just get some more." Fine, if we can find some more, well, that's a great thing. But I certainly can't understand why we can't conserve also. That doesn't make any sense to me.

What I'm hoping on our place is that we would continue to build up some of the paddocks, and get things so it's easier to move cattle around on there. I don't think that's much of a problem. My son has some interest, and my daughter is twenty-one and has also expressed some interest in coming back to the farm. It will be kind of fun in five years to see if they'll be either farming with us or nearby. That would be kind of fun. I don't know if that will happen, but if they'd like to, I think it could be interesting.

I don't know how I came by my conservation principles. . . . I'm going to be forty-five in October, so I suppose I was just on the end of the hippie age when I went to college—so maybe that was part of it. I grew up in northern Minnesota on a small, basically subsistence farm, and we were dirt-poor. We conserved everything—that might be part of it. My dad and mother are still up there and have retired. She was a schoolteacher. We were probably some of the last people to start using chemicals up there, spraying crops—maybe not because we had a conservation ethic but because we were leery of it, and the cost. That's part of it. . . . And then just my [educational] background. I got an ag education and ag mechanization degree, going to college in the seventies. There was a lot of stuff about environment and ecology and that sort of thing. I guess that's where it's kind of come from, you know. I've looked at it, and having a background, you don't believe everything everybody tells you. You have to be a little skeptical.

I take a look at some of the farm things when I was going to school, and it was always "Let's see what we're really doing here; what's the big picture?" And that's one of the things we looked at: if we're going to conserve, if we're going to grow crops, how are we going to do it for not just ten years or a hundred years—how about a thousand years? I don't know where I came up with that idea. We talked about stuff in China—some places there they've been farming continuously for nearly four thousand years. I think, too, being Norwegian (*Vance starts to laugh*) is a detriment to spending a lot. (*We are both laughing.*) My immigrant ancestors came here looking for land. They didn't have any in Norway. I think that's always been passed on: if you can get ahold of land, you cherish it, and you try to take care of it because there isn't much. That's the way it was in Norway. That isn't the way it is here, but I still hear that echoing, that sentiment. It comes through even when I talk with my parents, or talk with my grandparents, or even great-grandparents.

There's a farm on the base of this hill, that'll be on the right. That's the farm we're going to. They're grazers just like us. Dan and Bonnie Beard. They started grazing about the same time we did, though they've been farming longer than we have. They are really nice people. They do the rotational grazing. Dan and I were down to Nicaragua, teaching rotational grazing down there to farmers. We had a great time.

I shouldn't be surprised. One of the things I've learned is how much local farmers know about the rest of the world and how much they participate in it. The notion that farmers today are provincial or unsophisticated is far wide of the mark. I can't resist asking Vance when he was there.

Last October. I'm going back the end of August with my son.

I tell Vance I'm so envious I can hardly stand it. "I was down there in '86."

Really! What were you doing down there in '86? (*Vance pauses just a moment.*) Maybe I shouldn't ask. (*He laughs.*)

His laugh cannot quite hide his suspicion. I can't help but laugh too, knowing what he's wondering. I tell him, "I was not working for the CIA."

Well, that's good (*clearly relieved and very serious*). There were some bad things going down then.

"Oh, man, it was something," I agree, also serious now, remembering.

I mean, it was embarrassing being an American down there.

"Oh, it's embarrassing to be an American lots of places," I agree.

Yes, that's true. It was very embarrassing being an American down there. One of the things they told us before we went: don't talk about politics. We didn't, of course. But you couldn't help but hear things, you know. They would talk about the war, and they'd talk about losing people there. We couldn't go up in this one area because there were too many land mines still left over. And, Judas Priest, what in the devil were we doing there to begin with?

"The stories I heard in '86 would just chill you."

I believe it.

"The kind of viciousness that we put on them—"

Yes! And for what?

"For nothing."

Yes, absolutely nothing!

"The pretext was that those guys were going to march through Guatemala, Mexico, and Texas—"

And take over Washington! (*Vance finishes, his tone disgusted.*)

He points, and I pull the car into a farmyard, huge maples shading a picnic table that will soon be set with a typical farm spread—potato salad, green salad, grilled hamburgers, homemade buns, and dessert— all of it from the Beards' own farm or helpful neighbors. We drift across to a tractor and hayrack. The driver gets down from his green John Deere perch as Vance, too tall for my little Corolla, extricates himself from my car and swings out his big hand.

Well, Mr. Beard, are you up for a visit today?

A month or so later, I'm back at the Haugens' to talk with Bonnie. It's a cloudy day, threatening rain. I'd first met her at a field day held by another dairyman and grazer, Art Thicke, who lives a few miles east. She impressed me with her directness and a certain pragmatism. Larry Gates, a farmer and Department of Natural Resources hydrologist, introduced us and in a sentence told her I was writing a farm story. She looked a mite skeptical but agreed to talk with me.

Now we sit down in the kitchen. As in all farmhouses, it's the main room in the house. At the Haugens', the kitchen and the living room are all part of one big open space, fireplace at one end, and a guitar, mandolin, and violin hanging on the wall, with a music stand in one corner indicating that the instruments are not decorations. Bonnie starts some coffee, and I ask her how long they've known the Beards.

Well, that's kind of interesting—living as close as we are. We met at one of the grazing conferences in Wisconsin in '93 or '94. It's kind of silly, we're so busy—we're close and yet we don't see each other very often. Dan and Bonnie came up to our place to see our milking parlor while they were thinking about putting their parlor together. That was pretty interesting. When I show others our parlor, I always tell them they need to go see these other parlors—they have things in their parlors that we don't. Their parlors are better than ours because they learned from our mistakes. You saw theirs; it's really fine. Dan and Vance were together in Nicaragua last fall, talking with Nicaraguan farmers about rotational grazing. It was one of those experiences that deepens the bond. Let me see if we've got some peace coffee from Nicaragua.

We sip our coffee and Bonnie asks me how this project is going. I tell her one of the surprises has been that, for the most part, when I ask farmers about policy issues like the federal farm bill, "I get a response that goes something like 'Ah, the—expletive deleted—farm bill . . .' whether the speaker gets help from it or not. Then the subject promptly changes." Bonnie laughs, recognizing the experience. "The stories I'm hearing," I tell Bonnie, "are more personal, more focused around values than around public policy. I find that inspiring, because practically everybody I've talked to so far inadvertently reveals an idealistic or altruistic view of what they are trying to do. It's always about making a living, yes, but also more—about helping others beyond the family, or about working for the community."

That just gives me goose bumps because, Gary, we have to be pioneers all over again. And there are lots of pioneers around. We might not have to wash our clothes in a washtub or down by the stream, but we're foraging other waters. Instead of having to worry about taming the wilderness, or not having a doctor within sixty horseback miles, we have to worry about global markets, or how to make this farm work environmentally so it doesn't put itself out of business because it's drawing down our soil or water resources, or we have to worry about politicians who know nothing about farming, or consumer indifference to anything but the cheapest food—all those other threats out there.

Those threats are more serious than terrorism because they are more insidious, in my opinion. I don't wish for any more terrorist attacks, but if we get anything out of them, maybe it will get some people thinking about the food system being too centralized, and the benefits of buying more local produce, buying your food from farms that you know. But I think once you talk to people and they really understand a little bit more about the food chain and exactly the ramifications—it may be a little more expensive, depending on exactly what you're buying; it could be more expensive initially—but in the long run, it still keeps people in the community; you get good quality for your food, and then it's not a waste of food dollars. So I think once we get the public more aware and interested in acting accordingly . . .

Years and years ago, someone was reminding people that if you give people—he was speaking at one of the Bread for the World luncheons—if you give a person a gun, you might have a friend for a day, and if you teach them how to grow their bread or the grain, you're likely to have a friend for a lot longer. I think, socially, that's a lot of what our whole nation's problem is—you have to balance your security against your friends', but too often it's too fast and too easy to lean towards just the security and not thinking about others' well-being. . . . We might be in different countries, but we are still in the same world. We are still sharing the whole air and water and all of that. It's a difficult thing.

I guess it's a combination of everything—you get the people going more that way, and then that affects the elections. At the same time, we can't rely on the politicians, but we also need to work with them because we don't want them working against us either. We don't want them setting up policies that keep promoting more corn and the big companies. I think it's a real spiritual thing too.

Bringing it back to local—I think that's one of the reasons why it's so important in this particular area with our karst geology— that's one of the reasons why I'm so interested in working with more of a grass-based farm, even if you're not an intensive rotational grazer, just having more of the grasses there, just because we have our water erosion so fast and furious everywhere, and I think that's really important, going down to the aquifers as well as being close to the main rivers—

There is a knock at the door. Before Bonnie can get up to answer it, the door opens and a neighbor comes in. Bonnie introduces me to Judy and her grandson, a towhead who is tagging along shyly, one hand wrapped in his grandmother's skirt.

"Judy and Gary were dairy farmers for a lot of years, and they're not grazers like we are now," Bonnie introduces us.

"We were kind of . . . Well, we used pasture, but we weren't intensive grazers," Judy says, not defensively but thoughtfully, weighing Bonnie's comment.

"But you also didn't overuse your pasture," Bonnie prompts. She turns to me and says, "This is one of the best conventional farmers, right here. And she's been active in all the issues that have been going on around here." She explains to Judy that I'm working on a book about farming for the Experiment. I tell her that part of our effort is to involve the university in local projects that lead to sustainability. Judy says,

> I don't know; I've been kind of disappointed in the university. I don't think it's done its job. It's led a lot of people down the wrong path over the last several years. Maybe thirty years, I don't know. I guess what we've done, you'd call it "sustainable." Well, when we sold our farm we sold it to—actually they were farmers who were over on the next road. They have three together, two boys and the father. They have three hundred to four hundred cows. We had a dairy facility with a stanchion barn, and we built a free-stall barn for eighty. But it's really good for those kids. One was getting married and needed a place to put their extra cows because you can't put them all in one area. And we have pasture, because my husband didn't want to see our whole farm go into row crops and our hillsides be nothing but soybeans, you know. And they're making use of the hay, and they're good farmers. The kid, he's doing different things to the place. They have to do what works for them. The fields, at least, are not just one whole field of corn and beans.

I comment on some of the erosion I've seen in corn and bean fields this spring, sheet erosion and gully erosion moving tons of dirt.

"So what's the solution?" Judy asks. "Does the university have any solutions? I mean, there's that one guy, Gyles Randall, and I've heard Dick Levins speak. He spoke at our farmers' union annual meeting. He's very good. He does a lot of writing too. But do they listen to them? We talk about . . . keeping our agriculture more local. The university talks, but they don't actually get anything done."

Bonnie picks up on Judy's point: "I think there's a bunch of farmers that could really do it, and a bunch of them are all set to . . . and could effectively feed everybody locally, which I think would be a great thing. But the big hurdle we have there is the consumer markets. There are all kinds of people who would just as soon go to Wal-Mart and buy their stuff there, as opposed to patronizing the local people. Even the local grocery store, when I asked them where their meat comes from, they weren't sure. They buy it from a supplier, and who knows where the supplier gets it from?"

"You have no idea," Judy responds. "Like with the latest meat recall.

How many people got meat from their freezer that came from ConAgra, but it's not really ConAgra, it's really somebody else."

Bonnie nods agreement, adding, "And even as far as the grains go, the sustainable ag women's group is sending a notice around right now. They're concerned about some of the test plots where there are plants that are being grown for therapeutic antibiotic stuff—"

Judy interrupts for an instant, "I'm really worried about that."

"They're really concerned about those fields not being well contained," Bonnie continues, "and it starts cross-pollinating, so all of a sudden you might be eating cornflakes with antibiotics in them that you don't want. That's a big problem. I do think there'd be a good place for that kind of plant use. But it needs to be much better controlled, maybe in a greenhouse. If you can get some help in the health industry with that kind of a thing, I think that sounds like a wonderful idea, but I don't think it sounds like it's being watched."

"Look what happened with Monsanto," Judy agrees, "just that little bit of StarLink."

StarLink corn, a genetically modified organism (GMO), was not approved for human consumption but was allowed as animal feed. It contained elements that could lead to allergic reactions in some humans. Part of the agreement in the Environmental Protection Agency's approval of StarLink for animals was that it be kept separate from other corn. Grown in the United States, it soon began to show up in foodstuffs throughout this country and in the European Union and even Japan. It was traced to contamination from its storage in Nebraska, where corn processed to use in foods was also stored. StarLink was supposed to be removed from the food chain but was still being found in our food aid exports in 2005.[1]

"And down in Mexico," Bonnie says, "the fields were eight hundred miles or farther away from that stuff, and it still got contaminated. And their corn was in zones that were clearly marked 'organic use only.' They should have been protected, but it didn't help."

Many scientists approve GMO crops because no immediate adverse effects on human health have been reported. But Bonnie and Judy have put their fingers on another serious issue with GMO crops, one that has not had as much press: the capacity GMO crops have to contaminate other crops, wreaking havoc on biodiversity, eventually leaving no pure strains. Then all seeds will be in the control of Monsanto and other transnational corporations. Once again, I find myself musing at the thoughtfulness of farm people who do not have access to information readily available to scientists, journalists, and policymakers, yet who understand issues in greater depth than they seem to.

Judy comments,

I guess the only thing I would ask is that—do you think that our only solution is for people to get together? Larry Gates says we've got to recover our respect for the land and for each other. But how do we do that? Most people are too far away from the land. We can't rely on the government to do anything. It just seems that everything is in a scandal. . . . You just don't know what to think. Because you look at the topsoil, yes, you do have to work with all of those things, because if you don't take care of it, it's gone. I don't know how long it takes to build up topsoil, but it takes a long, long time. Like here, the soil is rocky and everything. When you get down to rock, you can't raise anything on these sidehills. It's just sad, how much of it has been lost down the Mississippi.

Like with us now, we're kind of getting down to the end, but we still have the grains because we have our son-in-law who's doing our field-work. He has the big John Deeres and does custom work for people. But we keep him doing kind of what we want—the hay, the rotation—and he's agreeable to it. He doesn't really need the hay, but yet it's better for the land. You get the bigger equipment, and it's bigger than what we had, but what we've got is grain to sell. We do have pasture, and we do have the beef. . . .

We didn't ever raise many soybeans, but we raised some beans, and then we had hay. We used to raise a lot of oats; we still have oats in the bin over where we lived. We're trying to get that cleaned out. When we tried to sell them, they didn't want to buy them. But now there's a market, we're just going to get rid of them. We have chickens and turkeys. Actually, that's another thing too—I tried to talk my husband into raising more chickens, but then, of course, he's doing it, so it's easy for me! Maybe we can find a market for selling our chickens.

"We'll be happy to take some," Bonnie says.
"We've done that," Judy points out.

We used to do that for friends. I did talk to somebody up by Lewiston, and there's a place up there that processes them and then you can sell them. But then, that's the thing—if you raise a whole bunch and then you don't have a market for them, you're left with a whole bunch of chickens, and the turkeys . . . are the same way. If you go to sell a bunch of turkeys in the fall for Thanksgiving, you might get some calls and you might not. The marketing is tough. . . . The marketing is the hardest—you're competing with the Big Cheap [Wal-Mart].

Judy's grandson is getting restless now, tired of hearing the old folks talk, so they head for the door, the boy shouting a bright goodbye.

Bonnie pours more coffee and turns back to an earlier point.

We were talking about losing respect and a spirituality for the land—but it is so involved. And to do it right, you need to start with the kids when they're little. And then I also think of triage. Like when people are in a nasty accident and they do triage, there are some victims that they know are too far beyond it and the thing to do is to work on this one instead because she might live. Should we treat the environment and the society that way? How do we treat our society and environment?

You were talking earlier about finding some people that are altruistic—and yes, we have some of that on our farm, but it's a constant juggle between where you want to be and working with the daily struggles.

"*The way I understood Vance—when you got this place, the hillsides over there were eroded, and not in good shape, and it had been corn and beans for years. It seems like the thing you are struggling with now is how do we get it to come back, so that we can make a decent place out of it. I can't think of much that's more altruistic than that, given the direction of farm programs for the last forty years, which has been pretty much the other way.*"

When we go on a tour, I'll show you that one hillside that we've been working on. The soil thing has been coming up a number of times in the last couple of weeks. Just yesterday again. I think even if you're doing row cropping and you add all the proper phosphorus, K, and urea and nitrogen, you know, unless you're really adding some organic matter back in with it, I think you're still going backwards. Your soil tests may say you've got the proper components in there, but what you really have is the proper chemicals. You don't have anything that keeps the soil alive. Even the best crop guys, unless they're adding organic matter back in—and I'm not a soil specialist, this is just my thinking, undocumented, unstudied—but it just seems to me like that's part of why everything goes backwards. It seems like so much that's going backward.

I remind Bonnie, "That's exactly what Ralph Lentz was saying at Art's field day." Art and Jean Thicke dairy with Ayrshires in one of the steepest bluff land areas of southeast Minnesota. "Remember? Ralph held up that dried-out cow pie and there was stuff sticking out of it all over the place. It was still full of O matter. It's clearly a lot healthier than not having it there." I also mention a story I've heard. "The details may not be quite right, but it seems that a few years ago, DNR [the Department of Natural Resources] *tried to check the amount of runoff on Art's place, partly because the hills are so steep. They had a tank truck, and they just dumped the equivalent of three or four inches of rainfall on one of his fields. They said they couldn't find any water running off—it went straight down into the soil. Art got a little flack from a Wisconsin guy about not having the right kind of grass, and then I found myself thinking, I wonder if 'kind' is really what's critical here, or if it's more about biomass, carbon/nitrogen sequestration* [when the organic matter in the soil retains water, it also holds carbon and nitrogen, "sequestering" essential nutrients for use in growing plants and keeping it out of our rivers], *and holding the soil in place. I mean you have to feed the cows too, but—"*

That's a fun story—I hadn't heard of that particular one, but I think you're right. It's not necessarily the kind. I think that's what a lot of grazers are finally thinking, and we're hearing it from the different grazing groups. It's not only the kind; it's what works for your farm, and I think in some of these cases we may need to utilize the chemicals to get something started growing so we've got the feed, so if we run the cows out there to build up the fertility, you've got something to feed them as you're working on it. That's what we had to do. . . .

You know, we had a rainy year last year. All spring, we had plenty of rain, and Art and Jean's pond never rose an inch. Finally, the ground was just saturated, and then we had that four-inch rain in June on top of it all. I called, and Jean answered the phone, and I asked, "Did your pond rise last night?" She said, "Oh, yes, it came up about an inch." It came up just an inch!

"It really is a cycle, isn't it? It all has to work together or it doesn't work at all."

Yes, I think that's right. Cycles are normal. Last year when you were here, it was wet; this year when you're here, it's dry. We have day and night; we have seasons; the moons come and go; we wake, we sleep; we work, we rest; you know. Food grows, it doesn't grow. That just kind of fits in with all kinds of things, and sometimes we just have to put the food cycles and the farm cycles in with that. It's a cycle either way. It just depends on if we're tipping the cycle to the benefit or if we're tipping the cycle to the disadvantage. It's cycles, no matter what. It just depends on which direction you're going.

I am reminded of Gary Snyder's poem about Lew Welch, a friend of Gary's who disappeared and was never seen again. In the poem, Lou comes back to say, "Teach the children about the cycles. / The life cycles. All the other cycles. / That's what it's all about, and it's all forgot." I'd tell this to Bonnie, but I can't remember the words just right and don't want to botch them up. Instead, I tell her, "I've run into more than one farmer this year that says, 'Well, I don't make all my decisions on the basis of money. If I did, I'd be working in town and making a lot more.' I've also run into folks that say, 'I just want to prove it can be done. They tell me it can't.' So everyone has their own way. And I keep running into folks who just say, 'Well, this works for me—I'm not saying it will work for anybody else. Everybody has to find their own way of doing it.'"

That particular attitude is coming through the grazers a lot more clearly in the last couple of years. To start with, we didn't have many examples to look at. A couple of generations ago, they knew more about it, but they were not rotating as fast. They didn't have to because they had more pasture acres. They had fewer cattle numbers and they were diversified. But now, all of a sudden, there isn't as much out there. Then there was a little more of an attitude: you need to do this, you need to do that—and maybe for starters, you'd better think about this area, that area. But I think the grazers that are really making it are the ones who looked at that and said, "OK, will that apply?" and if so, OK. And if not, "What do I need to do different to make it fit on this farm?"

For instance, here is one of the thoughts at that time: when you got your farm, you could just let the quack grass, or whatever, grow. They got part of this right, because they said, "Whatever grows best on your farm, you let that grow." You should get enough quack grass, wheatlage, whatever. OK, now that may be fine if you have some kind of old pasture or old hay field established to start with. But Vance said, "We just don't dare do that." We had so little pasture then, he said, "I don't know that we can rely just on that." It's really good that he thought of that and said, "We should seed down at least some." So we did seed down about sixty acres with some grass and clovers. Had we not, I think we would have lost the farm the first year.

This was mostly just corn and soybeans—there wasn't enough residue of anything left in the soil there to really grow anything. There was no O matter left after all those chemicals. In fact, there's part of the valley where—well, there were things that grew, but not much . . . It just didn't grow as well as anything else. Not that many other places grew well at all. That was a constant battle for a while. Now it's not as good as I want to see it, but it has improved. I wonder if part of the reason why is because it's lower in the valley where things have fallen out. . . . I'm wondering if there were all kinds of chemical residue from whatever it washed down and just took that much longer to break down as it sat there. I don't know.

"Is that still the case? Are you still getting a lot of runoff from farms above you?"

Yes. And most of them use chemicals. We have seven dams on this farm, and there's one that holds water. The biggest one is at the end of the farm. That one is two or three acres, depending. It overflowed in '97—that was the first year—then again in 2000. Supposedly, it's big enough and built so that it probably would overflow once every twenty-five years. It's already done it twice since we've been here.

Another thing that I think, as far as the water, whenever it goes places, if we could get more of the land in grass—at least in specified places; if not in fields, at least put in more waterways, at the very least—in all these fields. And bigger buffers would keep the water from getting to the lakes and streams as fast, which would keep it from going through the rivers as fast, which would also save us some of the flood problems. I don't think people should be allowed to continue to build on floodplains in the first place. Wouldn't that also help for flood disasters? Think of all the money we spend on that. If we could come back to the roots, so to speak (*we both laugh at her pun*), back to better roots in those places, it just seems like that could help alleviate a lot of problems.

"What puzzles me is that we knew all that back in the thirties—when the Soil Conservation Service was getting cranked up and their guys were going around. Then all of a sudden, we get these huge, record-breaking floods in the nineties, and everybody's looking at each other and saying, 'Whoa, what . . . ?' As if it should be a surprise!"

Well, Gary, in the end it is a surprise, and that's because we've lost some of that memory, and part of that people contact, and we don't have the stories from the other generations as much—you know, the connectedness.

"We don't have the institutions that connect us anymore either," I add. "The Grange doesn't mean as much anymore, and the church doesn't mean that much. We don't stay around there after service, just hang out and swap stories. We tend to go home and watch television. And we don't often get to work together with an older generation that can pass good stories on."

A friend came to us years ago and said, "Help me figure out what else I can do, because I think I'm going to have to sell my cows because my knees bother me so much. He was milking about fifty cows, and his knees bothered him so much he would take aspirin at the beginning of milking, in the middle of milking, and at the end of milking. The end result was, with neighbor help, the grazing group . . . put in his parlor, and I think it cost him $4,300—and that included the steaks for the thank-you party. He does not have a flush system in there—it's only six units—so he takes out one pail of manure every milking. The point is, he's still able to milk, he likes milking, and now his knees and back don't hurt so much. Instead of having two people to milk all the time, he's at one. So his wife went to town and got a job. This works very well for him.

We were talking about the university people. . . . I think that for a long time the grazers have been disappointed that some of the university people didn't see the relevance and the importance of intensive grazing and sustainability to begin with. Honestly, I also believe that it's good they don't grab every new idea real fast. On the other hand, I can understand why people are frustrated and not wanting to listen to the university quite as fast—because I've heard farmers say, "Well, that doesn't make sense. Years ago they told us we would make better beans if we kept our cows in and did all this out in the field and brought the feed to them." It's too bad that we had to go through that step, but we did because we believed their advice, and at that time, because the cows were just standing in the barn or the barnyard, the pastures were getting depleted and nobody understood rotational grazing. That was some of the better economic advice at that time, but maybe then we went another step too far. We didn't remember to keep the blend of the two, and so the rural cropping has gotten to be too much the focus, and a little out of hand . . . from the farming community all the way up through the university system. The pendulum needs to swing back some. I think there's a place for row crops. . . . There's places where it's acceptable.

We keep talking about how important all the connections are, and yes, we can appreciate the work up there [at the university], but it's so valuable for them to come out and see some of the people they're supposed to be helping with that research. But they need to be out here, once a year anyway. Something like that. But it's tough to get them to realize that this may be as important, if not more important, than a biologists' conference up in Toronto.

I was very pleased when we did the streamside grazing workshop last year. We invited a lot of university people, both from Minnesota and Wisconsin, as well as a number of legislative components. A lot of them came. It was great! I was very glad they made the time to come, and it was a lot of fun to show them some of what was going on here. It was even more fun when they had an interest[ed] ear and had intelligent and serious questions. That was really worthwhile. Even if we gripe about the university systems, we need to remember they are people too. . . .

Even here, as far as the systems and the university, the Minnesota Pollution Control Agency [MPCA], the Soil Conservation office go . . . new pollution control rules were going to be made, and we needed to do something in our barnyard. Vance said, "Let's sign up for this and get it done right. We can get it done with their blessing right away—rather than do something and then have it wrong." It seemed like a wise idea, but partway through, in the midst of it, we were thinking that maybe we shouldn't have started, because when we signed up, we were originally approved for forty thousand dollars' worth. They were going to do one drainage tile out there, which we were not pleased about. It was going to be for that—it was going to be for waterline, and seeding of pastures, as well as some kind of manure management around the barn area, all those components for that money. When they did the ratings and comparison of other things, they only had so much money—so ours got cut down twenty-six thousand dollars' worth. We just figured we'd do what we can do.

At that time, we had 130 or 140 cows. That told MPCA, in their computer program, that we're dealing with this much manure in our yard. They were programmed for people who kept their cows confined to the barn or a dirt exercise lot right next to it. They took no account of grazers at all, had no computer programs for that, so that meant we had to come up with this kind of manure management system, even though we had only 25 percent of that. So, four years later, they had learned enough, there was enough attitude change, they said, "OK, if you're a grazer, we understand—you don't have that much manure—we don't need to do that much here." So they did change that.

Originally, what we were hoping to do was some kind of picket dam. There was grass stuff, so that would filter out the liquids, and with the picket dam, we were going to scrape the solids and do the haul whenever we needed to. With that kind of manure that they said we had, that would not be acceptable, so they got into some other plans—a number of scenarios. One plan with the manure system alone was going to be forty thousand dollars. And the rain gutters we had to put on the building to keep the clean water out of the dirty water. The rain gutter thing was going to be eight thousand to eleven thousand dollars, because they had to be commercial. We still ended up with a tremendously expensive plan. I think it's working, and what we ended up with is fine. We abandoned one feedlot. Rather than do rain gutter stuff, we put up a roof structure; we extended our drive-by feeding. It still ended up being too expensive.

"But that's a case of you educating the agencies and the university instead of their helping you learn things," I protest. "Thank God they're willing to learn. Nevertheless, you'd like to think you could call them up and they would come down and help you instead of your having to help them figure things out."

If we could educate the public enough to make them understand, when they accept huge monies from Cargill, it may look like it's saving the university money, but in the long run, where is Cargill getting the money from? The taxpayers. It would still make more sense to not take Cargill's funding, even if you have to raise taxes a little bit. It would still be a better future. And people would be more willing to pay those taxes if they got some real service instead of lip service. Maybe they could understand that we are actually thinking more about long-term effects on the rural community, and not just next year's dollar. So how do we help you tell the story so you can do that?

"One of the reasons I was interested in this, and the board and I talked about it quite a bit, is that I happen to be a real believer in the power of stories. So part of the thing that gnaws on me is whether or not I can tell this story in a way that really would have some effect on folks."

So they understand we're real people and they don't remember us only when we're in a crisis.

"The other thing is, we really want to tell this story in cities. They need to hear it as much as anybody else."

As well as the schools—and don't forget the colleges. Sometimes they will listen.

We leave the house, go down to a four-wheeler waiting in the lane, and mount up. Bonnie drives and I perch on the back. We are embedded in engine noise as we bump downhill, a bit past the barn, and then head up the edge of the pasture, away from the buildings and the field that Vance showed me earlier. The rain begins almost immediately, hard, steady, almost straight down on this so far windless day.

Up there are the milk cows. (*Bonnie twists and points back.*) The north line is just in front of those trees. You can see the buildings are kind of centrally located.

At the top of the knoll, we stop and Bonnie cuts the engine.

This is where our cows are wintered. They come back here because they can get out of the wind. This is what we call our upper winter paddock. Those trees are what you see from down by the road. . . . So in the wintertime, the cows can get around that point of the woods, and be out of the wind. This wasn't grazed at all this spring; this is just what grew up. I'll have the cows graze through most of it, and then we'll clip it.

See those two fields that look like triangles? They're ours. And if you imagine that square of woods in between the triangles—that would be ours. And then there's a dam in between the triangles.

We swing around and head back down. We stop for a bit at the cluster of buildings, pulling into the barn. We are both soaked through, and the rain bangs down on the roof with steely pings. The big barn doors are open. Outside, the rain hits the ground in silver streaks, the drops bouncing high, tiny explosions like bombs cratering the puddles.

This is our barn. Some days it's cleaner, some days it's dirtier. That first winter, we took the stanchions out. Originally we were going to do a flat-barn parlor—eight stations. At that time, Swiss Valley had a program where they would help you finance the bulk tank and the milk equipment, and as long as it met their specifications, they'd give us a really good interest rate. By doing that, it gave us a little extra money, and we decided to put the pit parlor in right away. You can see the design; the holding parlor's in the back.

The last couple of weeks, it's been kind of funny. There have been different cows being the lead cows. Though they come up all right, the first couple of times they do, they're always a bit slower—it slows things down. Most of the time, they come in fine.

The first couple of years we milked without grain in here, and then we added the grain system. That was a good thing, because without it in here, we would feed out in these other bunks. At that time, they didn't all fit around one bunk. That meant partway through milking, you had to shift cows from going out that door to going out this door. And then you either had to deal with putting a little grain out every time, or, if you were going to put more than one feeding in the bunk, you'd run a wire around after a certain number of minutes to make sure nobody ate too much grain, because they can die from that. Those years we had Amish labor. They would milk, but they couldn't drive anything mechanical and we have no horses, so that meant if I was going to be gone, I had to make sure I juggled dealing with grain, as well as whatever else. By putting the grain in here instead, it cost us $5,100, but it was well worth it as far as labor savings.

This farm had two Harveststores [blue silos], and then we bought these bins. See that one pipe sticking out? That's where the slurry store was. So when we bought the farm, there was just the pad for it. That's why, now, our barn water goes down just by gravity flow. This roof stretch is new to us, just put up less than a year ago. That's part of the EQIP program [Environmental Quality Incentive Program], keeping clean water away from the dirty water and dirty water away from the clean water. I think that's really been a good thing for us.

Our plan is to calve seasonal because I like that the best . . . [even though,] economically, that's not the best way to go in the short term. I think it will be in the long run. In the short term, I think the most economic sense, probably, would be spring calving, fall calving. I think it makes good sense because I don't have the buildings and I don't want to have cattle inside all the time. But I don't want to freshen them in the cold winter months (*cows freshen when they begin to lactate after calving*). I've also had some freshen in the summer months, and I find I don't do well with those cattle that freshen from the end of June through July and early August. I don't know if it's because of the grass or because I'm tired from freshening in the spring—what it is—but I seldom do well with anything that freshens in those months. And I certainly don't like freshening in the winter. The cows stay out most of the time.

The older building—beyond this new one—when we moved here was four feet shorter, so Vance came up with a plan whereby we cut it in half and he raised it up four feet. Now we can get the tractor in. What we do is, on the worst days, which is usually the wind, or if it's icy, we'll give them a choice of staying in or going out. When it's really cold, as long as it's not windy, they'll go out. They would rather go back outside. . . . And they have the trees over there.

So our cows are outside except on the worst days. Occasionally, if we have a blizzard coming in, I might lock them in . . . just because I don't want to battle the drifts in the dark the next morning. I do get my four-wheeler hung up in the snow occasionally, but not too often. The heifers and dry cows are always out; they don't come in. I think they stay cleaner, even through freshening, when the cows freshen in the spring. I like for them to freshen outside. . . . If it's too muddy and I know one is getting close, I might bring her closer up, but depending on how I'm doing with the grass paddocks, I may have a number of different groups. I may have my milking group—probably two milking groups—and then I might have another group that looks like they're really close to freshening. If I can, I will keep them in an area that's closer to the building, just because I check on them a lot more often. I also run them through the barn after the main milk herd, at least once a day, so I can look at them. If it's heifers, they get used to coming through. They get used to the noise. I don't very often have anybody freshen in the other group that's farther out—where it might be, depending which field, that may or may not be a little muddier. If I can freshen them on the grass, I like that so much better than being stuck in the building, but if it gets too wet or too cold and windy, like March can be sometimes, I might have those close-ups in for a little bit. It just depends on how many are in the group and what the weather is. . . .

One of the things I've noticed about grazing is that pasture is so much healthier for the cows, not only when the grazing is good but during heat stress and drought as well. When it's really cold out, digesting hay heats cows up. So when it's really hot and droughty, I've found that it's more comfortable for cows to be on pasture instead of hay. If it's hot and the cows are fed on hay, they begin to pant. So that's another reason to keep them out. I have a friend who says that, in normal weather, when her horses are on pasture, they come to the stock tank for water three times a day. When it's hot and they're on hay, they're at the tank all day.

"So it's judgment calls all the time."

Yes. Vance gets really frustrated: "Why do we do things different all the time?" I don't try to be difficult, but I may be planning on doing this, and this week "this" works. But if next week, the cattle numbers in the different groups have shifted, then all of a sudden it makes sense to do it this other way. I can't blame him for being frustrated; he has a valid gripe. (*Bonnie begins to laugh. She has an infectious laugh, and I can't help but join in.*) But that's just the way it is (*right through her laughter*). He'll just have to live with it.

With all this work here last year, I really appreciated the EQIP program.

"Tell me more about that."

We did get cost sharing from both state and federal, but the EQIP is with the federal, the NRCS [Natural Resources Conservation Service] office. They cost-shared different amounts for different practices. This particular place they cost-shared at 75 percent. So I'm one of the people that benefited from the program, but it was because of our intermittent stream at the bottom here, which is mostly a dry run most of the time. (*Here her voice takes on a sing-song lilt, as if she is reciting an "I have been bad, but I promise not to do it again" speech for the teacher.*) I will not be polluting the waters of the state, because we have all this wonderful stuff here—not just the grass but that huge slurry thing that takes care of the wash water from the barn. (*She grins broadly.*) Even though I complain about the Minnesota Pollution Control Agency, or any pollution control agency, that's only when you get the people that deal with only the book smarts and not the practicality. I'm also very appreciative that our area has gotten some of these policies and laws in place because, let's face it, not all farmers are conscientious, and we do need to be able to put our hooks into some of those practices and pull them out.

Bonnie interrupts her thought to point out a cat, black and white, slinking like a hunter in the wet grass. They also have two black and white dogs, border collies that delight in chasing the four-wheeler as we drive, barking at the top of their lungs.

They do help me with the cows too, you know. (*Bonnie laughs. Haugens milk Holsteins.*) Everything on the place has to match. (*Bonnie laughs again, then returns to her earlier thought.*)

That doesn't mean that it's fun or easy, but I do think we need to do that. And I also think there are some cases where even the best of farmers, when they do those big expansions and have those huge lagoons and huge pits, even the best farmers can end up causing problems, just because the manure containment system has problems. So I'm glad there are policies and laws in place—even though I've grumbled about them.

When we bottle-fed our calves a couple years ago, we did make little pens in there. We did barrel-feed them too, for a couple of years. I think those first heifer calves that we raised, we raised on straight milk. . . . They always seemed more solid and vigorous than the milk replacer. I think because of the Johne's [a wasting disease that can infect cows, pronounced "yo-nees"], we'll stick with the replacer. . . .

This is the first year we've had corn down here (*pointing to another small field*). When this was farmed before, the normal field came clear down here to the bottom of the hill, so this is one hill that had the heck farmed out of it. . . . It obviously didn't grow very good grass either. So then a few falls, I'd have the dry cows and heifers, or I had some steers out there, and we'd feed them—roll the bales down the hill—and keep them out here as long as we could, depending on how the winter got to be. Then the next spring, it looked so much better. Last summer, a year ago June, in one forty-eight-hour period, the cows produced over seven thousand pounds of milk off of there. They produced more because they were out there for more than forty-eight hours, and they had a little bit of silage and a little bit of grain, but they got the bulk of their feed from that hillside, which wouldn't have produced anything for a few other people, you know.

Beyond the corn, you see the dirt road? That's the front of the dam that holds water.

"So Vance said you run your paddocks on the contour."

Yes. Yep, yep. Each high-tensile wire strip is about 150 yards wide, and then we subdivide it with the poly wires, of course. You can see those fence posts. . . . That wire was higher, but . . . we happened to put the lane not far from where the water naturally ran, and we didn't realize that the first couple of years, but we had a lot of erosion on the lane, to the extent that we decided we'd move the lane. So there's an old one over there, and I'm working on this new one here. Now we've got to get rocks. (*Bonnie laughs ruefully.*) So we too can be dumb!

At the end of this valley is where I was saying that I think a lot of stuff had been washed down there—we had a hard time getting that to grow much of anything.

It looks pretty good now.

Yeah, in fact, we had reed canary [a grass species with a root system that makes it especially valuable for holding water in steep country] that started this year, and it is doing much better.

On hot summer days, I will send the cows up to wherever they need to be, and then, at a certain time of the day, or depending on temperatures or whatever, . . . let them down. We have bigger paddocks from where you see the trees start down to the end of the valley, so the cows have plenty of shade. We rotate through them as need be. We have shade available when they need it.

We have a musk thistle problem, and it certainly has increased in the last couple of years, so we try to do a little spraying and a little clipping—and we do a lot of sputtering. Let's go up the hill. . . .

The rain has let up, so Bonnie cranks up the four-wheeler and we start out for the pasture where the cows are. Fifty yards from the barn, the rain starts again. Harder. We pass a little swale where the grass is deep.

I scared up some baby pheasants here last night.

We move higher and stop at the hilltop near the cows. From here we have a panoramic view of the farm. Thunder rolls and echoes all around. The rain picks up again, lightning punctuates the gray sky, and across the fields, all the farm buildings are ghostly shadows behind the scrim of rain.

Talking about karst geology, see that different grass there, where that little dip is? There's a spring area there. Around on this hill, there's a spring area where, years ago, some farmer must have decided to access the water, because there's an old tank and piping there. Straight across the hill in that area—where we often have our calves—I told you there was wild rice in there, there's a spring box up there. Farther along there, remember where the reed canary was, just below the corn? There's a patch of trees we call the willows. That's very springy in there. What we see is all about the same level, a few dips in there, and right behind the house we have more springs.

"You're essentially talking about the same level of rock, bedrock, though."

It seems that way. The long-term plan is to figure out how exactly to develop some of these springs for natural water in here.

"That would help, wouldn't it?"

Not only would it be less demand on the well, but you'd have variation in the length of time of the year when you could have cattle out here. You can see, between the valley and up there, that's an area that wouldn't have been farmed with any kind of row crops. Part of it was, but not all of it, otherwise there wouldn't have been trees on that hillside. But there, too, I have fed animals there, usually in the fall—either early spring or late fall of the year. I'm working hard at getting the fertility up. That's one of the things we do talk about. Until we get more organic matter in this soil, it's not going to be good, productive. . . . We'll see. Now you can see a little more of that hillside.

Bonnie fires up the vehicle again and we lurch off, the four-wheeler slewing on the wet grass. "When I think about what that must have looked like just ten or fifteen years ago, that's pretty impressive!"

It's good for me to be reminded how tough it was to get started in the first place. We want this to have more legumes in it.

We reach a low spot, the "intermittent creek."

Now you know it's been dry this year. If it had been even average wet, there might be a little trickle running through here.

We swing around a tight corner in the lane and pull up where we can see long rows of hay wrapped in white plastic wrap. This is new technology since I knew anything about farming, but I have seen it in many places. I ask if the plastic is a pain to use; I have an image of some machine stuffing hay into a white plastic sleeve, the way a sausage maker fills sausage.

No, we have a machine that just wraps it and wraps it. Some friends use bags, like stuffing sausage, yes, but you have to have a pretty clear idea of exactly how much hay you're going to have; otherwise you fill one half full, hit the end of the hay, and waste all the rest of the sleeve. So we decided to wrap. Wrapping has worked out pretty good for us, and we find the cows waste less of a bale of baleage. Even a big round bale like that, there's still going to be waste. I suppose if you put the pencil to it, the question is, Do you do the round bales and pay for a building—so you don't have to worry about losing waste there—as opposed to the plastic? However, it's much easier to get the baleage up, because when you're baling, you don't need the hay as dry. So, in the spring and early summer, when it often rains more, you don't need it as dry to put it up. I think we're going to waste less of the feedstuff. In the long run, we're going to waste less using baleage over trying to get dry hay, and [we'll have] better quality with the baleage rather than trying to get your dry hay.

We've pretty much seen the farm. I mention that Vance had said he thought one or two of the kids were interested in pursuing farming.

One of the things, a couple of years ago when I was having a really hard time personally, and things would get frustrating on the farm, and I was getting more and more fascinated with more and more things, I talked to the kids and I said, "You're working too hard too, aren't you? Maybe we should just sell the whole thing." And they would just come and they would say, "Mom, settle down. Settle down. We're OK, we're not working too hard. We don't want you to do this. We know you won't be happy without your cows, just settle down." Indeed, while two of them were in high school and one of them in grade school, all three of them would come and say that, and I think that if we do work too hard here sometimes, and we do, we've got to be doing something right, because all three of those kids have that interest in working and doing things on the farm—not every day, of course, because they're kids—and each one of them has some interest in doing some kind of farming afterwards. They're young enough that we don't know what their future will be, of course, but I think just the idea that they weren't really anxious to get out of here and never come back [is a good sign, since] there are some dairy farms where the kids just can't wait to get out of there.

I think about how satisfying that recognition must be to a parent, but I also compare it to the indicator of success many folks would look for—how wealthy their children were getting. I look again at Bonnie. She is soaked through, her short hair plastered down and water dripping into her eyes, off her nose and chin. She drives down the lane, parks the four-wheeler in the barn, dismounts, begins to wipe it down. Watching her, I remember cowboys I knew in Montana who always took care of their horses before they took care of themselves, and who knew their cows and their country better than I know my backyard.

CHAPTER 4

Farming Is a Spiritual Responsibility

MIKE RUPPRECHT

It's April, but you couldn't tell that from the weather. This year, it is still winter. I drive through the milky translucence of a winter day with snow falling onto the snow on the roofs of barns, falling onto the snow on the trees and on the ground, turning the whole landscape into a pale, mostly white, minimalist painting. Maybe the artist today wanted to paint light but didn't want to overdo it, or perhaps just ran out of color.

I am headed for Earth Be Glad Farm near Lewiston, Minnesota, to talk with Mike Rupprecht, a beef grower who also raises chickens. Mike comes out as I pull in to park behind the house. Mike is tall and slender with a quiet demeanor. The Rupprechts are a close family, and it won't cause a squabble if I tell you that his sister told me, "Mike is pretty quiet, a man of few words." We shake hands and go in to sit at the kitchen table to talk and sip tea. When I ask Mike about the future of agriculture, he responds simply.

I think we're doomed.

End of comment. I think for a moment the interview is over, that I should say thanks and leave. Mike's remark is stark enough, but when pressed, his reasons for that statement are pretty clear too. They are both social and environmental.

The key is going to be the consumers, and now the public is brainwashed by corporations. . . . I don't know how this earth will survive. We have been so hard on it. We cannot go on like this. Mother Nature always bats last.

Mike is a rotational grazer, which means his land, like the Haugens', is

fenced off in paddocks, and he allows the cattle to graze each one inten-
sively; then he moves them into another paddock. On Mike's rotation,
paddocks get a rest for thirty to forty days, till the grass comes back, and
then they are grazed again. Mike has mapped out his whole system and
has arranged it so his cattle can get to water no matter which paddock
they occupy. His hay and pasture hold the rain with the least possible
runoff.

I'm the fifth generation in this area. My great-great-grandfather,
Wilhelm Rupprecht, came from Germany and built a water-powered
mill for lumber and wheat. Ever since I was a kid, it's been fascinating
for me to go down there. I love to go down there and just think about
things, think about the history of this land. But bad farming resulted
in flooding and finally closed the mill. You can read about it in *Pio-*
neers Forever, a book by Marvin Simon.

It's true that Mike speaks softly. Perhaps because he is very thoughtful,
his sentences tend to be short. But he looks me directly in the eye, and there
is both power and passion in his quiet demeanor. He clearly cares, not just
about this place but about agriculture, and ultimately about the whole
earth, and the world we are leaving for our children and grandchildren.

It was a crime to let the topsoil erode. We're not going to go on
very long if our soils are depleted. That's why I have so much pasture
on my farm. We raise Angus beef. Also chickens; we have eighty laying hens. The whole key is that they need to be on grass. What's killing us isn't beef; it's the cholesterol from all the hydrogenated oily
foods, and the sugar.

Mike asserts that profit is not his primary goal.

We have 160 acres, and we rent a hundred or so from Mom. We
want to prove it can be done—that you can make a good living on a
small farm. And not just that we can do it, but that others can too. . . .
It has to work for more than just us. I only think about economics on
this farm from the standpoint of making a living. We have to make a
living or we'll lose the place. But I don't make all my decisions on the
basis of money. If I had to do that, I wouldn't farm.

The land here is continuing to improve. We use no chemicals, little
fuel, no till on our pastures, and reseed. Nature is telling us it's right
too—the songbirds are back. Now we are certified organic.

Despite that good news, Mike worries that his efforts, and those of his
colleagues who follow similar good practices, may not be enough. He be-
lieves not enough folks care, not enough follow good practice, not enough
really think about what they are doing to the land, the water, the future.

Farmers Talking about Farming

Mike and his family attend church regularly, he tells me, but he adds that he doesn't hear much in church about stewardship of the earth or taking care of God's gifts to us. It is not a common topic of conversation among other farmers in the congregation. Nevertheless, that church tie lends a spiritual dimension to his farming.

Knowing what happened to this valley, another part of this is a spiritual responsibility to take care of this gift that we have been given. I don't know exactly where it came from, but I know it's in my heart now. I can't change what's happening in the rest of the world, but I can control this farm. The message we all got, the message the European settlers all got, was dominion over the earth; conquer the wilds. Actually, we killed the Indians; we wanted their lands even though they were much better stewards than we. So it's more spiritual [for them]; they lived with nature very well. The message I like to get out of scripture, and you have to get back to the original texts, is that we are supposed to be stewards, take care of the earth. It only makes sense. I mean, don't we want to have a nice place for our children and grandchildren? Why should we be the last ones to use up the topsoil or the fossil fuel? What do we expect our grandchildren to live on? But I'm in the minority on that understanding within our churches. I mean, they probably understand it, but they don't place the importance on it. There's more importance placed on salvation and the Easter story. Someday we'll probably live on a perfect earth. To me, that would be heaven, you know? But there's more to it than just saying, "Believe in Jesus and you will be saved." Well, I need to do the best I can at whatever I'm responsible for in daily life and what needs to be pursued, and so spiritual life and sustainability, it's all connected for me.

"One thing that impresses me," I say, "is that the Old Testament prophets are very clear about what happens if the Israelites don't do right by their moral code. They said two things would happen: One is that God would send the Ammonites or the Philistines or somebody with a big sword, and they would cut everybody to pieces. The language is very graphic about that. The other thing was that God would turn this beautiful land into a desert where nothing would grow and everyone would starve."

So he'll turn the whole place into a desert as punishment? Or does God wish we wouldn't be screwing it up so bad and we would have kept it as a promised land, a land flowing with milk and honey? I think that part of this whole scriptural thing or biblical thing is that people are looking forward to a new heaven and a new earth, therefore we're not going to be here forever anyway. It doesn't say anything about "OK, what about our children (*laughing*) and grandchildren?" We don't know how long it's going to go on. So I know that's the view: just make sure everybody's heard the message about Jesus, and don't care about the earth. But the care of the earth is actually the care of humankind. So we need to care about our food system. So much of our health comes through our nutrition system. I was talking to a customer of ours in Rochester last week. I asked her, "Don't these doctors get this? Don't they want to eat healthy food, free of pesticides or that doesn't have drugs in it?" She said, "No, they think the answer's in the next new pill." I think she's right.

As I am backing the car out to leave Mike and Jennifer's place, Ralph Lentz's words echo in my head like a litany of praise: "On the good side, Gary, you can see it. There are a lot of farmers in this area who care. You can see it." Thinking about what I've heard from Mike, I'd have to say he's one who cares, and because I've had an opportunity to visit with others of the region's farmers, I know that Ralph is right: there are a lot of people who care.

CHAPTER 5

Timelines

RON SCHERBRING

The blufflands just north and west of Winona, Minnesota, rise above the Mississippi in steep hills. A few miles inland, the country is barely beginning to ease up a bit, has not yet relaxed into the rolling hills that appear just a few miles farther west.

Rollingstone is nestled in these steep, July-green hills so typical of karst topography. This town of about seven hundred has an impressive museum, a handsome high-spired church with carefully trimmed lawns, and a very neatly kept public park, complete with an immaculately groomed baseball diamond. The morning I was there, though they had built it, no one had yet come. The housing off Main Street is equally well kept and nicely painted. About half the buildings on Main Street, a block long, are brick, and they add an air of permanence, a dimension of character to the town, as if to say it was built to last despite the precarious terrain around it.

I'm early for my meeting with Ron Scherbring, so I go into Bonnie Rae's, the town's only café, and have some coffee. The menu sports an array of hamburgers (one-quarter to one-half pound), cheeseburgers, and diverse other sandwiches. Dinners range from $6.25 to $6.95, depending on whether you get three pieces of fish, hamburger steak, chicken strips, an eight-ounce New York strip, a six-ounce rib eye, or batter-fried shrimp. You can also get a side of "cheese bombs" for $2.95. A sign posted near the icebox announces, "Desserts are Homemade." Today's desserts are cookies, "fresh homemade pie"—apple and banana cream—and Snickers cheesecake. I'm happy to settle in here; it's my kind of place, and I'm comfortable here. Though everyone who comes in takes a second, quizzical look at the stranger, no one pays any further attention. They don't seem to mind my just waiting, reading, making notes, drinking coffee. I am a chain drinker, and the waitress is attentive.

I recognize Ron when he comes in, though I've never seen him before. He has an air of authority about him, one I expected from our phone conversations and because it fits his role as president of the Southeast Minnesota Ag Alliance. He's also the only guy who's come in looking as if he's looking for someone. Ron is a regular here in Bonnie Rae's. Everyone in Rollingstone is a regular at Bonnie Rae's. Ron introduces me to Dave Wardwell, the owner and the man behind the cash register. Bonnie Wardwell works the tables and is the one being generous with the coffee. Both serve as "chief cook and bottle washer," depending on the traffic, each doing whatever is necessary to keep things moving. Ron asks Dave if we can sit someplace quiet, and he seats us in another room, one we have all to ourselves.

We settle in over coffee, and I tell Ron what I'm up to—trying to tell the farm story of southeast Minnesota and beyond, not writing yet, but interviewing a variety of farmers, faculty, and extension agents, trying to educate myself. He listens carefully, then begins.

One of the things I've been doing lately is really looking back. I love timelines. When I look back at a timeline, I have a better sense of what's happening today. I was taught, but also believe, that history is the best lesson.

In one regard, Ron's historical view echoes H. C. Hinrichs's: the importance of the shift that took place in the early fifties. He offers a brief synopsis of our agricultural history.

We all talk about the thirties era, when we had the Depression, and we talk about the forties, when the war was taking place. And then we lose track of it and we jump over to the seventies. The fifties were really the transformation age, where we really lost the horses in agricultural power. We did have some tractors before that, but we really separated ourselves then. By the end of the fifties and early sixties, we pretty much went from horsepower to tractor power.

That midcentury point marked a kind of watershed in Hinrichs's view, and it does in Ron's as well.

There was a group of people that really began to know what they could do when they moved from one setting to the other setting in tractor power. So we started out with fifty or sixty horse tractors in the early sixties, and we ended up with 160 horsepower. . . . A lot of those people really fell in love and embraced a concept that we could be a production era. Not everyone bought into the newer systems. It was really hard for some to change.

Perhaps the real change in the last half of the twentieth century lay not in technology, important as that was, or in the sociology of farming, but in the mind-set of farmers.

To me, two kinds of farmers emerged. We went through the farm crisis. In my opinion, when I'd look and talk to a lot of the people that I work with, there are two types of farmers. There were farmers who really held fast and hard to the traditional way of farming and thinking. We were beginning to be told in the early eighties and by the nineties that we were businesspeople, we were businessmen and -women involved in production of some sort. There's a group of people that really embrace that, but very quietly. They embrace the fact that, yes, they needed to learn business skills and they needed to do those things that were not quite traditional.

"And the second kind," I prompt. "What about the ones who decided they didn't need to be businesspeople?"

Well, they more or less just probably didn't . . . think that change was necessary. I think there are a lot more in business. I think some of them have even modified somewhere in between. They'll do some business tactics, but it has its limitations. That has really helped me understand who's involved, who would make good colleagues. Some of the folks I work with, and that are the most fun, are the ones who say, "Yeah, it's OK to change, we can do things different."

Ron was one of those who embraced the idea that agriculture is a business. He did not reject the idea that farming is a way of life, but he recognized that it had to become more business oriented if current practitioners were to survive. Farmers in an Experiment meeting in Lanesboro in 2002 agreed. When one urban participant suggested that farming was a "way of life," all the farmers nodded in accord, but their response was deeply qualified: "Well, yes . . . but it sure is a business too."

One of the things that my brother experienced was getting out of agriculture because it wasn't profitable. He knew that in the hog business, he was going to compete with larger and larger hog operations. He didn't think his skills were there . . . but to admit that in an agricultural setting was terrible. Now, if he was running a printing business and wanted to get into selling shoes or something else, yeah, great. Nobody would care. So we have this stigma that would instill people to be quite quiet. So I think that's where the real difference is in different philosophies in agriculture.

Ron took considerable pleasure in the shift. He seems to enjoy the challenge of figuring out how to make the farm pay, and he makes it clear that he thinks the future lies in farming as a very serious, growing business. There is a kind of exuberance in his words.

The nineties were, to me, a fun period of time because we could take those business philosophies, those business principles, and apply technology to them. I did not believe that in one point in time, I would own a laptop computer. My business thoughts throughout the day and week and months would be a mixture of what I wanted and believed to do, along with what my computer said was right and feasible to do. That takes place now on a regular basis. It took, for a lot of people, basically the nineties to get through that and figure it out and embrace it. We're still working on that.

So I think . . . it gives you a background how everything . . . kind of fits into place, as we think about it. I think the industry is challenging. I love the agricultural industry. I think it's the most exciting place anyone can be. And one of the reasons is there is so much uncharted territory out there that we need to really come to grips with, and not so much from a philosophical standpoint, but a real standpoint. Is a larger farm contaminating the earth and the world? Is profit bad in agriculture? Is profit the only thing we should embrace in agriculture? Of course, you have both ends of the spectrum. The environment . . . it's just tremendous, the effect that we will have and leave in the environment. And it's constantly changing. You can write those environmental regs, and then new concepts and ideas come out and you have to rewrite them. Not only rewrite them, but then reeducate the mass of the people to embrace them.

Ron is among the many farmers I've met in Minnesota who not only farm but work hard for their community too.

I think everyone has a tremendous amount of responsibility, not only to themselves and to their operations, but again to the community and to society, to embrace a system that does work and that they're comfortable living with, because you have to live with yourself. But I think people like yourself talking about these stories and telling about things, it's . . . essential to do. I don't think we're going to get very far by pointing fingers or with a dialogue that is really negative. We don't have the luxury of society participating in that kind of work. Although that's the one that always gets the media attention and it's easiest to write, I'm really kind of against that kind of stuff.

"I'm glad to hear you say that," I tell Ron, trying to explain my role (and understand it myself) in this farm story project. I'm glad to hear it because part of my job is to tell as comprehensive and accurate a story as possible, so I really need to talk to commodity growers and agribusiness farmers as well as sustainable folks and organic farmers. And one of the things that strikes me about the small farms that are still making it is that, whether or not it is business that's driving them, their undertaking is as entrepreneurial as one can imagine. "I mean," I say to Ron, "they're just scrambling to find markets, to find transportation. . . . I like that distinction between the folks who say it's all right to change." I have yet to meet somebody who doesn't have his or her own personal twist on how to make a farm work. "Everybody is doing something just a little bit different, and some of it is really exciting! One of the reasons I love this project is that I think there are some stories from farms in the region that are appropriate for the whole country."

That's absolutely correct.

I talked with Dick Levins, an ag economist at the University of Minnesota. He said, "One of the things that farmers need is for folks to like farmers again. I can write all the ag stuff I want, but it's always going to be technical, and it's always going to be dry. We have to figure out some way to put some heart into the farm story." I tell this to Ron. "I can't do these stories and not like the people who are telling them, whether you agree with them or not." Ron picks up on that.

You know, that is powerful, and I agree with it, both Levins and you. My wife comes from Chicago, so she does not have a major background in agriculture. Her grandmother did. And it was interesting as we developed our courtship and she got introduced to agriculture. She would go back to Chicago—she was in college here in Winona at St. Teresa's—and talk about her boyfriend and that he was in farming, and the comments that were made toward agriculture were basically somewhat negative. . . . They weren't the warm-feeling "oh, that's a wonderful career." They were very negative. And rightfully so. I think it's partly because of history. When people left the farm in the past, they did it under severely, very, very hardship circumstances.

"So people hung on too long and they suffered for it."

Yes, that was the way of life. It was a basic primal instinct to be able to raise your own food and raise your own cattle and do that, and if you failed at the basic job of sustaining life, what did that say about you? So, yeah, they did hang on too long. Either that or they worked too hard and had strokes or heart attacks. So when people left the farm, it was quite tragic, in most cases. And they just plain didn't like the work. The work was hard and long. Even today, that is what most people believe agriculture is all about. The challenging thing is, there's a bunch of us—we don't live that life. I don't live that life of hardship. I did. But I made a choice, my life was not going to be tales of gloom and doom and stuff like that. But agriculture has not been looked upon as the occupation where that can be your first choice; it's usually your last.

There are numerous farmer-organized associations throughout the Midwest: the statewide Sustainable Farming Association, for instance, with numerous regional offspring, and the Practical Farmers of Iowa. The Land Stewardship Project has farmer members, as do the National Farmers Union and the American Farm Bureau. The Southeast Minnesota Ag Alliance seems unique. I ask Ron how the Ag Alliance came to be.

Well, [it] was formed when about eighty people got together. And we basically sat around and talked about the idea. We were tired of negativism. We were tired of everybody looking at the industry in quite a negative way. We understood that there were so many different types of agriculture. And they *all* should be positive. We didn't have an issue with organic farming; we don't have an issue with small operations; we don't have issues with large farms, or whatever. We don't like to think of them as "family farms"; we like to think of it as families being able to operate agricultural entities. That's a little different, because with "family farm," you're basically referring to the land. We're out to foster more than that, and to reduce stress on farmers.

Ron takes pride in his practical, reality-oriented approach to agriculture. Nevertheless, he is eager to get right into principles and an underlying philosophy, and he articulates them with an assurance that can come only from long consideration and careful thought.

We did identify three things that we thought we needed to hold pretty fast to. The first one is that agriculture is part of a business, not a social event, not a government event. It should stay in private industry and it should be on a solid business basis for its existence. One of the things we need in business is that people need to be able to do the essential things so that they can grow with their businesses. . . . Now with modernization, you could be seventy and still want to actively operate your farm, but maybe you want to make it so that a son or daughter could even be part of agriculture too. So we want to grow. We want a profit. We think people should not live like paupers. It's OK to have some money; it's not a bad thing to make a few bucks. Now, we're not saying that all farmers should just get choking rich and live like corporate executives . . . especially lately! (*Ron concludes with a laugh.*)

Though growth and technology are now readily sought and utilized, Ron notes that there are other serious issues for farmers.

The other things that we in the Ag Alliance feel very strongly about are that we need to be in tune with the environment. We need to save soils; we need to talk openly about all of these things. We have recognized there's just not one thing that's going to do it. Agriculture and the industrial age are developed by many concepts, but really focus on the facts, and the fact is we need to save the soil. We do not need to pollute the streams. We think that with the technology that's out there and with a little education, that when people on the farms spend that money, whether it's cutting the trees up the stream bank and having grass so erosion stops, whether it's somebody going through your Farm Service office or [the Natural Resources] Conservation Service and requesting their technical support in developing waterways or re-stripping the farms, it's really essential that we talk about that and that we work with that.

Ron likes the diversity of approaches to farming that are represented in our area, though he and others feel some criticism of their operations coming from farmers who take a different approach. I tell him part of my purpose is to include everyone in this farm story—commodities, organics, sustainable farmers, large operators and small. Ron agrees with that purpose.

In a later interview with Gene Speltz, a dairyman friend and neighbor of Ron's and another member of the Ag Alliance, I mention this necessity the Experiment feels. Gene replies, "You bet. We need everybody —we need the whole range. I think that's one important thing that we need to realize—we do need all these people, but we still all need to work together. And I think that's where a little bit of conflict comes in. The organic says, 'You're not doing your job out here.' Well, regulations say we are. Everybody wants cheap food—and they're getting cheap food. We have to put chemicals on to keep the weeds down, to keep the bugs out. If we didn't do that, our corn would only be this high, and it would proba-

bly take twice as many acres to feed our cows, which would pull away from the selling of the crops and the food that feeds the country so well. Then maybe the price of food would just double.

"There's a balance there that they need to realize—we're trying to be as efficient as we can. We want as little acres as we can to feed our cows, which is normal, and that's what we're trying to do. We're going by regulations—we can't go against them. We know it's better for us too. There's no doubt. . . . But we've been efficient. . . . We've been doing real well here," Gene concludes.

I ask Ron what the program of the Ag Alliance is.

Educating means providing our citizens with information. Sometimes we realize that some of the information that's provided out there is slighted; it's not whole. And that is done so that somebody can get across a particular viewpoint, or particular ideology—and that could be political; it could be nonpolitical. But . . . see, the American public is very sharp. If they get the whole story, once they get all of it, they can decide who's farming up- and downhill, stuff like that. But we need to tell the whole story—that yes, there's a place for all of us. So, out of that, we said, "OK, how are we going to do this?" Well, we're going to do it through education. And our main focus is taking those materials that currently exist, putting them in a format so that the local people that are in our local communities in southeast Minnesota can understand the real facts. How much agriculture is leaving the area? You do have a facts sheet. And it's from Winona County. This is where the averages are; this is where it's done. When farmers tell other farmers their story, they have a lot more validity. But it's just that we need farmers to start talking. When I see two farmers at the local café or filling station and they're talking about farm issues, I often wonder why that conversation shouldn't be positive. And we're starting to see where some of it is.

This is an area where Ron has something in common with Peggy Thomas and Ralph Lentz, two farmers who align themselves with the sustainable agriculture movement in Minnesota. Both Peggy and Ralph are eager to describe the Sustainable Farming Association of Minnesota as a means of "talking farmer to farmer." Ralph explains that he means that to be more inclusive. "If we only talk to sustainable farmers, where are we? We're just talking to the choir. We need conventional farmers, organic farmers, everybody." That's one reason both Peggy and Ralph sponsor field days on their farms and work on field days with others. At *these events, farmers show other farmers what they're doing and how, sharing both information and ideas. Ron picks it up again.*

We've worked with ag in the classroom. We have an educational plan. We're trying to get farmers into the schools. We want to get rid of some of those old images. If you don't have your bib overalls on, are you a farmer or not? We just got done now with our fair. We had a booth at a local fair.

My story would be that both of my sons, so far, have chosen not to farm. . . . Those of us who are really into business . . . see and understand what it takes to at least hold our own, meaning financially not be going backwards, and what we have to physically do to challenge ourselves so that we can stay there and grow and protect the environment. You start to look at people like a good businessman when he starts hiring people as employees for his business. He can talk to them for five minutes and he'll know. I've had ten apply, and that one over there—you know he's going to do it. Writers, or anyone, they have that feel. There are not a lot of people yet, because we have moved so far and so fast. And part of this is because of the speed.

Here Ron echoes Eric Freyfogle in The New Agrarianism.

It's just the whole society, conceptually, hasn't been able to deal with it. Education systems—you can talk to anybody at the university, and they will tell you how fast their research has moved away. . . . That used to not be the case. But we've diversified to move that fast. My two sons, one graduated from Gustavus [Gustavus Adolphus College]; he graduated with a degree in physics, had a desire through school to work in astrophysics. He's now decided to work for HBCI [Hiawatha Broadband Communications]. He has a love for computers. He's greatly involved in Internet communications. He has a great job. My youngest one is in the theater. He loves to put on plays—to do the sound and lighting and all those kind of things. He goes to St. Mary's [St. Mary's University of Minnesota]. They just looked at farming as a very active, intensive business. For my wife and I, farming's just something we get off on. We really enjoy it.

"I like what you said about keeping the soil, because if we're not doing that, we're like a businessperson who's not keeping up the inventory. Soil is where it all starts, regardless of the kind of farming you're going to do. If you can't grow stuff anymore, you can't keep the farm."

Well, there's no need for soil to fill up our permanent wetlands, or for our soils to be filling up our riverbeds. That's just flat-out not acceptable. I don't think we need to do it at the expense of cropping. For example, in my operation, we live in a very hilly part, just four miles out of town here. And my dad was a big believer in conservation since the later fifties, starting with soil and water conservation. We continued to . . . work with that, but actually, my work, since Dad's conservation efforts, it was pretty much done, and I was merely involved in the manure management side of it. But with the latest technologies, we will not plow our hay ground up in the fall, we'll spray it with Roundup. If we're going to reduce tillage, we're just going to do it. On the hillsides, we'll put the alfalfa, the clover, back into it the following year. So we have a one-year rotation, instead of a two-year. . . .

We even use the weeds and the grasses to control soils. You can work with those kind of programs, and still expect 150 bushel yield, not put every chemical on it. We just use a simple chemical called Roundup. That may be controversial, maybe not. You just spray once. If you do it right and time it right, and do whatever tillage you need, use lots of organic fertilizer . . . Not only the cattle we have, but they spread it all over the land, and we even give some of it away to neighbors. And they begin to realize the value of that.

"Do you raise dairy cows or beef?"

Actually, it's dairy heifers. What we do is Scherbring Heifer Hotels, that's the name of our business. We were in the dairy business. I started in 1972, and in 1992, [I had] the accident—I was injured by a herd bull. . . . After I had one surgery, and another one transpired after that, . . . we made it a point, basically, to see if we could find something to do that was less strenuous, less stressful. That would be great. So my wife got a job in town, and we started raising dairy steers and a few heifers for some local folks. Then we started getting more involved with the heifers because the business for the heifers grew. So, in 1993–1994, we started building facilities to house heifers. We actually kind of hung our shingle out. We got rid of the dairy steers and we just started raising all heifers. We raise baby calves for eighteen other dairy farms right now.

"How long do you keep them?"

Well, we'll keep them all the way to pre-calving. Most of the animals are for six months, and then they usually will move on back to the farms or to other growers or to work into our other growing systems. . . . Most of them are Holsteins. There're some crossbreeds in there. Some dairy farms are starting to work with a little more crossbred.

"That's interesting to me, because my grandfather had Guernseys and Jerseys. In my lifetime, they've all but disappeared."

Well, the Jersey-Holstein cross has really come of age. A couple weeks ago I heard a nickname, Ho-Jos. And Ho-Jos are a big thing.

I remind Ron of an earlier comment. "You said the Ag Alliance was pretty idealistic. That's a thing I just keep running into. I've been impressed talking to folks. There's this idealistic streak in practically everybody I've talked to. And it's not anything that's real self-conscious. It just kind of shows up around the edges and reveals itself inadvertently."

There is. I understand that. I see it all the time. . . . I know a couple of guys who have actively gone out—they're very good with equipment—and they have developed a custom harvesting and planting segment to their farm. You would never know it. They don't have a lot of clientele; they don't advertise in the paper. But they do a wonderful job with it, and it would be the same for them too. What you're saying is that some of the successes are best kept secret. It's like the military—you can't get it out of anybody.

And the other thing that plays a part in this, that I know from myself: we're probably the last group of people in America that is so ingrained with the religious background. You see a lot of these people go to their churches, whichever denomination, it's irrelevant. Around here, they will go to their churches, and be a part of their schools. That's a great thing. Religion teaches you to be humble and respectful. So you don't go out there and crawl all over somebody else's failures. You like to see everybody come along, and that's something that needs to be reflected. You never bragged, you never flaunted money around. Humility was a virtue, and even though you thought you were doing good, you thanked the Lord and prayed, and you moved in those manners. And that's done in other businesses, but I'm just saying it's more of a pattern in agriculture.

The other thing is, we work with nature. I don't care who you are, you can be hailed out, or have tragedy—you won't be on your high horse too long. Yes, that's built into the system too.

We have long finished our lunch, I've drunk my customary amount of coffee—way too much—and it is time to wrap up.

I'll be very happy to participate in this. I will make a note to have you invited to our next meeting. I'm not sure of the exact time and date . . .

As we step into the other room to settle up and leave Bonnie Rae's, Ron sees Gene Speltz, another Ag Alliance member, having lunch with his daughter and her friend. Ron introduces us, and Gene agrees to talk to me about his farm. I stop at the cash register to pay my tab and tell Dave that I will be back to try that half-pound hamburger. He says, "Well, you'll find us right here," and laughs.

CHAPTER 6

The Absolute Last Thing I Ever Dreamed I'd Be Doing

LONNY AND SANDY DIETZ

Most of our city planners and many agriculture scientists would declare that the highland ridges above the Whitewater River in southeast Minnesota are "marginal lands." They're not suited for townhouse development or for growing commodity crops like corn and beans. But that's one reason Lonny and Sandy Dietz found them attractive. They had something else in mind, and they didn't want to be dependent, beholden to the federal government's farm bill that supplements the income of farmers who, for one reason or another, raise corn and soybeans.

After I find my way up steep gravel and pull into the farm at the end of the road, I learn that one thing about this place that appealed to the Dietzes was that it was situated here at the end of the road. That would limit development. Another was that it seemed well suited to their hopes for a diversified operation that would work on sustainable principles and that could offer a measure of self-sufficiency and an income from farm-to-market direct sales of healthy vegetables and meat. Another appealing feature, one not commonly sought, was that the land had been pretty much abused. Part of the Dietzes' purpose was to restore at least a small piece of southeast Minnesota to ecological health.

Lonny: Why get into this? I wanted to try to help. It kills you to see all these farms going under. We looked around, driving all over the country, studied plat maps and aerial photos, for four years before we bought this place. We thought the hill road would protect us here, and we could restore and protect the place, but development comes here too.

The farm had a history, as Lonny explains, of alternating good practice and poor. Its most recent owner had let the place run down.

When we moved in here, it was covered with brush. We moved 350 stumps from just around here. (*Lonny waves his arm in a quarter-circle sweep.*) The original owner had a very diverse farm and kept it up real well, but the next farmer just let it run down. (*Lonny points out a couple of shaggy goats standing in front of the shed.*) We had forty goats to start with. Sandy wanted to do some weaving.

We step up on the porch, enter the house, and sit down at the circular kitchen table. Sandy puts on coffee and joins us. Lonny mentions the goats again.

Sandy: I was pretty excited, but the market just went down. Now we're down to two—a couple of geriatric cases. (*She smiles.*)

Lonny: There's a couple of sows out there too. We usually have twenty or so.

Sandy: We sell them in halves or wholes.

I ask about marketing the vegetables.

Lonny: We have a small CSA [Community Supported Agriculture] going for us. Last year we averaged twenty-five members, mostly in Rochester, a few out here. We try to keep it on a personal note. We're doing a little different. Most direct sales or Community Supported Agriculture operations set up in one spot, and everybody comes to them, but we can deliver to offices in Rochester, where we have several buyers in the same building. We sell to Winona farmers' market, too, and some to the Twin Cities. It's always been a little different down here, the country is so steep. Customers try things [vegetables] they wouldn't have otherwise because of our diversity. We put a letter in their basket every week so they know what they've got and how to fix it. Next step is to get the vegetable production down [meaning "down pat," not down in amount of production], maybe expand into bigger fields—maybe ten acres of carrots, because there are some markets out there. But we want to get the small system down first.

We talk a bit about the difficulties of farming with care for the land, the animals, the humans on it, and all the related species who depend upon the natural systems around us. Both Lonny and Sandy are interested in social justice issues as well as farming and ecology.

Sandy: There's something wrong with a system where, to keep our level of life, we have to have poverty nations.

Lonny: We're starting to be a third world country in the ag business. John Ikerd [an agriculture economist] sure got that right. . . . The Netherlands are over here in northern Minnesota because our environmental rules are more lax than their own. So they invest in paper mills up north and hogs in Ohio, where they can have factory farms that they can't have at home.

What it comes down to is that you can choose to do harm or you can choose to do good. It's the hundred monkey story: There are a hundred monkeys on two islands isolated from each other. The monkeys on one island learn some good thing, and when enough have learned it, the monkeys on the other island learn it too, without ever having contact with the other monkeys. You reach a certain number, and the tide turns.

Sandy: It takes each individual to make these changes.

Lonny: We've lost the art of observation. . . . We don't see natural systems the way we did when we were hunter-gatherers.

Lonny and Sandy watch their landscape carefully, looking for signs. They describe some of the ways in which they are now trying to restore their ridges to health, the things they have tried to do to grow healthy food. But the first thing that they mention is a marker of success they take real pleasure in.

Lonny: We've got a couple of eagles, red tail hawks, owls, songbirds. We've restored four acres of native grasses. Now we want to get some forbs in. We just started doing greenhouse. We had strawberries the first year, but they didn't work so well. Now we do early crops. Six acres of vegetables outside: lettuces; potatoes; some broccoli; cabbages; a lot more cooking greens, like kale, chard; shell, snap, and snow peas; green beans; beets; radishes; onions. We can't find organic onion sets, so we have to start those from seeds. We try to carry some potatoes over for seed, but the volume we go through makes it impractical to carry many over.

We have thirteen acres in set-aside. On the steep hills, we'll eventually do some rotational grazing of a few beef. And eventually we'll have fifty acres of row crops. We did it on custom, mainly small grains. Those are organic too. But it is hard to find a custom guy who will clean his equipment carefully enough to protect the organic. . . . We bought the place in '91, but it was all in CRP [Conservation Reserve Program] the first six years.

Lonny had worked for thirty years as an automotive engineer.

This is the absolute last thing I ever dreamed I'd be doing. Last year, I quit my job, so now we have to learn fast.

We're into this to prove it can be done. So many people told us it can't be done. [But] Mark [Ritchie, director of the Institute for Agriculture and Trade Policy], Dick [Levins, ag economist at the University of Minnesota], John [Ikerd, professor emeritus of agricultural economics at the University of Missouri]—they give you hope. You're running into walls everywhere; so many people are saying it can't be done. The idea out there is that to succeed, you have to grow. But when prices are low, you have to stay small. Extension preaches that you have to get bigger, so now, instead of five hundred farms, there are fifty, and we don't need the extension agents anymore.

One problem is that the emphasis is still on monocropping. Ag classes in the high school are all geared toward agribusiness. There is no hands-on farming. As Ralph Lentz says, they do "animal science" now instead of "animal husbandry."

Lonny goes back to his discussion of their general plan.

We've got our wildlife area now, row crops, trees planted, prairie restoration. Set up a whole farm ecosystem. It's all necessary. First thing was prairie, then retention ponds, then trees. We're trying to build live soil. Compost teas. Create "live food" full of enzymes and such, so we can get a healthier crop from healthier soil. Get more off less land, so there is room for wildlife.

Then Lonny takes the challenge right back to the city planners and the ag scientists I mentioned at the beginning of this story.

What's to stop somebody from making a living off marginal land? It just gives you more options. (*Lonny smiles.*)

Like numerous others I've met, Lonny and Sandy both give time and energy to causes beyond the farm. They have been active in various farm-support activities. They both attended the early meetings of the Southeast Minnesota Food Network, and they still participate. The Food Network helps local producers market and transport their food in a coordinated effort to make the region more food sufficient.

Sandy: The Food Network is exciting.

They agree that the Food Network is important for local food producers but also comment how important it is for the community to have access to food that is healthier than most of the food that is available in supermarkets. Even if it is not processed, much of it is transported over long distances, from sources we do not know—perhaps Mexico, where foods are subject to pesticides and herbicides that are not allowed here.

One year, Lonny was a board member of the regional Sustainable Farming Association (SFA). Now he has become the president of the statewide association.

Lonny: SFA can help the whole process. So many people are getting out of farming week after week. I want to make a difference, get the word out there that there is this hope, and I think that is the avenue for me to do that. Otherwise, it was just a weak moment.

Lonny and Sandy have two daughters and a son.

Lonny: [Our son] does chickens, ducks, and turkeys, a few laying hens. He really likes the live things. (*Lonny laughs.*) Has no use for vegetables. He wants to get into pheasants and quail too, for restaurants.

We go back outside and duck into the hoop houses, with white vinyl taut over the plastic hoops. Though it is cold outside—it was below zero last night—it is warm in these glassless greenhouses.

Sandy: In the hoop houses, it is eight degrees warmer under the row cover. It was fourteen below last night, but in the greenhouse—the one without a heater—it is sixty degrees above at eleven A.M. In the other, it's seventy-two degrees. The hoops have a single layer of plastic. You can have it about four or five degrees warmer if you use two layers of plastic, but then you lose 10 percent of your sunlight, so it's hard to figure what's more profitable.

Lonny: We learned something about single-layer last fall. The two you're looking at used to be sixty feet long. Now they are thirty. The wind did that. The new hoop house is longer and wider—and sturdier.

Lonny has that characteristic smile as he tells the story, as if to say, well, yes, those things happen, but we're learning. I marvel at the midwinter starts for a wide variety of vegetables, at the idealism and altruism of this couple tackling marginal land and working so hard to make it healthy as well as profitable, and at the systematic way they have of experimenting with their whole ecosystem. I head back down the hill thinking about the motivation that drives them and others I know, hoping as much as they do that it will all work out.

CHAPTER 7

I Just Felt It Was the Right Thing to Do

DENNIS RABE

It seems as though Dennis and Sue Rabe (pronounced "Ray-bee") have tried it all—conventional farming, high-production farming, value-added products, and direct marketing to individuals and to farmers' markets—and now are focused on pigs and beef cattle. They follow a rotational grazing pattern for the cattle and a Swedish deep-straw system for the hogs. Behind those changes and the evolution of their current methods lies a continuing desire to farm smarter, take better care of their animals, reduce inputs, have more time with their family, and not have to work so hard.

Sue, after farming for years, now teaches almost full-time, a matter of choice rather than necessity. "Farming is Dennis's thing," she tells me. "He just loves it. I don't! Three hours on the tractor and I'm bored silly! My work is different, and I do it because that is what I'm supposed to do with my life. So he gets to do what he loves, and I do too."

In the course of our conversation, what comes through from both of them is a profound desire to get it right: to achieve a balance on the farm that takes care of the land, the animals, and the family, and that generates enough income to provide for their needs and perhaps a bit more besides. Dennis and I settle in over good coffee across the counter from one another in the Rabes' attractive kitchen.

> As our children started getting older, we found ourselves getting older too! And working longer and longer hours . . . When we started looking at that, that's when we really started wondering about our system.

It's not something where you wake up overnight and say, "Hey, this is wrong." You grow into it. I farmed pretty conventionally for ten, twelve years and started questioning some of the things. I know, I suppose the late eighties, I was told that if I didn't market at least a thousand hogs, I couldn't be a player. That's really the minimum! We had a system where we had the crates, the hot nursery, the cold nursery, and we were farrowing once a month and really pushing out hogs. . . . Then I was pasture farrowing to add even more hog inventory, and I had rented two finishing barns, and when I got done at the end of the year, I had an eighty-thousand-dollar feed bill that I paid. There was nothing left for me! So why go to all that work?

And that's what I see with factory farming or volume farming. They're always looking at, if you make a buck a hog, for instance, all you have to do is sell forty-five thousand. I didn't want that; I just didn't want it. So I started looking around. What could I do? How could I change some things? Maybe we're kind of in the minority of the people that want a change and actually have a life. . . . But Ralph Lentz was kind of seeing some of the same things that I was seeing.

Ralph, who farms a few miles away, is a cattleman, a rotational grazer who is always paying attention to his soil, his grass, his creek. He taught agriculture in the local high school for a generation or more, and he clearly had an impact on his students—his name comes up in conversations around the region more often than any other. Dennis, like other students, has remained a friend of Ralph's through the years. He often judges his work according to what he thinks Ralph would think, still wondering what grade Ralph would give him for his work on this chore or that. Ralph likes to talk about his teaching and does so with considerable modesty. A few years ago, Ralph's heart stopped. The two emergency medical crewmen who showed up were former students. Ralph grins and swears that he heard—though the doctors say he couldn't hear anything—one of them say to the other, "I don't know, this guy didn't give us very good grades. You think we should start him up again?" Ralph's laugh affirms that they must have, perhaps still hoping for better grades. Dennis was the first one to visit Ralph in the hospital.

We were real different—Ralph was retired and there were no mortgages there, but he was kind of watching me, with a family and mortgages to pay, and [we] couldn't keep up. So that's why I've kind of looked at some other methods.

"But you pretty much devised your own scheme for doing it."

Well, there were a lot of things that went along that were all related. I learned about pasture farrowing on my own; I wanted to expand at that time. Actually, there were times it worked unbelievably well for me. One year, the sows that I took off of pasture farrowed; they were gilts, and there were three groups that came back into my crate system that weaned over ten pigs per crate. But the thing that really killed me was the market all the time. Three pretty good years, and then, all of a sudden, you have a real bad one.

Ninety-eight is when it went down, and you know, it hasn't recovered since. And it's not going to, because 80–90 percent of the hogs are contracted. That bad year in '98 put me in a position where my system was worn out, my crates were worn out, and I looked at myself and I looked at my age. . . . I didn't want to be in my fifties and washing crates! It took me four days a month! When the price of pork went down to eight cents a pound, I just said I wouldn't do it anymore. I started looking, and that's when I came up with the Niman's market. So it wasn't anything specific. . . . It just kind of evolved.

"Tell me more about the Niman's market."

The Niman's market is an antibiotic-free market, and there's a certain protocol. One of the things that is different is that the Animal Welfare Institute sanctions it. . . . Your breeding stock has got to be on straw or pasture. They all farrow on straw or pasture. There's no use of crates, unless you get an animal that's very unruly, then you just use a crate for a minimum amount of time, two or three days, until she settles down after farrowing. They will allow something like that. Most of us just farrow, and if you get an animal like that, she goes to town anyway. Ninety-eight was when I first signed up and started, and now, the last three years, I've been selling through them steadily.

"They're a private corporation?"

"Niman's" is Niman Ranch Pork. It actually started with the Niman's Gourmet Meats in Oakland. They started with beef and pork—both organic, all organic. It goes to high-end, upscale restaurants mainly, and it also goes to small grocery stores now too. They raised their own pork, but they weren't satisfied with it. They tried some of Paul Willis's. Paul is the one that got this started. He's a farmer in Iowa. It started with one shifting to him and working with him until there's two hundred Niman's farmers now.

"But you're not a contractor?"

No, we are not a contractor. No.

"How does that work? You just raise them and ship them and they buy them?"

We have a forty-cent floor. We're never lower than forty, which is higher than any of your contracts. The other thing is, they want heavy hogs, 260–280 pounds, so that's kind of an advantage because it's easier to put that last twenty pounds on than it is to raise a lot more hogs. And they have to be antibiotic free, of course. You can't feed any meat and bone meal or any animal by-products to them.

"So there's no danger of something like mad cow disease getting into hogs."

Right. And when interior Iowa gets forty cents, we get a six-cent premium; we're at forty-six cents, so the floor protects us. Like now, hogs on the conventional market are twenty-seven cents. Last week they shipped just fifteen hogs, and there was over six hundred dollars' difference between the conventional market [and] Niman's. So that floor is just huge!

"So one of the reasons they can make it work is because restaurants and everybody else are willing to pay for the quality."

They're paying for the quality, and they're raising what they feel is as environmentally safe a product as you can make it. There's a lot of deep-straw systems out there. Most of them will copy the Swedish system, where you're raising the pigs on deep straw. You don't have pits, you don't have liquid manure, because it's all absorbed in the straw, and we spread it back out on the land. It works out pretty well.

"So that's your mulch, your ground cover, and some nutrients to boot."

Yes. It holds most of the nutrients. What I like about it is it's just a much easier way to raise hogs, and they're more content and quiet. And I have way less work! (*Dennis laughs and shakes his head as if he can't believe his luck.*)

"Sounds like a great combination! Not only that, but you're making more money per pound."

There are advantages and disadvantages to both systems—and I'll recognize that. When you go antibiotic free, your feed efficiency [goes] down—meaning it takes more feed to make a pound of gain—but I'll take the higher pound, as long as I got the forty-cent floor.

"That more than pays for the feed."

Especially [in] the last few years, when feed has been so cheap.

"Is it going to stay that way?"

Oh yeah. We're going to see a little bump now, because basically the

U.S. has had a poor grain crop this year [2002]. Corn and bean farmers are going to make a little money, but you know that everybody is
going to grow as much or more as they can—if they make any money
at all next year. If we get a normal year, that price will be under two
dollars in no time.

"Will that be mostly corn and beans?"

Yes. The government program too, that's all they talked about was
the corn and soybean base.

"I went to that FSA [Farm Security Administration] *meeting in
Mazeppa where they were going to explain the new farm bill, and—"*

You did, huh? No conservation was there!

*"The moderator said right at the beginning, 'If you're doing grazing or
vegetables, forget it!' It wasn't until the end of the question-and-answer
period when some guy sitting across the table from me asked, 'What if we
want to follow some more conservation-minded practices?' The moderator said, 'Well, there's a new provision in the bill about conservation. Any
other questions?' Just dismissed it! That was the only mention of conservation the whole evening! It was entirely devoted to 'How do you fill out
the forms so they will show the most acres for corn and beans* [called "the
base"—on which all federal payments are calculated]*?' And a tiny
pinch of time for oats."*

Oats was two and a half cents, $1.20 for most farmers, at sixty bush-
el an acre. There was no conservation. I went to Wabasha and they
didn't even mention it! Not once! What it stated was, the more bushels
of corn and soybeans that qualify, the bigger your direct payment will
be, and disregard any conservation whatsoever.

*"I was talking with George Polk, the soils man who lives in Rochester.
He mentioned oats and how good they are for the soil. I said, 'Why doesn't
anybody raise them?' He said, 'Well, you can't make any money on them.'
So whether it's good for the soil or not doesn't matter."*

The other thing is, when you take livestock off the farm, there's no
way to use oats. You have to sell those oats in the conventional market.
I raise oats and dump two to three hundred pounds of oats in every
ton of feed. Well, that way I'm selling my oats for the same price as my
corn or maybe even higher. So that's a heck of an option! The other
thing is the straw. A lot of times, the straw will make up the difference.
Actually, it'd be a better crop than corn or soybeans. But you can't always count on that.

"So how many acres of oats do you have?"

We run around forty to forty-five.

"That's enough to mix with everything else?"

Actually, in the last few years, I've been gaining a bin every year. So this year the price of oats was higher, and I sold my oats. It wasn't high, but it was better than it's been for the last couple of years. It's like everything else; it's just a juggling game on the farm, and I had a poor farrowing December, so there are fewer hogs. I needed to have some money, so the oats went to town because I had plenty of oats in the granary. Sometimes that's just the way it is!

"That's got to be a pretty serious kind of juggling."

Oh, I'm always juggling, you know. You always have to figure what your best option is.

"I guess that's one thing that's impressed me when talking to folks around here. It isn't often very explicit, but it's just a continuous 'make a choice' day after day after day after day."

Yeah. Yeah, you don't want to lose too many of them. (*Dennis begins to laugh.*) Actually, there's a couple of big losses out here. I'll show you a couple of my mistakes.

"Are you still doing the value-added stuff you were doing?"

Sue and I talked about that, and we decided to let that go. There are a number of reasons. First off, we have the home 160 paid for, and then we became grandparents, so there are other priorities in our lives. The kids, the youngest is now twenty-one, and they're all pretty much out on their own, so there isn't any help here. The other thing is my wife had gone back to school and is teaching 50–90 percent in the RN program. Well, then we got some benefits, and that health coverage was just huge! So I couldn't find a reason to try and just kill myself.

You know, the value-added is great! You make more money per animal, but the time you put in it is really tough! I always thought we'd be able to develop a market where we could be delivering more pounds, but it takes a very special farmer to go ahead and get that done. I think Dave Minar might be successful at it. But he did have a few more things involved than just that. He had the location. We sold at the farmers' market for seven years. . . . When you start talking value-added, on a small basis it works terrifically well. But if you start trying to do that 100 percent, a lot of these people are driving themselves into the ground. I watch some of these vegetable people, and I just—

There is a knock on the door. Dennis answers, steps out on the back deck to visit a few minutes, then comes back in.

Just a seed salesman. I made a decision about four years ago that I wasn't going to buy any what I call "expensive seed," and his wasn't expensive, but it didn't perform that well for me. But I haven't seen a big yield difference in a $50 bag versus $120 bag. I figured, well, the $50 bag, I can pay for my planting, plus have $20 in my pocket. And hailed-on corn (*we both start to laugh*)—it doesn't matter how much you spend for that seed!

"*For sure . . . So do you have lots of corn and beans?*"

Just what I need to finish out about six hundred hogs a year. That's what I try for, that and about twenty steers. . . . It's right around that fifty-five acres of corn and forty acres of beans.

"*And that all gets ground up along with the oats?*"

Most of the time. This summer, I didn't even realize it, but I had extra beans on hand—no, in storage. I'm on a program where I take them to a guy, he stores them for us free, and then we pay a set fee for extrusion and he brings some back. Well, I had more beans, and I called him up and he said, "Why don't you sell about three hundred bushels?" And I hit just the right day! I got $5.80. So once in a while you sell a little, but it isn't that profitable overall. A lot of times I'd sell some hay. I don't try to grow extra crops; it seems I get burned every time.

I'd been down on the Mississippi at Red Wing, watching the barges load up with grain. Cargill was shipping corn from a previous year, holding it back for a better price. Dennis knows how the farm bill underwrites Cargill's expenses and sells surplus grain to the transnational corporations at a cut rate—grain that our taxes paid agribusiness commodity producers to raise. Because they get it so cheaply, Cargill and the other traders can sell it in the foreign markets of poor countries at prices below local farmers' costs, squeezing them out of business. It's called agri-dumping.

Right. Your question is, who is going to benefit the most from this farm program? The chemical and the seed companies, and the person that buys the corn—it will set them up for six years. . . . And there's no way to educate the people fast enough to stop it. First off, they don't want to be educated. Everything is going pretty smooth, and there's always food.

"*What do you see happening in the next twenty-five years in farming?*" *I ask.*

I don't see anything good. I see it getting bigger. . . . Way fewer farmers—a lot more erosion.

"*One of the things that's scariest to me is over-pumping of ground water for irrigation—which has supported an awful lot of agriculture.*

And that's happening not only here, but China, India, other places are all drawing down aquifers faster than they can replenish themselves. I just think, What are we doing to ourselves? It seems bizarre to me."

Well, I guess if we use it up fast enough, we'll be done.

"But man, is that a bleak future! And we're not exempt. We don't have any more food security than anyone else in the world, at this point."

It's interesting too—I was listening to an economist. He was talking about food security, and he was totally against any help toward the farmer, and his thought was, We can import our food cheaper anyway, so why raise it?

"So then we're dependent upon everybody else in the world for our food, and that will be good?"

Yes. It puts you in the position where—just like when we have trouble with the oil countries. All of a sudden we're really looking and scratching and trying to wake up and . . . why would anyone say such a stupid thing?

We have a long way to go. I remember in the eighties too—I got the *National Hog Farmer,* and I read about all the disasters they had with their manure systems then. Fish kills for miles! I thought, That won't come here. But here it is. Right now. Just the fish kills you hear about, they're only about 10 percent. It just makes me sick when you think we're not supposed to eat the fish. . . .

My dad went to ag school in the thirties or maybe early forties. They taught all of these different things then. But I think what really threw us over the boat in agriculture—and we all thought it was wonderful—was the chemicals. And once we'd gone from the chemicals that help weed control, then we got into all these monocultures—and we got to the point where we could control just about anything, anytime, like now with Roundup Ready.

You don't have to be a farmer to farm anymore. An example: I knew this guy in Bellechester who rented 160 acres—it was all corn, it wasn't plowed. So he hired someone to no-till beans in, which is conservation minded. Then he hired the elevator to spray the things, then he hired somebody else to combine the beans, and then he hauled them to Red Wing. And he has the equipment to haul it away. He made about eighty dollars an acre—it was about ten to twelve thousand dollars. But all he did was coordinate. It was nothing but telephone farming. Telephone farming.

"Yes. Gates [a farmer and Department of Natural Resources hydrologist we both know] *says we're on a two-track system: One track is corn*

and beans. He says you can't call those guys farmers; they're 'producers.' The other track is still farming, still diversified, raising animals and feed and vegetables instead of commodities. It seems to me, from what I've read about poultry, that's about the scariest contract anybody's doing. Those guys have you hook, line, and sinker. They can tell you exactly what to do and when to do it, and if you don't, you lose your contract. Plus, you lose the investments you've made in the buildings they required you to build to get the contract in the first place. And you can't get another contract or sell the buildings, because they put the word out that you are uncooperative or inefficient, sort of like the old Hollywood blacklist. So no other contractor will take you on, and you're bankrupt."

Yes, and I think your hog contracts are the exact same way now. If your hogs don't come in on certain specifications, they'll come right out and tell you what to do with your buildings and how to change your breeding stock, and even what feed to use.

"So we've got that track, and then we have farmers still, like yourself. As far as I can see, contracts lead to indentured servitude. You don't have any freedom; you don't have any control; you just do what somebody else tells you to do. I can't imagine spending a whole life doing what somebody else tells me."

That's really true in the hog industry— I know two fellows that in '98, their ledgers went up over a million dollars at Hormel. They are not independent farmers anymore—they're just enslaved farmers to Hormel to try and get this paid off.

Two of the things I see, though: If it all becomes corporate farming, they're going to make a profit. But if they can see they're losing money by all this erosion, it might be corrected. The other thing is, if we finally get to the point where we have enough eroded land, they may come back and say we need to have the smaller farmers.

"I guess that's one of the few things we can be hopeful about. You can only look at destruction so long until you decide you have to change things."

Dennis laughs. Can you imagine having corn and beans on Art's farm?

Art's farm is on very steep land. "I love the jokes I heard from guys standing around listening to Art that winter field day. One guy said, 'Man, you couldn't get me on a tractor anywhere near this place!' And Art's little quip about coming out one morning and the tractor's gone. Looked all over, and found it on top of the tree at the bottom of the hill."

They have found a number of things in the woods, though. Art said that one year he lost a round bale—and they were seventy-five dollars apiece.

"*Man, that is steep country! That's one of the good things, I guess, about karst topography. It does limit what you can do with corn and beans.*"

Yes, but you go over in the west, Florence area, and that's all corn and soybeans—it's just unreal!

"*I've seen corn and bean fields this spring where the rows ran right up and down the hill, and you could see the mud moving right off to the bottom of the field.*"

Yes. Every cornfield is just like a flood plain. This summer we had somewhere between twenty-six and thirty inches of rain. It pounded hard, and then it was smooth, and then you get three to six inches of water on it—it has to move. I saw water coming out of places I've never seen before.

I hand Dennis a photo of a cornfield's erosion that I took just last week along Highway 50. "The corn isn't really good corn, but it's six feet high, and even as high as the cornfields are now, you can still see erosion like that. And that rain that we had five, ten days ago, it looked just like this at the bottom of the cornfield, except the corn is still standing up. I don't know how many tons of dirt have moved off of that hillside this spring, but I wish I had a camera big enough to pick up the two or three other places just like this in the same field."

But you know, you were looking for this, and the average consumer, out for a drive, doesn't see this at all. Even if you handed them this picture, they'd say, "Oh, that's a nice picture, but what is it?" They don't know—they absolutely do not know. You can see gullies a foot or more deep, cut right down.

"*Those gullies are really deep.*"

And that's not even the whole part—the sheet erosion across the whole field is unreal!

"*Absolutely. If we had another six or eight inches of rain, these things would have been two feet deep and you would have had real serious gullies.*"

It hadn't rained that much. We try real hard—and I had some gullies like this, even with all the conservation. It didn't leave the farm, but . . . I can't imagine, if the whole thing would have been in corn or beans, what would have happened. That was a test this year.

"*You know, the other thing that puzzles me is what we've forgotten. When your dad went to college, the Soil Conservation Service was just*" *getting cranked up. I can remember those guys going around and talking to people. We knew then how to avoid floods, how to strip crops, how to contour, how to keep pasture, how to do all the things you need to do to keep the water on the ground instead of running off. And then it seems like we either decided we didn't have to do that anymore or we could ignore what we knew, or we just forgot—I don't know what!"*

Yes, there's just no interest, and unlike the thirties, the government is not going to make anybody do anything about conservation. To me, one of the simplest ways we could do something—like in our county, you should have at least a three-crop rotation, meaning you'd have to have some forage, or at least oats. Why couldn't they do that if you want government payments?

"They could."

But they won't. It's the almighty power dollar! That's what bugs me the most!

"Actually, sitting through that meeting the other night, it seemed to me there wasn't all that much difference, in spite of all the ballyhoo between the previous farm bill and this one. They aren't all that different."

I didn't see any. The only thing I saw was the previous one. Yes, it has direct payments and gives you flexibility—all that did was build everybody's base. And now, we have the direct payment and we also have that counter payment when the price goes low. Well, they're guaranteed $2.60 for corn and $5.80 for beans. There's no way these guys aren't going to go after it.

"No, and the whole point of the meeting was to help guys figure out how to get more and more acres on the form."

Wabasha County, they figure 161 bushels is our average. You take that times the $2.60, what I understand is the guaranteed price—that's $418. Now, if you have 161 bushels, you get the twenty-eight cents on top of that, that's another $45. I talked to a crop guy the other day, and he said it looks like he will average $500 with his corn under this new program. I'd grow it too for $500 an acre, if I was just interested in money.

"Well, a lot of folks are. That's the other part of it. Behind the farm bill, there's a whole mind-set: This land is here to be used for our profit, our benefit. You can do anything to it that you want because, somehow or other, it's going to take care of itself. So that's what we do. We invent stuff for the farm bill, make a few folks rich, and whittle away on the pool of resources that we have. I don't know what we think is going to fix it up. It seems clear that it's easy to fix, actually. A little ground cover goes a long

way in stopping that kind of erosion. A little manure goes a long way to-ward healthy soil. It's not as if there was some great mystery about how to make it work better. But it seems the folks in power don't want it to work better."

Like the agribusinesses too . . . I asked Pioneer Seed Company one time, when I was a salesman, why didn't they fork more money into grasses? No interest. Or oats? No interest. "Not enough money there; we'll let the university do that research."

"I can remember my grandfather—I don't know where this memory comes from because he died when I was about eight—but I can remember him and his cronies. Two big memories I have: One, the neighbors used to come to his place and they'd all sit down in front of a Philco radio—Grandpa had the only radio around, one of those old cathedral-window-shaped radios—and listen to the radio and talk. The other one is how often they talked about the new hybrid seeds—who was for them and who was against them. I don't recall anymore, but I do remember long conversations about whether hybrids were a good idea or not. I can remember my granddad standing up for a hybrid that I haven't heard of since his day. It was called McNally's Hybrid Seeds. He had a sign, just like the ones you see now. They probably went under in '39 or '40, I don't know. But it was a huge issue for those Iowa farmers then, and now it's all we've got."

If you watch things . . . it's real interesting on how the power dollar—they'll make everything work. You know how there was a big fight about the hormones in the milk—nobody says anything any-more. It's not even mentioned. Nobody pays any attention. You have a few that will buy organic milk. We buy milk that's not supposed to have anything in it. But that took a lot of dollars to get that through. The same thing is going to happen with your genetically modified crops.

"The only thing that's saved us from ourselves so far is Europe—Latin America, even. The whole section in central Mexico that was set off for hundreds of purebred corn varieties and no GMOs—now it's infected with GMO pollen. It all blows much further than Monsanto said it would. Seems like it's going to sweep the world."

There are two sides to it, though. Your Roundup Ready beans—when you see somebody no-tilling them in—there's an antierosion piece there that's good. And I recognize that.

Well, should we go outside and take a look?

We put on the already muddy boots we'd left in the back hallway, step

*out the back door, cross the road, and head out toward the barns and the
surrounding fields. The clouds are heavy, and though it is not raining
now, it soon will be. We walk down to look at his barn and pens. I ask
Dennis, "So, what kind of rotation do you actually run?"*

Well, it boils down to a lot of different ones. . . . We usually have
about a minimum of three, up to seven or eight years of forage. Then
it will go corn, soybeans, and then oats, and then hay again, depending
on where it is. There are a couple of fields—this one is real flat, and we
run that as a corn-soybean rotation. Too close to the barn, it gets too
much manure. It's so productive, I don't know if I can put it into hay.
I've had it in hay already, and that's productive too, but it's just kind of
handy. When we go out there, you'll see a bunch of strips, and that's all
highly erodable, so we keep that in forage as much as possible. The
forage—rotational grazing—builds the land up for a couple years or
more. . . .

To make this operation real simple: the cattle pay for the mortgage
and hogs pay the month by month.

"How many cows?"

We have sixty-five cows. I'm kind of figuring to wind up around
eighty, and to use the beef cycle whenever it's profitable. I'll build
breeding stock when the prices are low.

*We walk down to a barn, and Dennis describes his farrowing system.
He moves mothers and small pigs from one deep-straw pen to another as
they grow bigger. After they are weaned, the new pigs move out into pasture until they are ready for sale.*

This barn was originally set up as a confinement system, and now
we run it on a straw system. And this is the result over here. They farrow in there, and then this is what you would call the deep-bedding
system—where they take care of their young kind of on their own.
And you know (*Dennis laughs again at his good fortune*), it's just so
darn easy. They start throwing the tenth of August and I put in three
bales, and other than that, I haven't done any cleaning or anything for
them. The feed is a self-feeder, and they're happy and the little ones are
happy. My average number of pigs is less than with crates, but not that
significantly.

"What is the difference?"

We're running right around that eight, nine mark.

"And you were getting ten out of the crates?"

Not consistently, though. There are times where eight or nine will
be down to six with this system, but with crates, I had six too. It's real

hard to compare. I think in the long run, as you keep your breeding stock that does well under your system, I think that can be improved. You know, they just stay in nice shape, and . . . it's interesting to watch them. Watch their tails—they're pretty content and happy and just kind of fun. I like that! The other thing is, what equipment do I have invested, other than the pole building? I don't have any crates, and all the fancy stuff is nonexistent. You look at that—the fancy stuff costs money. A nice farrowing barn now, with crates on a pit, is over three thousand dollars a crate.

When I thought about the best way to farm, I said to myself, Well, I'm going to farm the way I want, and if I don't make it, I'll be a consultant like everybody else. (*Dennis laughs and heads into the barn and the farrowing pens.*)

You want to real consistently see healthy livestock, otherwise there's something wrong with your system. In all the confinement systems, you have to have a lot of vaccinations and they have to feed antibiotics; otherwise, they can't keep them alive. That system can't work, in the long run. This way they don't have the stress at farrowing time that a first-time animal has. When I want to wean them, all I do is bring the mothers outside, and they'll run them across to that pen over there (*he points across the way*), and the little ones all stay here—and they usually aren't stressed. It seems to work awfully well.

The one thing I have is a lot of flexibility. I can do it this way. I have sows out on the pasture. In the wintertime, they will stay right in the barn, but I like to have them in a group like this the best, outside. I went down to Paul Willis's, and that's kind of the way he works it, year around, and it's a little bit warmer in Iowa. I'm a believer. You look at some of these. Just look at the udders they're carrying on them.

Well, let's go take a look in the barn. (*Dennis laughs.*) We'll make you walk through the dirt today.

We walk down to the barn, and Dennis shows me how he moves the sows and pigs from one space to the next as the pigs grow, and pasture where they roam outside.

This gal and the gal in the next pen here will be put together. They will end up out in the pasture; these all will. This one over here, she just farrowed this morning. Last night she was in the hut, and this morning she decided she'd like to lay right out here. She's got ten little pigs. Ralph doesn't believe me, but I can look at their eyes and I can tell pretty much when they're ready to farrow. You hang around them too damn much, I guess, when you can do that.

"Why doesn't he believe that? He'll tell you all kinds of crazy things about what he can see in his cows." We're on fun ground here, for we both love Ralph. Dennis laughs affectionately.

I know it! That's just the way he is. But it is true—the more you hang out with cattle or hogs, the more you become a part of them; you understand them. That's why I like farrowing this way—they enjoy life. . . . I had Ralph in my farrowing barns with crates, and they were pretty darn content in there too. But the thing of it is, what I hated is when you took them back out of the crate, so many of them would hurt themselves because the muscles just weren't there. They would fight—anytime you mix sows, they're going to fight. You hear that snap, and they would break a leg or a hip—they're done then. I haven't lost a sow that way for three or four years, [with] the system I use now. . . . It's a nice, easy way to raise them. (*Dennis walks among the pigs, talking to them.*) I like to farrow them in here, give them their freedom, but I still have a little control. Out in the pasture, I don't have any control. Before they get about five or six days old, we will castrate them, and I may take one sow out—sometimes I'll take four out at once, with little ones—but they all have gotten a chance to get going. I think that's the important thing. You'll see some that are out in the pasture, and you'll see quite a difference. All I'm doing is copying the Swedish system and modifying it in my own barn.

Inside the barn, there is deep straw scattered across the floor. It seems clean and crisp. "How did the Swedes get onto it?"

That was interesting. The people made it illegal to have pits, and the people said they don't want that type of high-production confinement pork with antibiotics. They don't want antibiotics in their pork. All of a sudden, the farmers said, "Hey, we have to change and learn how to produce the product they want." To make a long story short, this will all be cleaned out next week, and we whitewash it and give it a new coat. I haven't cleaned it since early this spring, but I don't have to.

"This straw and everything in it will all go out on the field?"

Yes. . . . A lot of times, pigs will take one area and stay there. That's where the manure is, and you'll get a lot of stuff like this, where it isn't even dirty. I put that in my pole barns and use if for the cattle, so I kind of reuse it, in a way.

As we round the side of the barn, Dennis points up to the barn roof that is partially covered with sheet metal.

I'm doing this myself. It's well over $10,000 to do the barn, to have it hired. I know that the metal, just the sheet metal, will be right around that $3,300 mark. With my own labor, if I can figure out how to get up there, look at the savings that I can do. They're 16½-foot sheets. (*Dennis begins to laugh, and he points again at the corrugated metal, bright and gleaming even on this drab and cloudy day.*)

I think about the weight, and about how awkward those sheets are, even in the lightest breeze. "How do you manhandle those up there all by yourself?" I ask, thinking that I would not want that job, not even on the most windless day we ever had.

I slide them up there on that roof (*pointing to a section of barn that is lower*), then I slide them over, and then I walk the length and I have a ladder—see the ladder? It's got a hook on it so I can work on the ladder, and if I get the first one fairly straight, the rest of them will be fairly straight! (*Dennis is laughing again.*) And when Ralph comes out, he won't crawl up there to give me a grade anyway. I know where my mistakes are. I can afford a few mistakes this way, can't I?

"*Right.*"

I worked at that for three days, and I got that far in fifteen hours. Today my knees are hurting—I need an Advil! A lot of guys put the wood purling on, and I decided to go through the cedar shakes with screws, so it's all screwed down, rather than nailed down.

Dennis is laughing, thinking of the winds that come fast, and those that occasionally whip into the tornados this country is famous for.

You can lose a roof in this country pretty easily. I think it will work out good.

We head for a gate on our way to another barn. Inside this barn, Dennis again walks among the pigs, talking to the animals as he goes.

Come on, you guys, don't act so scared; you know better than that.

Dennis worries about the health of his animals, but over the years, he has learned how to raise them without antibiotics.

I haven't fed any antibiotics in probably ten years. That doesn't mean I haven't lost some pigs while I was learning. But I felt it was just the right thing to do. My thought on it was—and I kept track—that it used to cost between $2.30 and $3.20 to feed antibiotics for that feed efficiency. Well, corn was less than $2.00, so why not just feed them an extra bushel of corn and get rid of the darn stuff? It didn't make any difference. So I just felt I looked at the whole situation, and then when we were selling a lot of meat direct, I felt that was the right thing to do. All of that idea, so to speak, led me into this Niman's market. I had done some of the research to get there.

"*Where did you come by this value system of yours to start with? It seems like you're trying to take care of everything all the way through: the land, the livestock, rotations . . ." Dennis doesn't even think twice about his answer. It comes promptly, easily.*

My dad was that way—I think it was learned from my folks. In my teaching, I taught conservation. I was always pro-conservation and livestock. I always thought that had to be a part of the farm, to make anything work. I tried to look at the whole picture. It just seems like it's the right way to go—not necessarily the way it's going, but it seems more right to me. It's a way of lowering my inputs for raising the crops that I do have. Though the last three years, I can't say that for sure. There were two years in a row that we had 150–170 bushel corn, and our cost was about $1.30 a bushel. But my inputs are so low! And that's using a hundred-dollar land rent included. It wasn't because I fertilized so heavily or bought the most expensive corn. . . . It's because I'm using a system that is just going around. I put the manure down, and that gives you the corn that you're going to need for the next year, without a lot of fertilizer.

I've experimented with corn silage in particular. I have a planter, but I don't use it. It ruins my quality of life, machinery. I decided a long time ago. We have two guys; they're neighbors, and they just love machinery. They always had their corn planted before I got started anyway, and for ten dollars an acre they come in and plant it for me. It's a one-day shot. I look at that as being pretty darn cheap! What do they say on the average? Fifty-five, sixty acres. At six hundred dollars, can I afford a twenty-thousand-dollar planter? No, but you can afford a six-thousand-dollar planter that you're repairing all the time—and I just didn't want anything to do with it. So, that quality of life—I do a lot of decisions that a lot of people won't do, and I custom hire a lot of stuff. But you know, this corn out here is going to go 160–170 bushel, without too much trouble. Once in a while, I'll run into a snag where everything goes wrong and I've got corn up there that's probably going to go 110–120, and it should be my best corn, but because of my system, trying to prevent runoff and soil erosion, it held so much water, it was just so wet until after the Fourth of July. Not much you can do about that one!

I was going to tell you my mistakes. (*Dennis grins.*) I bought twenty-six Berk [Berkshire] gilts—went into the purebred Berk end of it, with the idea of selling it to Japan. The market and everything is there, but the Berk gilts were just lousy! I took some of them, and now they're coming around the second time, and they've been real good mothers, but I crossed them. Their pigs weighed about 1–1½ lbs. more—but I spent an awful lot of money on those damn gilts! (*Dennis is again wearing that grin, laughing at the joke on himself.*)

The other thing is, I have two silos—one I never use. And I've got a hot nursery I never use. You add that up—that's thirty-two thousand dollars right there.

"*More or less wasted.*"

I paid high tuition for my education. (*Dennis laughs ruefully.*)

From our position between barns, we have a good view of the cattle pastures, the ridge where the corn begins, and, if we turn, the cornfield behind us.

This piece of ground is too wet to farm (*pointing down a little swale*).

You can see the cattails down there, and there's no real way to drain it. The neighbors dump their tile line right at my line up there, which really doesn't help. I grazed this with the cows early in the spring, then the bulls went in here for six weeks, then I turned around and gave it ten days of rest, and then I had my breeding heifers in here for a short time on a rotation basis, and now the sows and little ones in the fall. So I'm making good use of it.

"*It doesn't look hard used.*"

Sometimes in the fall it will be just black, looks like no grass at all. They'll go in and root it all up. They're chasing some kind of grubs or something underneath. Next spring, by the first week in June, it will be all nice and green again. We never seed it; we never touch it—it just kind of keeps coming back. We don't exactly understand that. The pigs have the greatest time in the fall! A month from now, they'll really be running around; they will all go into the cornfield. When the combine comes, they just chase it around. If I could ever get a video of them chasing the combine down, that is so funny!

If you look up here, there's 240 acres here. There's 180 of it in pasture and forage, and it's all being stockpiled now for feed in October, November, and December. I'm not rotating my cows like I would normally rotate them. They're down there (*pointing*). I've got two pieces that will be to corn, but I'm keeping the cows on them during the month of September, when the rest of my ground needs a chance to build up their roots for the winter. But just think of all the manure that 160 head are putting down on that piece of ground! That works out real well. . . . Otherwise, to have some ground that you are going to tear up each year to go back to corn, you're just harvesting the nutrients.

"Well, you're getting the O matter as you go. That's something I guess I had to learn, that you can have soil that may be nutritious for plants—like corn and beans—but it's not necessarily healthy soil."

Yeah.

"And I think that's another thing the general public doesn't understand either. I had a friend say one time, 'Well, chemical fertilizers provide exactly the same nutrition as organic fertilizers, so what's the deal?' But they don't do anything by way of organic matter, and they don't create healthy soil, but dead soil with no organic matter at all. So there is a huge difference. They're not equivalent, no matter how you cut it."

No. But the general public doesn't understand that. You start talking organic matter or bacteria in the soil and they look out there and see all these nice crops—and they are nice—but long term, can it stay that way? I don't know, and I don't think anybody does.

"Except that if you were going to make a bet, you wouldn't bet on that long term."

No. You can't keep killing stuff and expecting it to keep coming back.

We turn to head for the pickup and Dennis smiles, looking up again.

I can't get over how that roof makes that building look brand-new!

We board Dennis's pickup. The rain has done its work, and Dennis shifts into four-wheel drive. As we drive, Dennis points out features he wants to be sure I notice.

This is the water system I put in for my grazing. Otherwise, all the cattle have to come home all the time. Always was wet around the drinking area, so we went with central tanks. (*He points to another paddock.*) Then I have some tanks out here—

"So they can get to them from either side."

Yep. I always tell my kids I put those in (*Dennis grins*) because I want to party till the cows come home—and now they don't come home anymore. (*He laughs.*)

I know Dennis is interested in diversity, and I ask, "So what's in this pasture besides clover?"

There's orchard grass, got some reed canary, alsike . . . I'll try anything.

"Is your alfalfa a mix too?"

Yes. See, we're not going to get over there, but three years ago, we wintered the cows in that bean field. . . . That's light soil, and it grows great crops after you winter them like that. The cows leave so much manure in the field, I don't have to fertilize. I set the bales so the cows move across the paddock, covering the whole thing over the winter. That manure is distributed as evenly as you could do it with a spreader.

The piece that's one paddock over and in between the beans here, we grazed it twice and then I baled it twice, and I've taken three tons of feed off of it and I still have a big grazing left on it. That's my experimental plot, where I have reed canary in it—to see how that fits into my grazing system, and it looks like it might. The only thing is, reed canary will take over everything. There will be a point where you wouldn't want that much. I'm trying to find the mixture of grasses that does so well on this farm, and I don't know what it is yet.

"Do you move those round bales occasionally or just leave them there?"

Those I baled second crop and left them right out there. I don't bring them home and then take them back out. . . . You can see them up there. . . . There's 150 of them, and when the beans go off, then I'll put them out. We'll winter on this half of the bean field. And then that piece will go to corn silage, and what we'll do there is . . . wait until late May to plant it. I'll use eighty-eight pounds of actual N [nitrogen] but no starter, and I'll get over twenty-five or thirty tons of corn silage that way. But the wintering—you have so much of the nutrients there. It's simple, it's easy, I'm not working hard doing it, but it's making me pretty good money, when you think of it. And I don't really know how long—are we getting six years of benefit from that manure, or ten? I can't tell you.

Well, there's one thing you do know, if you have a system—a livestock system—if you have 120 bushels of corn to the acre, that will finish out twelve hogs. Twelve hogs is twelve hundred dollars that goes back to that acre some way. That's what I keep thinking. So I mean, we average 150 bushel corn; why would I ever want to worry about 180 or 200? It doesn't seem right to worry about all that extra input.

Dennis sounds a bit like Vance Haugen here, but even more like the old Roman senator Cicero, who once asked what good a surveyor was who could tell him exactly how many acres he had "but cannot tell me how much is enough."

I'll tell you, the worst about this whole thing—I like the way I'm farming now, and I can do it financially because I've got things paid down—but you take the young person that wants to start farming, it's not going to work. Land around here, right now, is going between $2,500 and $3,200. Well, you have to find a gold mine on it to make it pay for itself.

"That sure is the story I hear all over the country! When I was in Montana, they used to say, 'You can marry a ranch, inherit a ranch, or you can go to Hollywood, become a star, and come back and buy one.' That's about it. And in those years, taxes were eating folks up, maybe even worse than they are now."

We dismount from the pickup and walk over to the paddock where the cattle are. As we come up on the cows, they begin to move and to bellow louder.

They're yelling at you. (*Dennis laughs.*)

"Ah, they think you have something for them."

These're all pretty much Angus. There's a little more marketing opportunity with them. Feeders still believe the Angus is still the most efficient. Angus heifers, bred heifers, are always in demand. (*The cattle are eager, and the calves are nosey.*) Calves kind of think they can sneak beneath the wire. (*Dennis grins.*)

As we watch the cattle, he brushes the tips of some grasses.

We were talking about the different kinds of grasses. . . . This is mainly orchard [orchard grass]—but the beauty of a grazing system is the dandelions. A lot of weeds can be eaten, and they like them. Dandelions, for example, are very high in mineral. I like this one that's underneath here too, this fine stuff, that's a fescue. That's what I look for. . . . Now this is typical hay ground stuff right here. With the grass in it, it dries so much easier than straight alfalfa, but you probably sacrifice some tonnage. . . . But you know, they kind of ruined alfalfa by growing too much of it. Now they have the weevil or something that everybody around here is spraying a couple times a year. Well, I don't even think about it.

"Ralph doesn't either. He's got his mixed with some grasses that keep the bugs away."

Yes. Well, the bugs will attack this, but there's plenty of grass there, and I still get enough feed, so I don't worry about it. You just can't worry about everything! It's just like insurance—if you bought all the damn insurance they want you to buy, you'd be broke.

"So they get all that hay, and all that feed."

Yeah, and two days later, it's all gone too. I give them new feed every day, though. But with this system, like in the winter system, you put your bales on top of the hill—right where you need most of the nutrients. Your cows will follow the bales and leave their manure right where you want it. Next summer the nutrients are all back, and the grass is high again. You can't beat a system like that, can you?

"So you calve in the spring, or all year?"

We calve in May and June. I have it set up so the cows and the heifers are all on pasture. The heifers get a little special attention. They'll have a pasture right by the building site, and they'll also get bred to a bull that throws real small calves the first time around. Two years' time, and Ralph has used him twice too, and we haven't had a heifer that's had a problem. One of the things that I do that's different than a lot of grazers: in the springtime, the cows all calve out in the pasture. There's a couple problems there: They're always on their own, which isn't really a problem, but if you give them a lot of pasture, they milk too good and get scours. . . . If you give them a lot of spring pasture, they eat too good and sometimes, especially by mid-June, they get a big calf. What we do is slow it way down. Actually, by slowing it down you will create a different environment, and then the white clovers will come through in those paddocks.

"I noticed some when we were out there before."

Mm hmm, but I found out that I can pretty much control any scour trouble with calves. I think I treated one this year and none last year, and none the year before. So it's working out real well for me. That goes against a lot of the producers. They'll sit there and calve in March, or even in February, and they'll have them in the darn mud lot, and then they end up with so much trouble to keep the calf alive. They get a bigger calf this time of fall, but dead calves are hard to sell!

The cows are grazing, moving slowly, but eating as if they enjoy it. The air is loud with snuffles, the tearing of grass, the rasp of hay pulled from the bale. Dennis talks to his cows too.

You guys must be hungry!

"They sure took to this, didn't they?"

Oh, I know it. Pastures aren't just pastures. They're a birthing area and reduce our cost for medication, and I guess the only thing I don't like about our system is we'll get a few late calves, and they always don't seem to do as well. They're born in the hot summer. Marketing doesn't bother me, because we market heavy feeders for steers; we've been averaging between $500 and $650 each on those. That's worked out well because you've got just about nothing but forage in them. The good heifers go back into the herd or sell as breeding stock. The poorer ones are finished out.

I always try to figure out how I can make the system easier. And that's not what agriculture these days is all about. You're supposed to confine everything and implant it or, like with hogs, you're supposed to have spotless farrowing crates, and a spotless nursery, and then another growing unit . . . We don't need any of that! Most of my cows are fairly young yet. The cows are reasonably good priced. . . . I'll let the older ones go now and replace them with good heifers. Anytime a cull goes, that's breeding stock, that goes to the farm payment. I don't even try to figure that in my cash flow. The first six months of this year went great (*Dennis begins to laugh again*), and then the one tractor just died so I had to buy a new one, and then the pickup—I got so sick of trying to make that one run, I traded that. A lot of other expenses fell in—

"It happens—"

The way I look at it, you think, Oh, Jesus, you know . . . but between my wife and myself, we already have twenty-five thousand dollars paid off this year's principle, and I gave my daughter six thousand dollars for a wedding—well, what more can I ask for than that? Even if you have to borrow sometimes unexpectedly, it's still not too bad.

"Well, you have to be doing something right, because it sure seems like the system is working."

Well, it will but it won't—if somebody comes in and tries to buy three-thousand-dollar land, it won't! I got all of our land for a thousand dollars. Once that first 160 was paid for, it's a lot easier making payments on the next 80! Some day, when all of the farm is paid for, I think I'll probably increase the cow numbers to 100 or 110 or something like that, and try just total grass—no feeding down. Do what Ralph is doing and see how it works out. But for now, I'll just keep watching him! You can do your own experiments, but you also want to watch the guys that are experimenting, because you can learn a lot more from them.

You want to hang on to this? (*Dennis hands me a paddock wire.*) I have to move these two feeders yet.

He moves out in the rain and rolls a couple of hoops for bales into a new paddock. He moves the iron fence posts, takes the wire I'm holding, and moves it too, organizing a new space for the cows to feed, moving them off a paddock that they have already grazed and into this one.

Look at all this nice feed. . . . But the cattle need to have something else besides just lush feed. They will go after that hay, even though sometimes it will be the poorest, even woody-looking stuff. But they will go after it midsummer. We all think we know what they need—

"But they do."

I think they have a pretty good idea.

Farmers in the upper Midwest often get three cuttings of hay during a summer. If the weather is right, it grows back after the first cut and can be cut again, and often a third time. Dennis and a few others I've talked with don't cut it the third time, even though they could. The surplus can be sold, but if the hay is left, it provides good ground cover, returns some nutrients to the soil, and saves on labor and fuel. "So you always just cut hay twice and third cut goes to grazing?" I ask Dennis.

Yes. That's another quality of life decision, and right now, I don't think I need to have that extra feed. I have a guy who will take all the hay I make—but sometimes it isn't worth it! It's just a decision I make—I'm not going to cut third crop, so I'm doing some crop improvement, and the hay stands last a lot longer! They don't see the manure, they don't see the cattle, but I have a hay stand that's on its fourth year and it looks real good. Well, if you give it a rest now, it's going to yield two and a half, three tons that first cutting. (*Dennis grins.*) I shouldn't tell you this, but after second crop hay is done, which is usually around August 1, I have between six weeks and two months where there will be 160 head of cattle and 600-plus hogs—and fifteen minutes of chores in the morning. (*By the time he finishes the sentence, we are both laughing.*) It's the systems that I set up that allow me to do that.

"I won't tell," I say.

Ah, I don't care.

"I may write about it, but I won't tell."

Dennis just laughs.

It's time for me to go. "Thanks, this has sure been a treat!"

Yeah, for me too. If you have any questions, just give me a call.

As we walk across his field to get to my car, Dennis tells me about his shortcut back to Red Wing. He waves as he starts back across the field toward the cattle, and I head north on the gravel. As I drive, I think about what I've heard here, feeling better than I have for a while, sitting at home reading documents about the World Trade Organization and NAFTA. And I know why.

I think about a phrase that cropped up several times this morning: "the right thing to do." I've heard Dennis use it before today, and he used it more than once or twice while talking specifically about his farm on today's tour. It's not exactly un-self-conscious, but it is so offhand that I know it is not meant to call any attention to itself or to Dennis. It is just the truth: "I think this is the right way," he says. "I've thought about it, and it seems like the right thing," he says. I don't know all the meanings that phrase might have for Dennis, but I've heard it in enough different situations that I know he isn't talking just about the right thing economically. "I could probably make more money if I farmed in other ways," he says, "but I think this is the right way." Though he doesn't say it directly, Dennis uses that statement in ways, and in settings, that reveal a kind of ethical choice he's making. It is the right thing for the soil or for the pasture, the right thing for his animals—meaning they are happier and healthier. And I think about his pleasure in seeing them that way, his interest in just watching his pigs, thinking about what they are trying to tell him, what he might learn from them, how he can look into their eyes and tell if they are ready to farrow. I hear again the way he talks to his cows when he moves among them: whether they are newborn calves or nearly finished, it's that same even-toned but often teasing voice, often accompanied by his lopsided grin, that I've also heard him use with Ralph and other friends.

Ultimately, I suspect, Dennis believes these are the right things to do in order to be the kind of person he sees himself as, an image of his own integrity that he wants to uphold. "They were going to pay me $1,800 for that, but I told them I wanted to forfeit that. . . . I did it for less than the grant." It's an image that matches the one Jefferson had of an independent yeoman farmer. Now, as I write this at home, I can see Dennis wince and squirm at such high-faluting ascriptions of character. But if I weren't sure I was right about this, I wouldn't put it down. "Sorry, Dennis," I say to him in my mind, "but we all know you're no saint, so don't worry about it."

I've not heard other farmers use exactly that phrase as often as Dennis does, but I've heard similar expressions from others often enough. Somewhere in his introduction to field days on his place, Art Thicke uses the same words each time: "We're interested in healthy soil, healthy animals, healthy humans." I think of Lonny Dietz telling me a year ago, "We've got our wildlife area now, row crops, trees planted, prairie restoration. Set up a whole-farm ecosystem. It's all necessary." He clearly thinks this seems like the right thing to do. So I never have to wonder why it is that I leave these farms newly buoyed, hope restored, happier than when I came. So many are trying to do the right thing on this place. How could I not be inspired by that?

Part III

Farming in America: Who Cares?

CHAPTER 8

They Say Eating Is a Moral Issue

BILL McMILLIN

When I drive in, Bill McMillin asks if I wanted to take the Mule for a quick tour. I say sure, so we start out on the four-wheeler down his lane, with mowed grass lawn to left and right, cross the highway that divides his property, then bump along his neat contour strip. From where we stop, we can see his contour strips and two other farms. It isn't his strips he wants to show me. He points across the big coulee to a field of corn, not his, with rows running right up the hill. Every row has a gully in it—some of the channels much wider and deeper than others—and the spring mud is pooled in the low places, drowning all the vegetation and running off downcountry to drain both topsoil and chemicals into the streams that eventually flow into the Mississippi, ultimately feeding the toxic bloom that kills fish in the Gulf of Mexico.

Bill begins considering his own farm, in view of these other places.

To go back, our farm, I think I got it right, was homesteaded in 1850. So you go back to the early 1800s, and you can imagine what this was all like. And then you kind of think about what's happened since then. We don't know a lot about what happened in those early years, but you can probably go back with some certainty to when they started the conservation districts. At some point, the government programs came on and the government got involved in agriculture. Since then, you can just see how government policy is affecting what's happening to the land.

Like this farm over here (*pointing at a cornfield across from our vantage point, on the edge of his contour strip*). When I was a teenager, it was in the soil bank, so it was all planted in grass. So if you look back, these headlands around here . . . the last twenty years they've been planted just like that, the corn or beans right up and down the hill, and there have been ditches eroding just like that. Every one of those rows has a ditch in it. You can't see it very well, but every row has a ditch in it. Yet the government keeps paying. But [the owner] doesn't farm the land himself; he rents it to a guy who raises a lot of crops. You got to wonder about having some discretion about who you rent to.

Bill has an expansive, comprehensive view of the connections between things, a sense of the broader ranges of ecology beyond the immediate area.

I don't know the science exactly, and the numbers just amaze me, and I see what's happening every day. When I farm next to this, I see this. And now the hydrologists, they're figuring out that it's having less land in grass and alfalfa that's breeding some of these severe thunderstorms we've been having. So it's all tied together. All the systems I hear about are circular, and they either feed each other in a healthy way, or else they ultimately destroy themselves. That seems so apparent. I can't understand why someone would buy into a system that is going to defeat itself. Probably a lot of it has to do with the system we were taught growing up. Who knows? My system is probably not sustainable unless we can do something about receiving a fair price for our product. That's part of it too. If we have to keep increasing the number of cows all the time—a 10 or 15 percent increase per year just to maintain . . . [our] standard of living—that's not sustainable either. Maybe if you've got flat land and stuff, you can do it without the erosion, but you still have the pest problems, and you still have the purchased inputs, so in the end it's maybe not as destructive as this we're looking at, but it's not sustainable either.

Bill is a careful observer of what is happening to the land. He's been looking at this soil, these creeks and rivers, these hardwood groves and rocky outcrops for years. He has a timeline in his mind.

To go back to what I was talking about before, somebody drew a line on a map and split this hillside into two farms. Each farm has its own history. Like a set of identical twins that are placed in different foster homes shortly after birth. They each have their own experiences that shape them. Some of their experiences may not be good, and they leave them with deep scars. This place is different from the one next to it.

Bill gestures to another farm a bit farther away, pointing out the split between the farms. They have fared differently over the years, depending on who owned or rented the place, alternating between good practice and poor practice.

So there was this line here, in time, and then you think about government programs and the SCS [Soil Conservation Service], and see how it's affected the land, and see how the changes of ownership have affected the land—like I said, the one guy had it all in the soil bank, and then the next guy put it into strips. . . . Then there was a farmer who planted it all into grass, and he wanted to be a cowboy and he was going to raise cattle there, and that was good for the land. But there wasn't enough money in it for him to do it, so he put up the hog buildings you see there. Then the whole farm was plowed up for row crops. At one point, there was a guy raising cattle, and then he wasn't making it, so then he found some investors and he built this pig setup, and at the time they built that, that was a fairly big pig setup. . . . That's the point it all got turned into corn and beans, so obviously the soil took a hit then, when that ownership changed. And then Carl Polhad bought it. He owned it for five or six years, and then he sold it to the present owner. He is a private owner, and he's a nice guy, but he's in it just to raise pigs, and he's doing a good job with the pigs, but the markets have been so low he couldn't compete with the big guys. Now he rents the land out, basically to whoever pays the most. I talked to him—he had some of it for sale a year ago—and my brother-in-law's looking for some hunting land, and I asked him, "Well, if you're going to sell it, could some of it be seeded down?" Well, no, he needed a contract to haul the manure on it, so you really couldn't seed it down either, so the confinement operation and the idea that it all has to be plowed up kind of go hand in hand. They have to have a contract on so many acres to haul the manure.

So there's no hope for the land in that system, not on that farm. The only way would be if somebody could contract with enough neighbors and then put some strips in there, but usually, most farms' manure about fits their number of acres. And it takes a lot of acres to clean out that storage over there. But I mean, there's a story there, and it's about how federal policy affects the land. And across the valley is this other neighbor with a different system for raising hogs and taking care of the land. His farmland is the twin with a happier experience.

The erosion we're looking at seems to give Bill physical pain. His reaction is one more of us should feel, whether we are farmers or city folks. This is the country equivalent of an urban drive-by shooting. There is unnecessary death loosed here, and, soon or late, it kills humans as well as fish. Without some depth of feeling on all our parts, the soil will wash away, the chemicals will continue downstream, and the government programs touted to be helpful will too often work out poorly or not at all. The programs are rarely based on farmers' actual experiences but on a bureaucratic or academic idea of what should work everywhere. Bill tries to affect that.

I've been bugging people about a very simple fix for those headlands. I mean, I've talked to the conservation people. I was on the soil and water district board for six years. I talked with the guys down there: "Can't we have a pilot program where the headlands are planted into grass?" And they say they're in compliance on the contour there, and because the headlands are only a small part of the field, they say, "Well, they're probably in compliance." . . . And their soil-loss formula . . . they changed it a couple years ago, and the current formula just doesn't reflect what's happening on the land.

So my idea was to pay them to seed down those headlands and just keep it that way, permanent cover. I haven't been able to get very far with that at all either, and it, it kind of bugs me. Such a simple program, I think it'd give real good returns for the dollars spent. The farmer wouldn't really lose anything. He'd be getting paid basically the income he might have off that, and he's saving the land and he's saving the water. So I went to the Minnesota Department of Agriculture, and they gave me the telephone number of a program in Washington. There was actually a program where you could seed down. On the contour you'd have a fifteen-foot grass strip, and then on the headlands you could have a fifty- or sixty-foot turnaround. Somebody in Washington changed that; instead of allowing the area office discretion to have a fifty- or sixty-foot strip, they limited it to fifteen feet for turnaround. You cannot turn around in fifteen feet! And I called them up, and they said, "Well, we have determined that fifteen feet is enough to filter the water." So see over there? Where the water's running down the hill? That fifteen feet over next to the fence—you can see the water doesn't go through that fifteen feet. And everybody in the offices knows about it, and they're all talking amongst themselves, but I think they're afraid to lay their jobs on the line and say, "You've got a big problem; somebody made a stupid mistake, and let's get it fixed."

Everybody up and down the road talks about that place over there, and even the guys at the soil and water office and the SCS, but it just seems like people are afraid to stick their neck out at all and try and make change, although I know of a couple of cases where they declared a farmer out of compliance. The farmer appealed, and the people above our local officials did not back up the locals, so it didn't do any good; the farmer was paid anyway. So it's not just one failure in the system. There are several failures. I think they see what's happening; they know it's wrong, but they just aren't able to make themselves work to change things.

I was wondering, you know, do we have a democracy, or do we just have the economy, the consumerism, basically a free market? That runs the country, I think. People don't make decisions; the market decides. At some point, we have to decide what kind of system is going to replace it, because this system can't last forever, just because of the way we consume. We're consuming all our resources. So the government encourages people with their policies to spend money, not to save money. It's all spend, spend, use stuff up. You just can't keep going like that. We keep having more people, and it's just like this pig farm over there: it drives itself into the ground; it's bent for destruction the way it's going. Well (*he grimaces*), that was my venting on that subject.

Bill starts the Mule up again, and we bump back over to his place, sitting on the Mule to talk, comfortable in the mid-June sun as we continue to visit. We are parked between the barn and the house, and Bill's thoughts go back to his place and issues he wants to work on here. I ask Bill how he sees his farm twenty years from now.

As for the future, I don't plan on milking cows many more years. If I found someone who was interested in working into it or wanted a couple of years learning experience, I would go that route. The runoff thing is something I will probably do regardless, because if I am going to use the silos, I will need to. But that's not a sure thing. I mean, I might decide to quit milking dairy cows and raise beef or heifers, and do it all on just grass. If I could figure out the way to do it with the stocking and keeping the animal numbers and everything, I'd like to see the whole farm grass. But I'm not certain; it's not easy to do it, keeping the numbers to the grass, 'cause as the seasons change, it takes more numbers, and that's where the silos come in handy. If you get a drought, you've got the stored feed, and it works real good that way. So the other option is maybe this side of the highway put all into grass, and the other side maybe raise some feed, forage, whatever. So I see it being a livestock farm, but I don't see it being a dairy farm.

My wife doesn't want to sell the farm; she's made that clear. I probably don't either if it came right down to it. It's home; I grew up here, but ... Years ago, like ten, fifteen years ago, I was thinking what I would like the future of this place to be. . . . It would be nice to have a place either for beginning farmers or to do experiments. At that time, I thought that when I got to the age that I am now or a few years beyond, that I would maybe have all the money I needed and be able to just donate it. Now (*laughing ruefully*) I'm not sure I'm going to have it. I mean, you don't know how much you need, really. You don't know about the Social Security, and you can't rely on that too much. So I'm not sure. But if something would happen to Bonnie and me, I would still see that as my first choice, to see the farm go towards something like that. I would want it to be a working farm that teaches or experiments with methods that are sustainable and that're going to protect the land and help young people get started. So I would see it going more that direction.

It seems agriculture changes more rapidly than many industries, including even the computer technology trade. I ask Bill if he sees some good changes happening among the farmers he knows.

One good change is getting away from chemicals. That's kind of where the SFA [Sustainable Farming Association] got started years ago. They've kind of moved away from that a little bit, but if there are that many interested in it, maybe they need to get back into that a little. Land Stewardship set up a program, the Stewardship Farming Program. So there were fourteen farms, and they did experimental projects. And that was learning how to cut back on chemical use and learning how to do intensive grazing, that type of thing. There was a lot of interest in it, and there were a lot more than the fourteen farmers who wanted to participate, so that's how the Sustainable Farming Association got started.

Art's got an impressive place, and boy, the numbers—I mean, it seems to be really working well for him, and he's going against a lot of the principles that a lot of the established grazers are going by. That's really interesting to see what he's doing, to see what happens over the years. . . .

I see more and more [people speaking out], even university people, and maybe they're just a small percentage, but I hear some of them speaking out anyway. A county agent had a column in the paper a month or so ago—and you know county agents are just people—and said that we have to start looking around now and see what's happening, or what we like now, what we love, is going to be gone before we know it, and it's going to be too late to change things back. So it's good to realize that some of them see what's happening and are willing to speak out a little about it. Some of them don't. It gets real frustrating when some of them just talk the company line, the way it seems. They're looking at it as "This is what's happening, and you can either be a part of it or you get out, and you can't change it." I look at it as "Where do you want to be?" and you take steps, no matter how small they are, to get where you want to be. You can make sure you're going that direction rather than taking steps that you think are wrong.... No matter how small the steps, then, at least if you're going in the right direction, you're making progress.

Well, the corn and beans thing, I mean, the inputs that are required for corn and beans, they're all kinds of limited resources, and we're losing the soil; we're paying the farmers, then we're exporting it overseas, usually at a loss; I mean, the government subsidizes exports. What is the sense in that? It's totally flawed. I don't know if Monsanto and Cargill and those people are writing the legislation, if they're so big that the legislators are afraid to do something that's going to hurt them. I mean, I debate too, a little bit. If we had a president that was really able to change things around, what would happen to our economy? I think it is strong enough that it would bounce back, but there might be some pretty lean years, and those people that have their money in retirement funds would probably really get hurt bad. It's sort of like making the transition from conventional farming to sustainable farming, where it's not as profitable for a year or two but then comes back. Trying to create an economy that works for the whole world and for the land and for us, you may have that same kind of trough for a few years, but after that, it seems like it would have to work better. I guess we need to try. We are obviously not headed in the right direction now.

If you can keep that outside influence out, you know, the big corporations . . . That's something else I struggle with a lot. I don't know how religious a person you are, but I struggle with our system and how it seems to be at complete odds with our religion here—the survival of the fittest versus what you're taught, whatever religion you are. They simply don't mix. I was at a Farm Bureau county convention in the mid-eighties. I joined to see where they were coming from. At the time, I was county NFO [National Farmers Organization] president, and trying to work with them and get some cooperation going. And that was the time when there was a law that a person losing the farm had right of first refusal to buy it back. Farm Bureau was opposed to that, and the churches came out in favor of it, and the county

Farm Bureau president made the statement that the churches should stay out of this. I said, "Isn't part of what churches do supposed to be to provide guidance, or provide a path on which to lead your life?" [Some people] claim to be religious, and yet they have this survival of the fittest policy that flies in the face of everything we've learned from religion.

Have you seen some of the stuff that Catholic Rural Life has put out? Boy, they've put out some pretty strong statements, oh yeah. They say that eating is a moral issue, that people have a right to healthy food, food produced in a sustainable manner. They put out a lot of brochures and literature both on the way that animals are raised and on food. I don't know how wide this is. This is the Catholic Rural Life in Minnesota and Wisconsin. I don't know how big an area this covers.

I am impressed by how often conversations with farmers veer to church or religion. Bill brings his own thoughtful attention to how religion might affect the way he lives, or agriculture as a way of life.

I was raised Catholic, yes, but am not so much a practicing Catholic now. I have a problem—my wife's Lutheran, and she's very Lutheran, and we talk about original sin. I think if you believe the story of the Bible, I think the world was basically a paradise until original sin, and original sin changed everything. My question is, Do the consequences fit the crime? You see everything that is happening in the world today, and if all that is happening [is] because of original sin, I don't think that's right. Even the U.S. Constitution has a little higher standards than that. So you go back to the Old Testament, and the God of the Old Testament was always leading the Israelites into battle. I don't know. But then there's the rest . . . from Jesus on, and I struggle with that. I don't have the answers. But I don't think it is as clear-cut as some people think it is. Some people don't like to question things. You know, if I was raised a Muslim, I would be a Muslim probably, or maybe the same kind of Muslim as I am a Catholic. People are what they are raised for the most part; most don't change very much. Makes you wonder.

I've been pretty involved in trying to get farm groups together and farmers working for collective bargaining, but there's a part of me that struggles a little bit with that too, because collective bargaining is, for the most part, size-neutral, and I don't want to do something that makes it so the big guys can just keep getting bigger and still keep forcing out the small guys just because of the numbers issue, the economy of size. So I talked with Dr. Levins, the ag economist at the university, quite a bit about that, and he says, "Well, you get enough guys working together enough, then you can limit yourself a bit, limit growth, and do that," but I am not convinced that it would work that way. So that's one kind of gray area or dark area in the whole collective bargaining thing that I am not sure about, but I don't know what else farmers can do. I mean, you either do that or you rely on the government, or you go down the tube.

So, somehow, we have to replace what we have with something better. And it's not going to happen overnight; there's got to be that transition, like we talked about before, where it's going to be difficult. But I think the talks really need to get started, and I'm sure they are [in] some places, but I don't know if there are enough people coming together to make things happen. You just can't keep this consumerism going where you just spend and throw stuff away and bury it or burn it and . . . You just can't do it. At some point, people have to realize that. You can't keep losing forty or sixty tons of topsoil every year, twenty years in a row. You have to stop and say, "That's enough."

And maybe [at] some point, you know, back to the religious thing, we're all maybe brothers, or we're all related, at some point on this planet—and maybe we're faced with it now and don't know it—the environmental crisis or . . . some other kind of crisis, some asteroid comes toward us from outer space . . . But maybe we have to work together as one, and work to solve our problems and think of this as our world; this [is] all we have, and we have to start working to save it.

We have been sitting on the Mule for this conversation, and it is time to move, stretch out the kinks, and say goodbye. Bill has a trucker coming to take some calves; I need to head for another meeting. I tell Bill thanks and get into the car.

I leave the farm thinking about Bill's last comment: "But maybe we have to work together as one . . ." Both our science and our religion agree with Bill that we are "all related, at some point on this planet," whether it be by DNA or faith or some profound sense of care for this place and love for all creatures that arises from a source we cannot name. I'm grateful that there are farmers, folks in any profession, who think that way, who know that we have to work together. That is not only a sign of hope. It also means that there is still plenty of good work, meaningful work, to do in this world, and we can rejoice, take refuge, even find solace in that.

CHAPTER 9

Farming Connects Us All

Every story has a context, akin to an ecosystem. That ecosystem includes its history, its environment, and its cultural and social context. This chapter, along with the chapters in part 4, provides our farm stories with their context in the state, the Midwest, America, and, in a small way, the rest of the agricultural world. Everything discussed, no matter how far it may seem from Minnesota, Kansas, Iowa, Texas, New Hampshire, Ohio, or California, is part of our farm story, influencing present circumstances and practice.

Ecosystem, then, is an appropriate name for such a context, for the first insight of sustainability is that everything is related. Further, our connections with the rest of the earth, with other species, and with plants and rocks and stars are not simply genetic or chemical or physiological. They are also temporal. So the ecosystem in which agriculture is embedded includes not only the present context of farming; it also includes a historical context, extending into the deep past. There are few contemporary farm issues that do not have ancient roots, sometimes in cultures foreign to our own. And every good farm practice put into use today creates a story that bodes well for the future. All those stories, past and present, impinge on our own. Farming offers us a comprehensive view of the larger culture that cuts across every concern we have in creating a sustainable society: poverty, the environment, racism, even finding and cultivating a meaningful life rather than a merely prosperous one. Agriculture is one of those great cultural intersections where environmental concerns and social justice issues, including the economy, come together and play themselves out for good or ill.

THE FALLING ARCH STORY

Ecologists today use the term "keystone species" to designate a critical animal or plant, like the center stone in an arch, without which a whole ecosystem would collapse. David Henry, an ecologist at Kluane National Park in the Yukon Territory, makes a very convincing argument that the keystone species in boreal forests, for example, is the arctic hare. Then— a Jack Benny pause—"unless it's the red squirrel," he closes, with a mischievous twinkle. The difficulty in pinning down the keystone species may be one of the reasons the idea remains controversial.

Even if the concept is controversial in ecology, I think there is a keystone in our cultural practice that is easier to identify. Agriculture, not the usual economic markers of our prosperity, is the keystone for the whole culture. Our best markers are not the gross domestic product and the gross national product; nor are they contemporary technology, industry, stock markets, or transnational corporations, most of which we could survive very well without. We have survived far longer without computers than we have with them, but we cannot survive long without a viable agriculture that produces healthy food. So the real marker of the state of the union may well be the health of our farms and our soil, the keystone essential to the whole culture. We'd like to think that Internet technology is the course of the future, but without healthy food and food security, which can come only from healthy soil, we have no future. The big-profit corporations like Monsanto, Cargill, Archer Daniels Midland, and ConAgra cannot provide us—far less the world—with healthy food or food security. Pesticides and herbicides and chemical fertilizers cannot provide us a food-secure future; the folks who make, own, and market the pesticides and herbicides, and the food they make possible, don't plant anything but test plots. Sustainable farming practices, however, may provide us both healthy food and food security, especially if they are combined with an alert consumer culture that understands the importance of healthy food, the dangers of processed foods based on industrial models, and where that healthy food comes from locally and how it is raised.

Ancient as it is, farming has always been a difficult and uncertain life. Some talk now of a farm crisis, as if farm crises come and go. Others insist that the crisis has been going on at least since World War II and that fluctuations in farm income are simply brief respites from the ongoing decline in the number of small farms and the increasing expenses and diminishing profits that provide the real pattern of farm life. The Farmer Summit, a group of farmers from the upper Midwest brought together by the Institute for Agriculture and Trade Policy at the University of Minnesota in 1998–2000, holds that the current difficulties facing farmers are but an extension of older crises. Indeed, there is evidence that the farm crisis has been going on from the earliest days of agriculture.

Lucretius, the old Roman poet and philosopher, perhaps understood the nature of limits better than we do. He composed his epic poem *On the Nature of Things* in approximately 100 B.C., but it sounds as if he is describing our current dilemma: "At length everything is brought to its utmost limits of growth by Nature. . . . This is reached when what is poured into the veins of life is no more than what flows and drains away. Here the growing time of everything must come to a halt." He adds, "Already the life-force is broken." The earth, in his view, was already past its prime. This was not a passing observation but something he had thought through. Lucretius believed that nature no longer produced enough food to nourish itself, let alone humans. The earth is like an old person, he thought, whose calories no longer provide the sustenance necessary to energize a hale and healthy body. Like an aging human, the earth did not seem able to lift nourishment into the "veins of life" to maintain its vitality. Some of Lucretius's comments about agriculture in his time make current conversations among farmers sound like echoes: "Already the ploughman of ripe years shakes his head with many a sigh that his heavy labors have gone for nothing: and, when he compares the present with the past, he often applauds his father's luck. . . . He grumbles that past generations, when men were old-fashioned and god-fearing, supported life easily enough on their small farms, though one man's holding was then far less than now."[1] Despite the apparent age of agriculture's problems, farmers have hung on and survived and kept food on our tables—at least some of our tables—for about ten thousand years. Today's crisis may seem more sophisticated and complex, but that is probably what farmers in Lucretius's day thought too.

Behind our efforts to understand these issues lies an assumption that we will survive these times too. Yet some people closer to the issues than I am are not so sure things will work out. In late 1999, I met with a U.S. Department of Agriculture agent in South Dakota. We sat in a truck stop filled with crowded shelves of trucker gear, empty paper cups, the clatter of dishes, and the drone of conversations. The agent told me he believed the difference between the farm crisis of the 1980s and our current circumstances was that "in the eighties you could see tiny holes, little lights at the end of the tunnel." Stirring sugar into truck-stop coffee so weak that, if you held it to the light, you could read the license plate numbers on cars at the pump outside, he continued, "If you could hang on and had a little luck, you could crawl toward one of those and find it." But not any longer: "This time it's just curtains. It's a dead end; there's no escape." Mike Rupprecht, the Minnesota cattleman who farms near Altura, knows the feeling. "I think we're doomed," he says. Mike Schuth, who has lived, farmed, gardened, and worked for agricultural agencies in Wabasha County all his life, when asked about the future of agriculture, says bluntly, "There is no hope. I don't see any."

Neither of these Mikes sees himself as a sign of hope that contradicts his gloomy view, but each of them is. And I have found many others in southeast Minnesota who are creating a future for agriculture, and for us all, by adopting good practice. Further, there are many small farms in the region that are making it: growing healthy food, keeping healthy animals, restoring soil health, finding niche markets, working the farmers' markets in our small towns and cities, and paying their mortgages as well. And many of those farmers are leading satisfying lives that, though they may be stressful and filled with hard work, are also rich with purpose and meaning. If some of the news about agriculture is depressing, we need to face it and understand why.

A STORY OF HARD NUMBERS

One can understand both Mikes' concerns, for the generalities that can be drawn from the statistics in our time are well known and deeply foreboding. I toured one Midwest county with a farmer who was born there. He pointed out the reality as we drove the back roads, past empty barns and useless silos: "This guy's not milking anymore. That guy isn't milking anymore. The one over there with five silos, he doesn't milk anymore." The USDA reported that total farm receipts for midwestern commodities in 2000 were the lowest since 1994. The total for 1997 was $207.6 billion. In 1998, it dropped to $196.8 billion, and in 1999, it fell below $192 billion. Crop receipts, including those for wheat, feed crops, and soybeans, were all down in amounts ranging from $2.5 billion to $15 billion. The only farm activity that was almost holding its own was livestock. In that area, receipts in 1997 were $96.5 billion. They fell to $94.5 billion in 1998 but came back a bit in 1999.[2]

For decades, urban incomes have risen and rural incomes have fallen. Even in recessions, when urban incomes fall more sharply than rural incomes, the gap remains. In an article in the USDA's *Rural Conditions and Trends*, optimistically titled "Rural Areas Show Signs of Revitalization," author Peggy J. Cook concluded that revitalization seemed to be occurring only in some rural counties and that, "despite a slight decline during 1993–1994, the percentage of rural people with poverty-level income in 1994 remains higher than in 1979 and 1989." Cook went on to acknowledge that, "during the 1990's, the rural-urban gap in real per capita annual income remained approximately $6,000 or greater while rural nonfarm jobs in 1994 paid $8,093 less than urban jobs." The Center for Rural Affairs reported in 2002 that "only one county among the poorest 50 counties is a metropolitan county, and many of the poorest 50 counties are very rural, agriculturally dependent counties. For the fifth year in a row (1996 to 2000 data), the rural Great Plains can lay claim to being the poorest region in the nation." And the USDA's 2002 agricultural census revealed that while the market value of agricultural goods was rising slowly, production costs were also rising, often at a higher rate.[3] Factor in the rising debt load, and at some point, the cost of production exceeds market value produced. The farm then becomes a small boat in a storm, with a hole in the hull and water lapping over the stern.

The data released by the U.S. Department of Commerce in November 2002, as reported in *Agri News*, were equally chilling. Those numbers showed that just twelve states received a larger share of personal income from farming than from unemployment payments in the second quarter of 2002. The year before, there were thirty-four such states. During the second quarter of 2002, seven states, including Minnesota and neighboring Wisconsin, had overall losses from their farms, with Minnesota showing a loss for the last two quarters. The year before, no states had shown losses. But more deeply disturbing were the numbers *Agri News* did not report, covering the data from 2001 through the second quarter of 2002. Minnesota began 2001 showing total farm proprietors' income at $271 million. The state's farm proprietors' income fell every subsequent quarter, to –$58 million in the first quarter of 2002 and to –$917 million in the second quarter. With the exception of the final quarter of 2001, non-farm income rose steadily, from $10.42 billion to more than $11 billion. Minnesota farmers' personal income for the same period also showed a decline, from $778 million to –$380 million, whereas non-farm income during the period rose steadily, from more than $163 billion to more than $169 billion.[4]

Many of the remaining farmers or their family members—mostly spouses and kids in high school—work off the farm more than two hundred days per year. Not all of those who work off the farm do it because they need the cash to keep the farm going. Teacher Sue Rabe, for example, says, "My work is different, and I do it because that is what I'm supposed to do with my life." But for others, working in town is a desperate attempt to keep the farm. Low prices for agricultural products and new demands for money for health insurance, equipment, and other necessities mean that the only way to sustain the family is to increase the size of the farm. As Peter Rosset, then codirector of Food First, a nonprofit organization that works on hunger issues, said, "It just takes more acres to produce the same income." The consequence is not simply that some farm family members are working in town; many families are leaving the farm altogether. As in our economic life in general today, the big get bigger and the small get shoved aside. In farming, as the number of acres planted in commodity crops increases, the number of farms declines. Jeff Gorfine, former chair of the Experiment board, put it as a series of questions as we rode through the country in his car: "You got a farm bill that's essentially business as usual for the next six to ten years. So aren't there going to be more people off the land . . . and isn't there going to be more of a land grab? I mean, aren't more people going to be pushed off because there's going to be more commodities?"

Perhaps such losses don't matter to others. Perhaps losing a farm doesn't mean much to anyone except the folks who lose one after generations in the same place. After all, the land will probably still be used in farming—unless it's near urban sprawl, in which case it may be taken by a developer bent on high-cost (and higher-price) homes.

But many do feel a sense of loss, as shown in a story about Harry Truman that Peter Forbes tells in his book *The Great Remembering*. Truman grew up on a farm. After he retired from public life and returned to Missouri, some developers prevailed upon him to say a few words at the dedication of their new shopping center, called Truman's Corner. When the time came, Truman, seventy-three years old and using a cane, mounted the lectern and discovered that the shopping center was built right on the old family farm. He began,

> It is a pleasure indeed to be present on an occasion like this. It gives a family a rather bad case of homesickness though. This farm has been in the family nearly 100 years . . . and then we've been right here on this farm since 1905 as a home residence. My brother and I have planted and plowed corn and wheat and oats all over this acreage here. It's now being turned into this wonderful business center. We certainly hope that it is a successful one, because while we would have liked very much to keep the farm as a home and have run it as a farm, we know very well that progress pays no attention to individuals. We don't want to stand in the way of progress, but that still doesn't keep us from being rather homesick for the places we knew when we were children, when we were three and five and two years old.

Truman rambled on a bit and then got warmed up again by his sense of loss. One can imagine the expressions on the faces of the developers whose new center was being honored as Truman thumped his cane on the floor and said, "And when I get false teeth and get bent over, you'll hear me pounding on the floor and telling the kids what a wonderful place this was before these birds ruined it."[5]

Nevertheless, the farms continue to go, their disappearance aided and abetted by increasing acres of commodities and federal policy. The Land Stewardship Project reports that acres planted in corn in Goodhue County, Minnesota, increased by approximately 20 percent between 1987 and 1997, while soybean acreage increased 80 percent. In the same time period, the number of farms went down approximately 19 percent. USDA figures show that, in the same ten-year period, the number of farms lost in Winona County was 197, while Dodge County lost 156.

Though the number of farms has declined steadily since the end of World War II, the 1980s saw a big hit. In Goodhue County, according to the USDA, the number of farms went from 2,849 in 1950 to 1,686 in 1987. Winona County suffered the same trend, with farm numbers going from 1,904 to 1,174. Dodge County took the biggest hit: from 1,712 farms in 1950 to 830 in 1987, a loss of more than 50 percent.[6]

The time frame from 1950 to 1987 generally correlates with a period when corporate industrialists, agricultural economists, university faculty members, policymakers, and government employees did their best to influence farm policy so that one-third of American farm families would be forced off the land and thus available for factory work in town. (See "The Role of Federal Policy," below.) The increase in commodity crops coupled with the loss of small farms is not going to end soon. Gyles Randall, codirector of the Southern Experiment Station in Waseca and a soils scientist at the Southern Research and Outreach Center, estimated, "In ten years, Waseca County will have ten farmers raising 170,000 acres of corn and soybeans." But I am not so sure. American taxpayers, under the false impression that all farmers, small or large, have their hands out for subsidies from the government, will not long continue to underwrite federal farm bills that sound like special giveaways that other workers do not receive. Newspapers reported at the end of 2003 that Vice President Dick Cheney had insisted that new farm supports be stricken from the draft of the Republican platform.[7]

Another problem with the system as it stands is that it depends entirely on chemical pesticides, herbicides, and fertilizers, all of which are short-term fixes that kill the microorganisms in the soil, rendering it essentially dead. Chemical mortuary services cannot disguise the death for long. Further, the higher production and lower chemical use promised by such products as Roundup Ready soybeans and genetically modified corn have not been borne out in studies. A final problem is that our agriculture continues to be hugely dependent on oil, a short-term resource. Oil prices have now ballooned, but they will eventually diminish because there will be too little oil left to support entire industries, and its value will dissipate.

Who would continue to buy into a system with such problems? And who will suffer the most? My guess is that it will be those who have prospered the most under the present system: the biggest agribusiness commodity producers. Who will survive? Not the industrial giants, but the smaller farmers who have kept input costs down, avoided the chemical rat race, and maintained a diverse operation with animals, ample pasture, and healthy soil: in short, those who produce healthy food. For now, they are often limited to local markets and community-supported agriculture. But if they can hang on— and they have survived to this point, even without federal support—the agricultural world will be theirs, for without cheap and abundant oil, everything in the culture will again become local.

WHY SHOULD WE CARE?

Why should the rest of us care if farmers are leaving the land in droves? Or if farmworkers live with poverty and ill health? Why should someone in Poughkeepsie or San Francisco care what happens to a small farm in Indiana or Kentucky? For the most part, America's farmers and farmworkers are out of sight and out of mind. True, they are more visible in some areas as tourist attractions, as in Pennsylvania. But too often, tourists look at farms on their weekend drives into the country as if they are visiting museums, not understanding much about the practice, politics, or economics of real farm life.

If we can raise enough food with fewer farmers, a common assumption, and if chemicals can increase production and decrease the rampaging pests, another common assumption, what difference does a decrease in the number of farmers make? Why should anyone in Rochester or the Twin Cities care? Gyles Randall reports, "I was asked that very question, 'Why should I care?' by a member of the audience after a talk I gave in Rochester, and it comes up in some form when I speak in the Twin Cities or other urban centers." There are important reasons to care and to be informed as best we can about our farms. The real problem is that much of the general public, and many members of Congress as well, put stock in a set of myths: that larger farms are more productive and efficient, that farming is about raising commodities, that chemicals can fix everything, that export markets increase farm income. The facts, as far as I can discern, indicate that all those notions are simply false.

Rather than lament federal farm policy, which is easy to do, we might raise questions that will help us see behind the myths: Can we have viable metropolitan and small-town lives with fewer farmers? Are larger farms really more productive? What are the impacts of monoculture and of chemical agriculture over the long haul, not only on the land but also on our communities? Recent research indicates that larger size isn't necessarily more productive or efficient and that commodity crops don't really pay without government subsidies. Further, reports on export values and farm income show that the greater the exports, the greater the gap between corporate export income and farmgate income (what farmers receive for commodities sold). So what are the real issues that farmers, and we, have to face?

TWIN MYTHS

One of the driving forces in agriculture is the apparent necessity for increased farm size. Productivity and efficiency are the main selling points for industry and the heart of the U.S. Chamber of Commerce's pleas for endless growth. And productivity, efficiency, and large scale are more realistic in the sense that it takes more acres, in this era of depressed farm prices and increased demand for new necessities like health insurance, to support a family. "Depressed prices, not increased productivity," says Food First's Peter Rosset, "is what cost us our farm families and smaller farms." But Rosset sees more to that picture. Indeed, we are just beginning to understand how to measure productivity. The traditional measure, still in use in some indexes, is yield, the production per acre or hectare of a single crop. The highest yields often come from monoculture, planting a single crop in a large field. But in recent years, the definition of productivity has shifted, even in such circles as the World Bank and the International Monetary Fund, and we have begun to recognize that yield is not as effective a measurement as total output. Measuring only single crop yield makes larger farms appear more productive than they actually are, says Rosset. Larger corporate farms tend to produce only one crop and little if anything else, getting as much acreage into corn or beans as possible, whereas smaller farmers, both in the United States and around the world, produce various grains, fruits, vegetables, and fodder, as well as animal products, including meat, cheese, and leather for hides.[8]

Total output, Rosset explains, provides a more accurate assessment of productivity, for it accounts for all that the farm produces instead of measuring a single crop. Examining USDA data and using total output as the measure, Rosset found that smaller farms out-produced larger farms; indeed, the smaller the acreage of the farm, the more productive it was. The 1992 U.S. agricultural census showed that the average four-acre farm grossed $7,424 per acre and netted $1,400 per acre. On average, farms over six thousand acres grossed $63 per acre and netted $12 per acre. Every level of increased size beyond four acres, Rosset found, reduced both the gross and the net income. Farms of "27 acres or less have more than ten times greater dollar output per acre than larger farms," reports Rosset. Among the reasons he cites for the higher productivity of smaller farms are greater land-use intensity, greater labor intensity, the use of "non-purchased inputs," and more efficient use of other resources, such as forests and aquatic resources. A synonym for land-use intensity and labor intensity is efficiency. Multiple cropping and more effective irrigation practices also contribute to the greater productivity of smaller farms.[9]

The most productive American county east of the Mississippi, according to Rosset and others, is Lancaster County, Pennsylvania. Its annual gross sales of agricultural products amount to $700 million. An additional $250 million comes from tourists who love visiting the beautiful, traditional small-farm landscapes. One factor that makes this possible is Lancaster County's large Amish and Mennonite population; the county "is dominated by these small farmers who eschew much modern technology and often even bank credit," says Rosset. He cites Dean C. Ludwig and Robert J. Anderson, who contend that Amish and Mennonite farm communities in Pennsylvania offer a model for local, "indigenous development": "Instead of placing emphasis on the highest or global level of competitive interaction," they argue, this model "starts at the bottom and places emphasis on the development of strong, independent, semiautonomous regions with unique identities." This is just the opposite of the direction taken by most of our agricultural corporations, which aim first at export markets, where more and more economists say the United States cannot compete. It is also contrary to our free trade globalization, which puts corporate agriculture above even our national sovereignty. Other nations, Brazil, for example, have greater natural soil resources and lower production costs. Studies show that each year of rising agricultural exports has shown a corresponding net decline in U.S. and Canadian farm prosperity. Shannon Storey, president of the Canadian National Farmers Union, says, "The traders and the corporations make more from it [farming] than the farmer does." *Western Producer* notes, "Behind Storey's complaint is the reality of a growing gap between trade growth and farm income. While the value of food trade increased 140 percent from 1989 to 1998, net cash farm income increased just five percent. . . . Farmers, it seems, have not been able to trade themselves into prosperity."[10]

The Pennsylvania Amish communities are notably self-reliant, yet "their economies are market oriented and highly successful." They trade extensively with the outside, are good stewards of the land, and, especially telling, "their members find a great deal of meaning and centeredness in their work. While their economies are market based, they are highly diverse and integrated rather than fragmented, cooperative rather than competitive, based on value added rather than on commodity products, and dedicated to reciprocity more than dominance." These farmers feature the local market first and turn to export markets only when they have produced enough to sustain the local community. All of this conflicts with much of the prevailing wisdom of our urban citizens and flies in the face of conventional agriculture as well. Ludwig and Anderson conclude that these farmers' overall practice does not follow conventional wisdom and, indeed, appears to reverse it.[11]

One difficulty for those outside such systems is that there is a disconnect in our economy between productivity and income. We do not reward the more productive farmer with an eye on strengthening the local or regional economy or on following the best conservation practices. Instead, via a succession of farm bills, we reward the less productive farmer who has an eye on global markets and prevent his or her gaining any rewards for good conservation efforts. Indeed, as Larry Gates, a farmer and Minnesota Department of Natural Resources hydrologist, points out, "It's not just that we don't support conservation with our farm bills; we *penalize* farmers who seek to conserve soil, prevent runoff, and follow good conservation practices." Thus the average farmer on 158 highly productive acres has an annual income of $8,690 and gets nothing from the farm bill, whereas the farmer who produces and nets far less per acre on his 6,000 acres makes more than $70,000 per year, much of it subsidized by federal farm legislation. There is not much doubt about which income is preferable, and therefore which farm size is preferable, for many producers. Neither is there much doubt about which size is more efficient or which is around for the long term.

In farm country, we have talked of economies of scale for years. What that generally means is that larger is not only more productive but also more efficient. With larger acres, reduced fencerows, and ample space for bigger equipment, we think we can create those economies of scale that mean greater efficiency. Neither the history nor the numbers bear this idea out. In a speech to the Wisconsin State Agricultural Society, Abraham Lincoln said, drawing, as he often did, on his own experience, "The ambition for broad acres leads to poor farming, even with men of energy. I scarcely ever knew a mammoth farm to sustain itself, much less return a profit on the outlay. I have more than once known a man to spend a respectable fortune on one; fail and leave it; and then some man of more modest aims get a small fraction of the ground and make a good living from it. Mammoth farms are like tools or weapons which are too heavy to be handled. Ere long they are thrown aside, at a great loss."[12]

In our time, the most widely accepted definition of efficiency has become total factor productivity. Through mechanization, larger farms may reduce labor costs, for instance, but there are other measurable factors that impinge on efficiency, including land, inputs (especially purchased inputs), capital, and debt ratios. According to Rosset, when economists looked at the total factor productivity of small farms in other parts of the world, including sub-Saharan Africa, Asia, Mexico, Colombia, and Honduras, they found data from the 1960s onward that revealed greater efficiencies on smaller farms than on larger farms. However, that pattern is less clear in the United States. Rosset speculates that, here, smaller farms are less efficient than they might be because they cannot "make full use of expensive equipment." Large farms, on the other hand, are less efficient because of "management and labor problems inherent in large operations." They may also be less efficient because of the nature of heavy equipment, which harvests large fields in less time but leaves more grain in the field than older methods. Peak efficiency in the United States, therefore, is likely to be found "on mid-size farms that have one or two hired laborers." Rosset cites Willis L. Peterson, who found that, when data from surveys were corrected for a number of biases, advantages of larger farms found by some scholars "disappear, while there is evidence of diseconomies as farm size increases." Rosset concludes, "Even in the United States there is no reason to believe that large farms are more efficient, and very large farms may in fact be quite *inefficient*" (Rosset's emphasis).[13]

There is another aspect to the efficiency myth. Profit ultimately comes not from efficiency but from economic power. That is the secret of the transnationals' success. Richard Levins, a former agricultural economist in the Department of Applied Economics at the University of Minnesota, points out, "Mergers and acquisitions have been the principal ways by which agribusiness has increased the size of its participants." The importance of those mergers and acquisitions is that "increased size, in turn, leads to greater economic power." Efficiency is not part of this equation. "Those processors having the greatest economic power will be more profitable than those having less economic power. Similarly, profits available to suppliers will be claimed by the most powerful among that group."[14] In fact, our system not only encourages power and inefficiency but rewards them.

Large-scale agricommerce has other costs that are not included in the economic analysis. The loss of biodiversity is one such loss, a factor of primary importance. It seems to be one of the legs on which commodity agribusiness stands, with its required pesticides, herbicides, genetically engineered seed, and sterile fencerows. Is it a rule of thumb that the larger the farm and the larger its corporate market, the more it tends toward monoculture? Or does it only seem that way? I asked Deborah Allan, a soils biologist at the University of Minnesota, about diversity. "If you were going to make a real simplistic list of good practice, just the easiest things anybody could do that would create a fine farm, what would be on it?" She pointed out some benefits of diversity that outweigh costs and losses. Her response came from working directly with farmers in southeast Minnesota. In part, she said,

> just keep trying to get as much diversity of materials as you can, by doing rotations or by using cover crops, but you want different kinds of plant materials, different quality, and you want fibrous, rooted grasses, like oats, in the rotations. Dennis [Rabe] uses that. And he'll say, "I don't get any economic benefit out of these oats, but I get such

huge benefits that it's worth every penny I spend to put them in there, just because of what they do for my farm." That's a great way to think!

So the grasses, the legumes in rotation for cover crops are giving you the nitrogen-rich materials. Diversity of input, of biological input, reducing tillage as much as you can, and keeping permanent cover on the landscape. . . . All the old stuff, the old conservation practices: contour strips . . .

So what is more productive and more efficient? Where does the best hope for the future lie? The loss of diversity defeats us all in the long run, but public policy helps us garner the short-term economic gain, essential to many farmers, by using techniques that destroy diversity. If creating and sustaining diversity is productive and efficient in the long haul, how do we create public policy that encourages that? There was a time in agriculture's history in America when soil conservation and farm policy went hand in hand. *Soils and Men*, a 1938 USDA yearbook, is very clear:

> The earth is the mother of us all—plants, animals and men. The phosphorus and calcium of the earth build our skeletons and nervous systems. Everything else our bodies need except air and sun comes from the earth.
>
> Nature treats the earth kindly. Man treats the earth harshly. He overplows the cropland, overgrazes the pastureland, and overcuts the timberland. He destroys millions of acres completely. He pours fertility year after year into the cities, which in turn pour what they do not use down the sewers into the rivers and the ocean. The flood problem insofar as it is man-made is chiefly the result of overplowing, overgrazing, and overcutting timber.
>
> This terribly destructive process is excusable in a young civilization. It is not excusable in the United States in the year 1938.[15]

It was possible once for federal policy and soil conservation to be mutually supportive. Is it possible in our time? The question of whether we have the intelligence and integrity to make that happen again is huge. It includes the question of what we want to stand for in America. Do we want to stand for a sustainable culture, or do we want to stand for a corporate empire, which inevitably has a short lifespan? Thus far, those questions remain unanswered, though a majority of our elected representatives appear to have chosen the latter.

REWARDING INEFFICIENCY

There is a certain irony in all this. One tenet of our economic system is that the market will reward the most productive and efficient worker or business. We have known for a couple of hundred years that this is not the case, but it makes currently profitable businesses look good, so we continue to say it. For more than fifty years in America, we have rewarded less productive and less efficient farm practice, crippled local economies, and stripped rural communities of their economy, vitality, resources, and social capital in the process. There is no doubt that small towns go the way our small farms go. The agricultural economists' rule of thumb is that for every seven farms we lose, a retail business closes its doors in an adjacent town. Whether this is folklore or fact, the decline in rural farm populations is inescapably connected to diminished rural communities, a notion we will look at in some detail in the next chapter.

There is also no doubt that, though corn and beans can be grown with reduced soil loss and fewer chemicals, large fields of corn and beans have led to marked soil losses and a concomitant pollution of streams and groundwater. Successive farm bills have encouraged such soil waste and have propped up farming practices that would otherwise bankrupt any practitioner. Large fields can be contoured and can include waterways. They can utilize strategies like no-till, intercropping, and rotations, and leave fencerows with abundant habitat for small game and birds. Often, as we have seen, they do not.

The data on crop production for corn and beans make the picture very clear. In Illinois, for example, in 2005, the market value of corn was $1.95 per bushel; the cost of production was $2.95 to $3.21 per bushel. Similarly, the average market value of soybeans was $5.50 per bushel; the cost of production was $6.39 to $7.35 per bushel. Moreover, costs of production were higher in Illinois in 2005 than in 2004 by 5 to 11 percent, depending on the region of the state where the commodities were grown. According to *Farm Economics Facts and Opinions,* "Higher crop (fertilizer, seed and pesticides) and fuel costs were the main contributors to the total cost increases, although interest and land costs also increased in 2005. . . . Generally speaking the same expenses that increased for corn also increased for soybeans."[16] This pattern in corn and soybean commodity production—costs of production that are higher than market values—has been in place for years and is one of the reasons that the farm bill price supports for corn and beans are essential to the survival of their growers. Because they receive federal payments, corn and bean growers continue to grow the same commodities.

Because both corn and soybeans are row crops, neither provides much ground cover during this region's season of heaviest rains. Erosion is apparent on the gentlest slopes. Since soil is the base of every culture's economy, such soil losses impoverish the entire culture—a cost we all pay equally, farmer or urban resident, rich or poor, and a cost that the development of bio-fuels, which requires more bean production, will only exacerbate. Will the fuel efficiency and cleaner air that bio-fuels promise outweigh their costs in energy, bean production, soil loss, and stream pollution? In 2004, Tad Patzek, a chemical engineer who teaches in the Department of Civil and Environmental Engineering at the University of California, Berkeley, conducted a study of energy inputs in ethanol production. He concluded that it cost more energy to make a gallon of ethanol than that gallon provided. The American Coalition for Ethanol complained that his estimates were based on outmoded ethanol technologies no longer used, exactly the same argument the industry used when a 2003 study by Cornell agronomist David Pimentel showed similar results.[17] No one knows for sure who is right—and not many have asked—and yet we plunge ahead.

Those dollar figures, grim as they are, do not factor in the additional losses caused by such inefficient practice: soil loss to erosion, the impact of chemicals, the pollution of regional streams, the toxic bloom in the Gulf of Mexico, and the loss of habitat for birds and small animals, such as meadowlarks and cottontails. So the notion that we have a system that rewards productivity and efficiency is simply a myth with which we delude ourselves, a cloak behind which we can nurse our denial that poor practice is as destructive and as inefficient and unrewarding as it really is. Meanwhile, we continue to reward poor practice and inefficiency with our tax dollars and our federal farm bills.

COLLATERAL DAMAGE

Whether one sees the current situation as new phenomenon or old, the numbers that come from the USDA, the Census Bureau, and the Department of Commerce, as well as anecdotal information, all point to the idea that farming is getting harder rather than easier, that world trade as it is now organized is harmful not only to independent producers but to regional trade centers and small towns, and destructive of our social and economic institutions. When war kills noncombatants, we speak of collateral damage. Our farm stories also reflect collateral damage, as both ecosystems and humans suffer. Both are often mere bystanders, wounded by industrial and trade movements too large and fast moving to contain. The connections between us and the environment make this a matter not of environmental rhetoric but of survival for our human species as well as others. The damage may be unintentional, but, as in war, the pain is real, and the casualties are more often civilians than soldiers.

A Story about Natural Systems

One of the indicators of the interrelatedness of everything and the cost of new inefficiencies in our industrial-scale agriculture is the impact that the heavy, less efficient harvesting equipment now used in the Midwest has on the lesser snow goose populations nesting in the Canadian Arctic and sub-Arctic. What does that have to do with farming in the Midwest? This traditional migratory waterfowl from the sub-Arctic used to winter on the southern Gulf Coast. The heavy equipment that now does the work of harvesting has left larger amounts of grain in the field, however, and that increased food supply is one of the reasons that the geese have begun to winter as far north as North Dakota and western Minnesota. The population of the lesser snow goose has escalated dramatically. *Natural History* explains, "The bounty of food from agricultural sources, combined with reduced hunting, cut goose mortality rates in half during the last twenty years." Other factors, such as warmer than normal winters, may also have contributed to the population increase. By at least one estimate, the lesser snow goose population is growing 5 percent per year.[18] This may be a natural fluctuation in population, but our Western science has not been in the area long enough to know. If it is part of a natural cycle, it seems clear that midwestern farm practice exacerbates it.

Back in the sub-Arctic, this larger population is eating itself out of existence and, in the process, destroying its own nesting grounds. In 1968, the "colony of snow geese nesting in the willow and lyme grass fringes of the coastal salt marsh at La Perouse Bay near Churchill, Manitoba, numbered fewer than 2,000 pairs." By the spring of 1990, there were 22,500 nesting pairs, which increased the colony's size an additional 90,000 goslings that summer. Under that population pressure and the loss of grasses, "inorganic salts move from underlying sediments to the bare surface, raising soil salinity to as much as three times that of sea water," write the authors of the *Natural History* article, Robert Rockwell, a professor of biology, and K. F. Abraham and Robert Jefferies, both botanists. Ultimately, the plants die, the shoreline dries, vertebrates die or leave, the feed for other shorebirds is lost, and "nothing is left but dry sticks."[19] Both the geese and their environment become uncounted collateral damage, never included in the profit and loss statements of our industrial-scale farming systems or our systems that contribute to global climate change.

Larry Gates tends to take a long view. In a conversation about the geese, he points out that the current phenomenon may simply be part of a longer cycle than biologists have studied. "Perhaps the population explosion and subsequent crash," he speculates, "is only a season in a long natural cycle. When the geese are reduced again, the tundra shores will come back, the snow goose population begins to rebuild, and the cycle starts again." For now, the snow geese do what U.S. farmers did from the eighteenth century till the 1930s—they move on when the old place wears out. Snow geese now go to less degraded breeding grounds to nest, just as we once moved to less degraded farmland. By the time all the suitable nesting places are used up, "large stretches of coastline will have been destroyed, and numerous other species will have lost their feeding grounds."[20] The decrease in efficiency in large farming practice clearly has ramifications that we did not anticipate—even broader impacts than those on the economy, the immediate environment, and farming, none of it recorded in our cost accounting. Collateral damage does not magically turn into collateral when you go to the bank for a loan.

The Worst Conditions of Any Workers

Here at home, the more that small farmers have to struggle, the less able they are to take care of those who work for them, while the large food processors use their economic clout to squeeze the summer workers,

136

many of them migrants, who wield the knives and operate the machinery in their plants. The National Agricultural Workers Survey found that, "over the period of the 1990's, with a strong economy and greater, increasingly widespread prosperity, farmworker wages have lost ground relative to those of workers in the private, nonfarm sector." Between 1989 and 1998, "the average nominal hourly wage of farmworkers has risen by only 18 percent (from $5.24 to $6.18), about one half of the 32 percent increase for nonagricultural workers." Over 60 percent of all farmworkers "had incomes below the poverty level." Thus the "median income of individual farmworkers has remained less than $7,500 per year," while that of farmworker families was under $10,000 per year. The Center for Urban and Regional Affairs at the University of Minnesota reported in 2001 that the average wage for the approximately twenty thousand to thirty-five thousand field workers in Minnesota was $5.50 per hour, and that "field jobs that paid more than $5.50 were rare." Most of the farmworkers were immigrants; 81 percent of them were foreign-born, and 95 percent of those were from Mexico.[21] An Oregonian who had left New York to get away from racism and violence told me that he was shocked at the overt racism expressed about migrant farmworkers in the West. "At least in New York, among educated folks, the talk about race would be disguised or discreet. Here they just say right out, 'Those damn Mexicans,' or whatever."

Yet it is no exaggeration to say that American agriculture depends on migrant labor. Without it, industrial agriculture would collapse, which is one reason agribusiness vigorously resists unionization, with its attendant risk of strikes. Despite legal efforts to improve the conditions under which migrant workers live and work, migrant workers' lives remain perilous. Baldemar Velasquez, founder of the Farm Labor Organizing Committee, says bluntly, "Migrant farmworkers suffer probably the worst living and working conditions of any workers in the United States." As recently as 1990, migrant farmworkers picking onions in Oregon earned about eight dollars per day, though owners had an elaborate system to prove they were paid the minimum wage. Onion workers there filled one-peck burlap bags or wooden crates with onions, working under a sun hot as that in the slave South, in a climate so dry that the constant wind sucks two liters of water per hour from the human body. These issues of special concern to the upper Midwest too. Though it is not a region strongly identified in the popular mind with migrant laborers, twenty thousand to thirty thousand migrant agricultural workers are employed in the Red River valley region of northwestern Minnesota and the eastern Dakotas, and still more work the fields and processing plants of central and southeastern Minnesota.[22]

Agriculture is, by any measure, a hazardous field. A 1988 study found that ag workers had the highest incidence of work-related deaths: Between 1977 and 1987, the number of combined deaths per 100,000 industrial workers was 11. For agricultural workers, it was 52.1. Mine workers were a very close second, with an average of 52. There is, however, a tremendous difference in organizational clout between high-risk farmers and miners—in 1987, our government spent $100.67 million dollars on mine safety and $0.49 million on farmworker safety. Workers in our food processing industries, many of whom are migrants, have the highest rates of job-related injury and trauma of any workforce in the country. In *Fast Food Nation: The Dark Side of the All-American Meal*, Eric Schlosser cites startling U.S. Department of Labor statistics showing that, in 1999, "the incidence of repeated trauma injuries in private industry was 27.3 per 10,000 workers; in the poultry industry [mostly migrant workers] the rate was 337.1." Schlosser continues, "In the meatpacking industry [also with a high percentage of migrants] it was 912.5."[23]

Migrant workers also have higher rates of many diseases than other farmworkers, and far higher rates than the general population. The National Center for Farmworker Health identified the following occupational hazards for farmworkers: accidental deaths and musculoskeletal injuries from heavy and repetitive lifting; respiratory illnesses and dermatitis from exposure to pesticides, dust, plant pollen, and mold; obstructive lung disease linked to livestock and grain work; skin disorders; eye injuries; infectious diseases caused by poor sanitation and poor drinking water; urinary tract infections caused by lack of access to toilet facilities; and possible cancers from pesticide exposure. Many pesticides pass through the skin very easily, and they can cross the placenta to impact the fetuses of pregnant women who work in the fields. Definite correlations between pesticide exposure and birth defects have been found among farmworkers, including facial clefts, spina bifida, anencephaly, and neural tube defects. Dina Schreinemachers of the U.S. Environmental Protection Agency studied rates of birth defects in wheat-growing counties in Minnesota, Montana, North Dakota, and South Dakota. In all, 85 percent of the acreage was treated with herbicides. Children who were conceived between April and June, when herbicides are applied, "had an increased chance of being diagnosed with circulatory/respiratory malformations" compared with children born the rest of the year. In addition, infant deaths from "congenital anomalies significantly increased in the high-wheat counties for males but not for females."[24] The tragedy is that many pesticide-related illnesses and deaths could be prevented by reducing on-farm pesticide use and by increasing bilingual pesticide safety education and training.

Such impacts are not limited to migrant workers. Sandy Dietz, who farms organically, told me,

> I walk with a friend whose husband has cancer. She knows it is from his handling of chemicals. She gets angry and has to get away, and so we walk. She told him that stuff would kill him. Doctors can name the stuff that did it, she says, but they do nothing. She finally had to get some counseling. But hospitals are dependent on the pharmaceutical companies. They have them in their back pocket. And they are the same companies that are producing the ag chemicals. It's just the food, the overprocessing. There is so much heart disease, and so few people controlling everything.

Sandy asks, "What kind of competition is that, that does so much harm?" Mike Schuth and his family have farmed the hills above Wabasha for generations. "Just think with the cancer rate and stuff," he says. "It's just the same pattern over and over again. And who's putting all these chemicals out here in the land? Monsanto. And the worst part, now they're down in South America doing it."

Perhaps we could weigh these human health losses against real gains for agriculture through the use of pesticides, a kind of grisly cost-benefit analysis. Measuring the loss of human lives against costly retooling is common in large industries—automobile production, for example— that assume a certain number of injuries and deaths as a normal part of doing business. But the tragedy of pesticides is exacerbated because the pesticides may be counterproductive. The herbicides generate "superweeds" that cannot be controlled by the herbicides we have, and pesticides may have toxic effects on those who use them. In a 2004 study, Charles Benbrook, former executive director of the National Academy of

Sciences' Board on Agriculture and Natural Resources and now an independent consultant, reported, "Weed scientists have warned for about a decade that heavy reliance on HT (herbicide-tolerant) crops such as corn, soybeans, and cotton, would trigger changes in weed communities and resistance, in turn forcing farmers to apply additional herbicides and/or increase herbicide rates of application." In the three or four years prior to Benbrook's study, that was already happening to Roundup Ready crops. In sum, Benbrook writes, "All GE [genetically engineered] crops planted since 1996 have increased corn, soybean and cotton pesticide use by 122.4 million pounds or 4 percent." In 2005, the use of human subjects, including children, in pesticide tests by corporations was revealed in a congressional report highly critical of their methodology. The bottom line of these health conditions is revealed in the statistics related to life expectancy and infant mortality. From New York to Oregon, the average life expectancy for Latino farmworkers in the United States is forty-nine, compared to seventy-three to seventy-nine for the rest of the population. Infant mortality for this same group is 25 percent higher than for the general population.[25]

These issues are worth exploring at some length because American agriculture is more dependent on migrant workers than it is on herbicides, pesticides, and chemical fertilizers. There is ample evidence in our agricultural experience that we can grow great food—even greater, healthier food—without those things. Yet we'd have little food at all without migrant help at every step of the way, from clearing the land to planting, cultivating, harvesting, and processing.

In southeast Minnesota, Centro Campesino, an organization that seeks to improve the conditions of migrant workers, often sees its members in housing with no potable water. Some housing options have no cooking facilities, prohibit families, and require married couples to live in separate trailers. If families wish to meet and eat together (a powerful ritual in the culture), they must do so in the city park. Another camp I have seen offers housing with broken windows and sagging doors. Even with the provision of such poor housing, there is not enough room for all the workers. Many cannot afford town housing, and local residents have been known to simply shut their doors in the faces of Latino couples who answer newspaper ads for rental space.[26] The only alternative, then, is to sleep in the car. Kathryn Gilje, an Anglo and one of the founders of Centro Campesino, has experience with migrant workers in Texas, New Mexico, Iowa, and Minnesota. "The housing here is worse than in Texas or New Mexico," she says of Minnesota. Apparently, "Minnesota nice" extends only so far.

Main Street in Owatonna, Minnesota, which lies between two rows of older buildings, mostly red brick, has a nice, historical feel to it. To the right is a central city park with tall maples and oaks, and on summer weekends, a farmers' market displays homegrown organic foods and locally produced, value-added canned goods, honey, and handicrafts. People from all over the area visit, walking from display to display, buying the week's vegetables.

Across the street, I make my first visit to Centro Campesino, going through a narrow entry to climb a wide stairway and turn into the offices. A half-dozen spaces are crowded with people and equipment. On workdays, the place is a buzz of soft voices, computer hum, and activity. On one wall is a framed collage of Cesar Chavez, Robert Kennedy, and others from the halcyon days of organizing in the sixties and seventies. In one of the tiny offices, Kathryn Gilje introduces me to Consuelo Reyes and Victor Contreras. As the months pass, I will get to know others on the staff as well. Though both Consuelo and Victor have good English language skills, they are more comfortable with Spanish. My Spanish is limited to the standard greetings and an occasional daring phrase to show my ignorance, like *un poco*, so Kathryn translates for us.

Kathryn and Victor started Centro Campesino in 1997 to organize Latino migrant agricultural workers in the Owatonna area. Both now work full-time for the organization. Consuelo was a volunteer "founding mother" of Centro Campesino; after being trained, she came to work in the office, just a few months before my visit. By now an old hand, Consuelo organizes workers and works with youth. She provides Friday and Tuesday afternoon programs in Mexican culture and leadership development for children in kindergarten through eighth grade, including programs on Spanish language, Mexican history, and current events. They call it Club Latino. Some Anglo kids also participate, which she appreciates; it seems to be a good sign of sympathy and support. Kathryn, Consuelo, and I sit down in a tiny office to talk, and I ask about Club Latino.

> We do this because we want the kids not to forget life in Mexico, and to learn about it, and to keep their language.

Consuelo sounds much like Eskimo and American Indian families I knew in Alaska, talking about their languages: they want their children to know both. But there is no bilingual program in Owatonna, and no prospect of one. Many of the children come to school as monolingual speakers of Spanish. Because the students are trying to learn content and concepts in a language they do not understand, the teachers sometimes think they are not very bright.

I worry about that. My littlest girl—I have two and one on the way—is very sensitive. The teacher asked the class one day to draw a picture of a slide. My daughter drew a nice picture of a slide with steps and a long curved slide. The teacher drew a big red X through it and said, reproving, "This is not what I asked for." It turned out that she had asked for a drawing of a sled, sliding on snow. . . . My little girl felt bad.

I ask if Consuelo did anything about it, an effort I knew would require a certain courage.

Yes, I went to talk to the teacher, and told her that she might have pointed out that it was a nice drawing of a slide, and that what she had really wanted was close to that, in the language, but different.

"Did it make a difference?"

(*Consuelo laughs.*) Well, she said she would try to work on it . . .

But it seemed clear that the concession was made grudgingly, and Consuelo did not expect things to improve. I learned they didn't until later.

It's better this year, now that she, my daughter, is in second grade and has more language.

Consuelo organizes the Club Latino program, but she also wears many other hats, as is common in many nonprofits, where folks tackle whatever needs to be done at the moment. Centro Campesino provides daycare for workers' children, and that is another of Consuelo's responsibilities. She also works on the legalization of immigration status for workers and with the Centro campaign to make worker policies at Chiquita Foods, the local processing plant, more humane. When not organizing, Consuelo is in the office, answering questions and passing out information to drop-ins.

Many people come here from southern Texas, where they are recruited and brought here. They are humble, honest people, looking for work. There are approximately 350 migrant workers in and around Owatonna.

Of those workers, 330 are members of Centro Campesino. There are also about sixty Anglo "friends" of Centro Campesino, who want to help, if only by contributing money to the cause. Centro's funding comes from both foundations and individuals. Memberships cost five dollars; friends pay thirty dollars.

The work has brought some good results. There is a carpet in the day-care center, some appliances in the kitchen, and a bathroom. It took some serious negotiations, but some changes have also been made to workers' housing, including hot water in individual housing units and a storm shelter, which is important in this country of destructive twisters. But another setback came the year after they won those concessions.

The canning company fired most of the workers who worked for change.

And wages remain very low. The starting wage for migrant workers is still $5.75, though union employees who do the same work receive a higher wage. Consuelo has been through the mill with migrant work, and she tells me about working for the Iowa Beef Producers plant.

I've lived that experience. They treat workers very badly. They are like slaves.

Kathryn has done some organizing work in Iowa too. She says, "Storm Lake is a scary town. We'd have a meeting in someone's house and the police would just keep circling around it as long as we were there." But it seems that things are not all that much better here in Minnesota. In my first conversation with her at Centro Campesino in 2002, Kathryn said, "We want to see changes in the housing. We've opened an office in Montgomery. Green Giant houses people in trailers there. They are divided by gender: seventeen to twenty-one men in a trailer, seventeen women in another trailer. Husbands and wives cannot be together. No kids are allowed. Workers are charged $3 per day per worker for the facility. There is no water and no kitchen. There are centralized bathrooms in a separate trailer. In Warsaw, a camp not controlled by Seneca, workers are charged $350 to $700 for a room about twelve by twelve [one side of a sort of duplex shack, it looks like in the picture I'm shown] with no bathroom and no cooking facilities. There are two toilets in another nearby shack, with the sewage running out on the ground about twenty-five feet in front of the trailer." Part of the dilemma is that trying to get the company to provide better housing can prompt retaliation, as Consuelo points out.

They shut the housing down, and then people have to sleep in cars. They have no place to go. . . . People ask us why we stay. But our workers come from the poorest parts of Texas and Mexico. They come hoping to make a little money so they can send it home to their families, and because they want their children to have a better life. They think they will find it here, and so, in spite of the difficulties, we stay. We have to sacrifice something for the sake of our children.

I comment on the way people in Mexico City walk in the park on Sunday afternoon, how they hold hands: men and women, men and men, women and women, and all the kids, and how they put their arms around one another.

Yes, the culture is more free, happier.

"More flexible," says Kathryn.

The workers who do not stay year around go home in the winter, and their children get to see that life in Mexico, see how families live and know that happiness. But we have to recognize that we will be staying here. Our children are born here. We have to keep working.

Victor Contreras has been involved in Centro Campesino from the beginning. Earlier, when I asked Kathryn how she learned Spanish so well, she smiled, pointed at Victor, and said, "From him." I ask Victor about migrant workers in the region, and his response is a neat lesson in both history and social justice.

Migrant people started coming into this region in 1910. This area shows up in the records from the 1920s. In the beginning, people came to work in the fields. Now they work in nurseries and processing plants. Now there are twenty-five thousand to thirty-five thousand migrant farmworkers in Minnesota.

"It's hard to get accurate data from the state," Kathryn explains.

Parents, kids, singles—they all come. They are mostly Hispanic and they come from the poorest areas of Texas and Mexico. They come with a vehicle, a necessity. Here they face a life that is difficult: very low salaries, very hard work. They work in the spring, picking rocks from the fields, weeding in the fields. They work four or five months processing peas, green beans, corn, and squash.

Victor reinforces what Consuelo said—"These are humble people"— then recites a litany of difficulties that Centro Campesino works to rectify.

Housing is an area where the corporation does not want to give attention. Our people get asked sometimes, by local folks, why they separate themselves. But the company places the housing outside of town. Diseases occur among the workers at a higher rate than the rest of the population. People barely survive, but the corporations prosper and grow worldwide. The crew leader, who is often the recruiter who goes to Texas or New Mexico to find people and promise them good wages, sometimes treats people very badly. We know from experience and from testimony. When you see large fields of corn, you can't imagine the numbers of Hispanic people working in them.

Victor's own experience began in Iowa, as did Consuelo's, working for Pioneer Seed. His story is not about himself so much as it is about his father and others:

When working in Iowa with my seventy-two-year-old father, detasseling corn, we saw a tornado coming. But the fields were very wet. I was trying to catch up with my father because I was worried about him, but it was very hard to run in the mud, and I noticed that the only people caught out like that, everyone trying to find shelter, they were all Hispanics out there at that time.

"For us," inserts Kathryn, "detasseling is a right of passage. Lots of kids do it briefly, for a summer or two, in high school or college—but if you are still detasseling at seventy-two, and no retirement . . ." She shrugs, palms up.

Victor describes the beginnings of Centro Campesino.

We know that farmworkers don't have easy access to unionize. Migrants do exactly the same work for less money. When the National Labor Relations Act was passed in the thirties, agricultural workers were categorically excluded from protection and forbidden to take part. We were looking for ways to struggle together. In 2000, we began a strong membership drive. Our people grew in leadership and values, and we confronted the company in that year. We won some changes in housing, things that had not been changed in forty to fifty years. You can drive around and see the old housing. The history is alive in the land. There is now hot water in housing. We won a little there. But in 2001, Chiquita would not hire the activists again. We put in a complaint to the National Labor Relations Board. Their attorneys thought it was an excellent case. Our attorneys thought it was an excellent case. People came from Texas and Mexico to testify. But in the end, the government didn't move. People stood up and told their story. We lost against the government and the corporation.

The Centro Campesino effort to get Chiquita to install some hot water, mentioned by both Consuelo and Victor, worked out well till the following year. "The next year, those who participated in the protests that led to the hot water were not rehired," Victor tells me. Further, the hot water is not potable; it must be boiled or treated for use in cooking.

Another protest led to a small increase in wages, an increase long over-due, but when workers came back the following year, the wages were lowered again to their original rate. Each struggle won seems to mark the beginning of the same struggle that must be undertaken again, making migrant workers the contemporary equivalent of Sisyphus, pushing the rock up the hill only to see it roll back down again.

Yet even here, one finds a hopeful spirit. I ask Consuelo, "How do you see the future?"

She becomes very quiet for a moment, then responds, "I just see a lot of struggle. . . . In these times it is more difficult now, since September 11."

"What do you do to keep up hope?"

"My hope is here, in the organization. The support we get from one another, the help we give each other, helps. People really work to help each other."

When I ask Victor how he thinks the future will be, his response is terse, realistic, but not in the least bitter or antagonistic.

"Hard. . . . There is no support from unions. Schools are not willing to help our children. The churches do nothing."

"How do you keep up your hope?" I ask him.

"Talking to people, seeing people react, being able to support people in their rights. I get the energy from my heart." He smiles, pointing at his chest.

A month or so later, I go back to take some pictures of housing. Consuelo takes me out to the local company housing, which has been much improved. It is being cleaned up in preparation for the arrival of this summer's migrants, who will soon appear in pickups and cars, some battered, others new and proudly shined. A concrete storm shelter, untried as yet, sits adjacent. The buildings are unadorned cement blocks, small duplexes lined up in rows. A high chain-link fence surrounds the place. There are swings, a teeter-totter, and some picnic tables in back. Consuelo shows me where the nursery will open in the fall. It has a rug on the floor, but there is no equipment, and there are no toys. But in one of the good stories about what happens to workers, this room will be transformed later.

In each tiny apartment, there is a two-burner gas hot plate and running water. The workers who come to live here for a few months will bring curtains to hang between the dining-cooking-seating area and the beds. There will be a picture of the Virgin of Guadalupe or of Jesus, photos of family, and a small shrine with a candle. Their belongings will transform the room from cold and severe to warm and homelike. But the toilets and showers are in a separate building across the alley, and the running water in the duplex must be boiled before it is safe to drink.

On the way back to town, I ask Consuelo about going up to Montgomery to take some pictures of the trailer houses there. I have grown accustomed to her courage, and the courage of the other Centro Campesino folks, in the face of transnational corporate powers. But now her eyes get big and she says, "Oh, no. That would not be good for you." For a few moments, that seems to be it. When I finally press for more, she adds, reluctantly, it seems, "They do not like to have people looking around. They would call the police. Maybe this summer, when there are many people. We could go then, to visit friends. That might be OK." Later that afternoon, I go to Montgomery without Consuelo and take a few pictures, feeling cowardly, looking over my shoulder as often as I do at scenes I want to shoot.

There are Anglos who are helpful and willing to put forth some effort to make good things happen. C. J. Taylor, an Anglo Texan who had never been to Minnesota, learned about the barren nursery. She prevailed upon her church members, family, and friends to contribute to it. Responding to a Centro wish list, she collected toys, furniture, playground equipment, an electric keyboard, cribs, beds, bassinets, soccer uniforms—so much stuff that she and her husband George had to load it all in a big U-Haul trailer. She drove the trailer from Salado, Texas, to Owatonna to deliver it in person and see the facilities for herself, and then returned to Texas to make a graphic report to her generous Austin Metropolitan Community Church, other friends, and family. Another Anglo couple was interested in helping with a new Centro Campesino housing project in the first stages of development. No land had been acquired, but Centro Campesino had its eye on some land that would cost one hundred thousand dollars, which they did not have yet. While listening to Kathryn's description of the project, Barbara and Erroll Wendland, also Texans, asked if Centro had an option on the land they hoped to purchase. Kathryn said yes. Erroll said, "We think you should exercise that option right away." Kathryn paused, unsure what he meant. Erroll continued, "We can't help as much as we might have a couple of years ago, but if you give us some time, we'll cover the cost of the land and you can get on with the housing. Meantime, we'll be happy to sign a note at the bank, if that helps."

These signs of hope, from all sides, indicate that the struggle is not only worthwhile regardless of the outcome but that the outcome may, at least in some cases, be better than anticipated. Maybe Centro's dreams for its workers will yet come true.

THE MIGRANT EXCHANGE RATE

Given the difficulties they face in our towns and cities, what do our migrant workers contribute to the lives, industries, and agriculture in southeast Minnesota? Fortunately, the Center for Rural Policy and Development in Mankato provides some hard numbers to answer that question in compelling economic terms. In the center's report *Estimating the Economic Impact of the Latino Workforce in South Central Minnesota*, author James J. Kielkopf concludes that migrant workers contribute more to the economy, and to our own economic gain, than many of us realize, and far more than they cost in government services. Even without the agricultural sector, in Mankato and the nine counties around it, the "total estimated value added to the Region Nine economy due to the Latino workforce is $484 million per year." Because there is no documentation of the ethnic makeup of agricultural workers in the region, it is difficult to put an exact number on their contribution in this sector. Nevertheless, the study shows numbers for some aspects of Latino migrant workers' impact on the regional economy. Latino workers involved in the area create 7,800 additional jobs for non-Latinos. The largest employers are in the food processing and packaging companies like those that Centro Campesino seeks to work with. The workforce in such plants is about 33 percent Latino. Because the products of a processing plant are largely for export outside the region, they bring increased wealth into the region, and the "effect on the local economy is multiplied because many other local industries serve it and depend upon it for survival." The presence of Latino workers and their families raises government expenditures in the area, reaching $48.3 million in 2000. More than $24 million of that came from state and local levels.[27]

There is a perception among some Anglos that Latino folks don't pay their way, that somehow they are a burden on the rest of society. Yet we all cost the government the public goods and services that we count on. And for our migrant residents, the government does not offer much by way of extra services. As reported by Consuelo, Owatonna does not offer bilingual education for Latino children or low-income housing for migrant families. Yet just in this region alone, migrant workers contribute $121 million in additional tax revenue, "$45 million of which is state and local tax revenue." Because their tax contribution is well over double the amount they cost in government services, "the Latino workforce's effect on taxes is to cause *lower* effective tax rates for the non-Latino residents of the region" (Kielkopf's emphasis).[28] In light of the numbers, it is clear that non-Latino citizens could afford to do more to ease the discomforts of those who come to work primarily for the benefit of others—including Chiquita.

Our migrant friends offer us additional important yet immeasurable contributions beyond economics. Many offer an openness to friendship that is all the more remarkable given the suspicion and even hostility they often encounter from others. In my experience, all it took was one smile, or one friendly gesture, and they were happy to see me again, show me their lives, serve coffee and food. Their colorful music, their guitars and maracas and melodic voices, and their rhythmic paintings and posters have become part of our local celebrations and parades. Their dances enliven our public events. Despite their desire for a better life and their willingness to risk everything to achieve it, they teach me how to be patient without losing my impatience for change. And it is not the least romantic, or metaphoric, to say that in the cuts and bruises, illnesses and accidents that our industrial and agricultural systems impose on them, they give us their bodies as well as their labor in exchange for the low pay we give grudgingly and reduce even further at the first opportunity.

A SERVITUDE STORY

Agricultural contracts are nothing new—contracts for "specialty grains, feeder livestock, fruits and vegetables, have been in use for some time"— but signs of their increasing use are all around us. Darren Hudson, assistant professor of economics at Mississippi State University, writes that in 1969, only 6 percent of all farms used contracts, amounting to 12 percent of the value of all agriculture. By 1993, 12 percent of all farms were contracting, and they raised approximately one-third of the total value of agricultural production. Since that time, the figures have varied only a little. Why the concern about contracts? Multiple issues face farmers who contract. All are related to a reduction in income, but a couple have a deeper, perhaps even more dangerous aspect. The following summarizes various reports on problems with contracts:

- Farming shifts from an owner-operated business to one in which the farmer becomes another "worker in an industrialized system controlled by someone else," as Neil E. Harl puts it. Those farmers who cherish autonomy may find that they have little to say about inputs, buildings, when to feed or harvest. A grower may invest in a particular kind of barn to house chickens, creating an enormous debt. If the contract is not renewed, the farmer, Hudson notes, may "thus be responsible for paying for assets that can no longer be used."
- The economic risks shift from corporate marketers to farm producers.
- Farmers are divorced from their markets, which are taken over by large production corporations.
- The winners in the contracting process are the input sellers and processors who hold production contracts that guarantee them a steady market for inputs and a steady supply of raw commodities for processing.
- In contracting, a greater percentage of the profit goes into the seed supplier's coffers, while the farmer's already reduced share shrinks even further.

Farmers nevertheless buy into contracts in order to have a guaranteed income and to reduce risk.[29]

Gyles Randall has studied farming as a soils scientist and an interested observer, not only in Minnesota but elsewhere in the United States and overseas. He described what can happen with contracting:

I remember out in New Jersey a couple of years ago . . . with the chicken and the broiler production being as intensive as it is out there, there was some movement by the integrator to put these custom operators on such a basis that you perform at such a level or you don't get a contract. So they were just jacking them up, and some of those growers felt that they were being penalized. They went into it for some security and a lack of risk, and after being in it for a couple of years, they're finding out that they've got a hell of a lot of risk involved in it. The company, they were pretty heavy-handed about it. I didn't get a lot of sympathy directed to my attention for those farms, or the way they handled the environment, or the ag economy, or the sociology of the area. Apparently they're kind of coming around to recognize that they've got to deal with these issues in a better manner. Capitalism had kind of gone too far in this case.

Both Harl and Randall might have added that contracting decreases the risk to the contractors while increasing their income. The basic concern is a growing imbalance in market power: input sellers and output processors gain access to steady markets and thus strengthen their economic power at the expense of the individual producer and, eventually, the consumer.

The power of food contracts with megamerchants such as Wal-Mart further illustrates how the system works. Randall pointed out that Wal-Mart is now the largest grocer in the world. What Wal-Mart does and how it does it affects its suppliers, local farmer producers, and its customers. Randall told about hearing the CEO of Land O'Lakes describe the bid process for a contract with Wal-Mart.

He outlined the difficulties in their business in getting the retail contract. He explained that Wal-Mart was out of Cheetos, and they needed cheese for the Cheetos. They put out this bid. They want your bid for cheese. There were about thirty-three different companies like Land O'Lakes and smaller that put in bids. After a month or so, Wal-Mart sent a letter back out to twelve of them saying, "You're in the running. Sharpen your pencils; we want your cheeses." And they sharpened their pencils and sent in their bids, and about two weeks later, they got a letter back. "You are one of three. Sharpen your pencils. Come on in." . . . It was really this kind of thing, right down to the hundredth of a cent on these mass volumes. Wal-Mart was able to squeeze them down because we have in this country this philosophy— you know—we'll run to Mankato to buy Cheerios because it is twenty cents a box cheaper than it is here in Waseca. We really are bent on this philosophy. So Wal-Mart screws down Land O'Lakes, then Land O'Lakes screws down the farmers, and they reposition their plants, and it affects all rural America, and the quality of life.

There are other problems, too. The control exercised by contractors over individual growers is greater than many realize, and their power is growing. Farming has always been an intellectual challenge requiring acute observation and high analytical skills. But now contractors can tell you when and how to plant your grain, what chemicals to use, when to harvest, and when to expend capital to rebuild your barn according to their or their architect's design. Dennis Rabe, redoing his own barn roof, would be in trouble. You follow the prescription or you lose your contract. In this system, you don't have to know anything about the land and its characteristics; you follow directions. It's paint by numbers for your land. As one farmer said, "Any fool can farm under those conditions." Whether you are foolish or wise, it is hard to prosper when you are merely following orders. As Ralph Lentz, the rotational grazer from Lake City, says, contract farmers are not farmers but producers.

Another analogy is that they are not farmers but franchise owners. In a 1995 article in the Pulitzer Prize–winning series "Boss Hog," the *Raleigh (N.C.) News and Observer* noted that hundreds of contract holders are "the franchise owners in a system that more closely resembles a fast-food chain than traditional agriculture." Indeed, consolidation of food processors and America's insatiable demand for fast food complicates production contracting, whether for animals or grain. North Carolina's hog population "has more than doubled in four years, and nearly all of that growth has occurred on farms controlled by the big companies. Meanwhile, independent farmers have left the business by the thousands." As in many other states, the land-grant university in North Carolina has long encouraged high production, and the state agriculture commissioner insists that there is no escape from the corporate takeover of agriculture. That is the future, in his view.[30]

The Northern Plains Sustainable Agriculture Society calls this movement the Tysonization of agriculture because it began in Arkansas with Tyson chickens; it has since been adapted to hogs and cattle. There is reason for alarm as the process expands. The agriculture society claims that Tyson's corporate profits rose "14 fold and the company's earnings per share from 1980 to 1994 ranked them No. 1 among Fortune 500 Companies." Yet despite the increases in profits, Tyson "refused to increase the price paid to growers during the following decade"—even though the cost of inputs required of farmers increased 50 percent during that time. "Chicken plant workers fared little better," and the environment suffered as well. Nitrate leeching and the use of cesspools for manure storage have polluted streams and caused numerous fish kills. In 2003, cattle raisers sued Dakota Dunes–based Tyson Fresh Meats, formerly Iowa Beef Producers, for monopoly practices that "violated the federal Packers and Stockyards Act of 1921 by using unfair, unjustly discriminatory or deceptive practices or devices; manipulating or controlling prices; creating a monopoly in cattle acquisition; or restraining commerce in cattle." Complaints occurred from 1994 to 2002.[31]

THE GENETICALLY MODIFIED MESS

Given our recent farm history, it is no wonder farmers have been scrambling, grasping for any straws that will keep them afloat. For many, one of the great hopes on the horizon in the mid-nineties lay in genetically modified (GM) crops. They promised resistance to insects, higher yields, and higher quality. "Corn hybrids genetically engineered to express *Bacillus thuringiensis* (Bt) toxins were developed in the 1980s and were commercially introduced in the mid-1990s," writes Benbrook. It wasn't until 1996 that enough acres had been planted to attract attention. By 2001, the number of acres had reached 70 million. In addition to Bt corn, Monsanto developed Roundup Ready soybeans, genetically modified to withstand Roundup, a Monsanto-registered trademark for an herbicide that is sprayed on the beans to control weeds. GM crops became such a lucrative pursuit that pesticide companies "simply bought all the major players in the corn and soybeans seed industries," Benbrook says. They also bought some smaller players. For $1 billion, Monsanto bought an obscure company called Delta and Pine Land, which owned the patent on a "terminator gene." "Engineered sterility is not uncommon," explain William and Ann Cunningham, authors of *Principles of Environmental Science.* But this gene-set, they point out, can be easily "moved from one species to another, and it can be packaged in every seed sold by the parent company." Saving seed is no longer fashionable in the United States, the European Union, and other developed countries, but, as the Cunninghams note, "it is customary and economically necessary in many poorer parts of the world." Indeed, they report that the inventor of the terminator gene, Melvin Oliver, said, "The technology primarily targets second and third world markets." Plants were not the only object of modification. Recombinant bovine growth hormone (rBGH), created by genetic engineering, is injected into dairy cows. Monsanto Canada introduced Nutrilac, a brand name for another genetically engineered product, recombinant bovine somatotropin (rBST). Both were intended to increase milk production by 10 to 15 percent.[32]

What we know about GM crops is that there is no known way to prevent their contaminating non-GM crops and entering our food supply. The world's greatest storehouse of diverse purebred corn species, located in central Mexico, has been infected with GM traits, meriting a cover story in the *Nation*. Though Mexico had outlawed the planting (but not the consumption) of GM corn in 1998, GM contamination showed up in Calpulalpan in 2002. The United Kingdom banned GM crops in November 2003 after a three-year study showed they destroyed pollinators. Greek cotton growers were dismayed to discover, in March 2000, that nine thousand acres of cotton would have to be destroyed because it was GM contaminated. They were more dismayed to find, in the spring of 2004, that 847 tons of GM-contaminated cotton and corn seed were on the market. It had come from Pioneer and Syngenta. In May 2005, "A ship originating from the United States was impounded . . . in Ireland after its cargo was found to include the experimental Bt10 GM maize not authorized for commercialization anywhere in the world." Another news release announced, "Illegal GM Maize Found in Japanese Imports." And, as we saw in chapter 3, StarLink corn, which is not approved for human consumption in the United States, has been found in food aid sent to Central America. We also know that the biotech corporations pay modest fines and generally do not alter their behavior; they continue to make their promises and insist, with overwhelming public relations dollars, that their products are not harmful. According to Jeffrey M. Smith, who has been studying GM-related issues for more than a decade, Monsanto is so powerful from campaign contributions, lobbying, and personal contacts that the rest of the biotech industry relies on its influence to pull them through the public doubts about their products and protect them from government scrutiny till they can produce something that may be beneficial.[33]

Monsanto and several other pharmaceutical companies had become famous long before. Monsanto acquired considerable public attention for producing PCBs (polychlorinated biphenyls), which have since been deemed carcinogenic and have been banned since the 1970s. Monsanto was also one of the companies that brought us Agent Orange with promises that it would defoliate Vietnam but not affect human health. It did defoliate Vietnam, but it also caused untold suffering among our veterans, and is still considered among the Vietnamese to cause birth defects and diseases. U.S. Department of Veterans Affairs Secretary Anthony J. Principi acknowledged in 2003, "Compelling evidence has emerged within the scientific community that exposure to herbicides such as Agent Orange is associated with CLL [chronic lymphocytic leukemia]." Other diseases linked to these herbicides include soft-tissue sarcoma, non-Hodgkin's lymphoma, and Hodgkin's disease.[34] What we have learned since Vietnam is that both herbicides and pesticides can affect our health.

Comparing the American discussions of genetically modified organisms (GMOs) with the British response reveals important differences. In addition to the scientific work, discussed below, that was undertaken in the UK and never matched in the United States, there were massive public opinion surveys and frequent flurries of debate in major papers like the *Independent* and the *Guardian*. The difference between the press coverage in the United States and that in the UK must have been obvious to even a casual observer. Long news columns, pro and con, helped inform the discussion of GMOs in the UK, whereas our own press has remained essentially silent on the issues. The British government, like ours, was eager to support the biotech industry and confident that the tests would show GM foods and crops to be perfectly safe, "no different from regular foods." In all likelihood, if the tests had come out clean, GM crops and foods would have come into the UK whether a majority of citizens wanted them or not. Nevertheless, Monsanto in the UK did not have an unquestioning press to support its efforts.[35]

One of the big selling points for GM ingredients in food was the claim that they were not harmful to human health. We are accustomed to assuming that the USDA, the Environmental Protection Agency, and the Food and Drug Administration conduct the tests that protect us from the dangers inherent in new sources of food and drugs. But, according to Smith, the USDA, under pressure from presidents both Democratic and Republican, approved the use of GMOs despite questions raised by its own scientists, in large part bypassing the regular testing requirements. Most of us have since been eating foods with GM ingredients, mostly unaware of what we are consuming. With little struggle and few facts at hand, Smith claims, Congress succumbed to industry pressure to prevent labels that identify GM ingredients or the food's country of origin. The system seems to work the same in Canada as well. There, the Health Protection Branch of Health Canada appointed five scientists to look at Nutrilac (rBST). The reason for the study, according to their 1998 report, was that "the New Drug Submission has been at BVD [Bureau of Veterinary Drugs] since February 19, 1990. Records indicate that the manufacturer of this product did not subject it to any of the normally required long-term toxicology experimentation and tests for human safety," in part because the chief of the Human Safety Division had never asked for such tests. "That no such tests should be necessary was due apparently to a mutually agreed upon assumption between Health Canada and the manufacturers of rBST products. Hence, the conflict; and the present 'gaps analysis' review by the rBST Internal Review Team." The report was negative toward both rBST and the department. In 1999, according to Smith, "Health Canada scientists told a new Canadian Senate Committee how they were threatened, harassed, and denied promotions in retaliation for their testimony the previous year."[36]

Similarly, in the United Kingdom, Smith reports, Arpad Pusztai, one of the most highly respected scientists in the world in his field, conducted an experiment in which he fed Monsanto's GM potatoes—destined for our fast-food french fries—to mice. Almost 20 percent of the mice died. The corporate response was not to look further into the potatoes' toxic effect and make them healthier but to launch a public relations crusade to destroy the career of the scientist. The story of the assault on Pusztai's integrity and his eventual vindication by other scientists was widely reported in the British media and helped raise the alert about GMOs there. It is a story that has since become a staple of anti-GM forces.[37]

In the United States, the decision of whether a new food is designated generally recognized as safe (GRAS) by the USDA has been left up to the company that makes it. A GRAS classification is important to industry because it means the USDA does not have to conduct safety testing. Thus, according to Smith, Monsanto certified that rBGH was harmless to human health, and the USDA approved it without further study. Calgene's genetically enhanced Flavr-Savr tomatoes made a brief appearance on the market; scary reports of tests indicating dangers resulted in their being pulled, writes Smith. In 1989, L-tryptophan, a GM dietary supplement produced by Japanese corporations, killed a number of people in the United States and caused exquisitely painful illnesses in thousands before the Center for Disease Control determined what was happening and the Food and Drug Administration pulled the drug from the market, Smith reports. Further tests in the UK, Europe, and the United States raised questions about potential allergens in GM foods. We do not know whether the dramatic rise in asthma and allergies is related to the allergen potential in the GM food we have been eating, because no one has done any studies to find out. In light of the thin science undertaken by industry and the lack of oversight by regulatory agencies, more than eight hundred scientists in eighty-four countries have now signed the "Open Letter from World Scientists to All Governments Concerning Genetically Modified Organisms (GMOs)," created in 1999, "which call[s] for a moratorium on the environmental release of genetically modified organisms (GMOs), a ban on patents on living processes, organisms, seeds, cell lines, and a comprehensive enquiry into the future of agriculture and food security."[38]

In fact, most of the scientific testing that has been done was undertaken overseas; the results, as we have seen, have led to the banning of GMOs in the European Union. In response, wrote Neil King in the *Wall Street Journal* in 2003, "U.S. Trade Representative Robert Zoellick in January blasted the EU for what he called its 'Luddite' and 'immoral' opposition to biotech crops, calling the ban 'a complete violation of WTO rules.'" The difference between our government's and corporations' views of GM foods and the EU's view has become so wide that Poul Nielson, a European development commissioner tired of Zoellick's criticism, said, "This is a strange discussion. Very strange. We are approaching a point where I would be tempted to say I would be proposing a deal to the Americans which would create a more normal situation. The deal would be this: If the Americans would stop lying about us, we would stop telling the truth about them." The *Independent* reported that when the UK's three-year study of GM crops' effects on the environment was released, EU Environment Minister Margot Wallstrom, in an undiplomatic moment, said, "Those Americans tried to lie to us!"[39]

One U.S. charge against European governments was that, in their ignorance, they were letting thousands of starving people in Africa die by hindering our development of GM crops that could feed the world. But people in developing countries starve not from lack of food but from lack of access to food. We now raise one and a half times the amount of food we need to feed the world, but hungry people can't pay for it, and corrupt governments don't deliver it. Even without genetic modifications and agribusiness, we have the capacity to feed millions more, now and in the future. Once again, it is a matter of scale and of will. Did we ever really intend to feed the hungry? There is not much evidence for it. Of all the grain that goes down the Mississippi on barges from the Midwest, only a tiny fraction goes to least developed countries, where hunger is the greatest. An Institute for Agriculture and Trade Policy study of the Mississippi River's grain traffic reported, "These grains are shipped to those who can best afford them, not to those most in need." The study compared shipments to twenty-eight wealthy countries and twenty-five countries designated by the UN Food and Agriculture Organization as "Category 5, the countries with the world's most serious malnutrition problems." The institute found that, for every ton of U.S. corn shipped to poor countries, 260 tons were shipped to wealthy countries. As for soybeans, none were shipped to poor countries; 17.8 tons went to wealthy countries. We raise food to sell to people who can pay for it. That's just good business. For the most part, the people who are starving are also broke. "But we know from several thousand years of observation," writes farmer Frederick Kirschenmann for the Northern Plains Sustainable Agriculture Society, "that small-scale, labor-intensive, local food production systems, wherein local people have access to production resources, are by far the most productive."[40]

U.S. scientists, too, have written reports, spoken at international conferences, and even testified in U.S. District Court about the difficulties of adequately testing GMOs. Richard Strohman, professor emeritus in the Department of Molecular and Cell Biology at the University of California at Berkeley, pointed out in *Safe Food News 2000* that Monsanto answers only the technical question of whether or not it can "move this gene and this characteristic from A to B." But, he said, "We know you can do that"; there are a "whole host of other questions."

Genes exist in networks, interactive networks which have a logic of their own. The technology point of view does not deal with these networks. It simply addresses genes in isolation. But genes do not exist in isolation. And the fact that the industry folks don't deal with these networks is what makes their science incomplete and dangerous. If you send these genetic structures out into the world, into hundreds of thousands of acres, you're going into the world with a premature application of a scientific principle.

We're in a crisis position where we know the weakness of the genetic concept, but we don't know how to incorporate it into a new, more complete understanding. Monsanto knows this. DuPont knows this. Novartis knows this. They all know what I know. But they don't want to look at it because it's too complicated and it's going to cost too much to figure out. The number of questions, the number of possibilities for what happens to a cell, to the whole organism when you insert a foreign gene, are almost incalculable. And the time it would take to assess the infinite possibilities that arise is beyond the capabilities of computers. But that's what you get when you are dealing with living systems.[41]

At the 2001 Toxicology Symposium at the University of Guelph, Joe Cummins, emeritus professor of genetics at the University of Western Ontario, concluded his presentation in clear language: "The bacterial genes used in GM crops have been found to have significant impacts on the individuals ingesting GM crops. The impacts include inflammation, arthritis and lymphoma promotion. The consequence of GM food genes being incorporated into the chromosomes of somatic cells of those consuming GM food and their unborn has been ignored by those charged with evaluating the hazards of GM crops."[42]

Finally, in his testimony before the U.S. District Court for the District of Columbia in the case of *Alliance of Bio-Integrity v. Donna Shalala*, Richard Lacey, a professor of medical microbiology and member of the Royal College of Pathologists in the UK, ran through a series of questions about GM safety commonly asked by scientists. Lacey, like Strohman, believes there are too many imponderables to check out entirely.

Whereas we can generally predict that food produced through conventional breeding will be safe, we cannot make a similar prediction in the case of genetically engineered food.

Therefore, the only way we can even begin to assure ourselves about the safety of a genetically engineered food-yielding organism is through carefully designed long-term feeding studies employing the whole food; and it would be necessary to test each distinct insertion of genetic material, regardless of whether the same set of genetic material in the same type of organism has previously been tested.

Even if the most rigorous types of testing were performed on each genetically engineered food, it might not be possible to establish that any is safe to a reasonable degree of certainty.[43]

The final story has yet to be told on this front, but American farmers have adopted GM corn and soybeans with astonishing rapidity. For a while, rBGH was fed to cows on the assumption that it increased milk production. It did. It also increased the incidence of mastitis, destroyed musculature, especially in the legs, and resulted in birth defects and reproductive disorders—risks that Monsanto's label now advises of. There are two problems with the increase in mastitis. One is that the condition must be treated with antibiotics, which may find their way into the milk and thus into humans, with unknown results. The other is that the potential profits from any increase in production are reduced by veterinary bills and losses from contaminated milk. Further, there are indications that rBGH itself can be absorbed into the milk we drink, and research wars are waging over whether that is enough to create problems in humans. Indications are that rBGH creates an insulin-like hormone that is present not only in cows but in humans as well. The hormone increases cell growth and is linked to breast cancer.[44]

There is other evidence now that genetic modifications are not as profitable as they were claimed to be, that over a few years they simply breed hardier insects and more determined weeds, and the cost of inputs goes up faster than the profits gained. Indeed, even the increased production seems to be spotty at best, and conventional corn often seems to do as well or better, at a lower cost and with more sustainable practices. In a recent paper, Charles Benbrook looks at the costs of Bt corn for farmers. He finds that

> the impact of the Bt corn premium on seed industry profits has been remarkable. In the case of industry-leader Pioneer Hi-Bred, the Bt corn premium boosted earnings from seed corn sales by 7.3 percent

over the 1998–2000 period. In terms of Pioneer Hi-Bred's after-tax income, the Bt corn and seed premium was almost 20 times greater, reflecting the loss of $100 million in 1999. Put another way, without the Bt corn premium, Pioneer Hi-Bred would have lost almost $200 million over this three-year period, or over 7 percent of total revenue from corn seed. . . . Bt corn had a similar effect on Pioneer and Monsanto revenues from corn seed sales. . . . Over the three-year period the Bt premium accounted for just over 9 percent of Monsanto seed corn sales. . . .

The financial impact and importance of Bt corn was greatest in the case of Syngenta. About one-half of total Syngenta corn seed sales were Bt varieties, more than twice the share of Pioneer and Monsanto. The Bt premium increased Syngenta corn seed revenues by over 18 percent in this three-year period.

Transnational corporations are making a fine profit. But what about the impact on farmer income of the premium prices that farmers pay seed corporations for Bt corn? "Every acre planted to Bt corn has increased farmer seed expenditures an average of $9.80 per acre, about a 35 percent jump," Benbrook reports. Further, based on USDA corn production, cost, and return data, Benbrook asserts,

the biggest jump in "Seed and Chemical" costs occurred between 1994 and 1996 and coincided with the emergence of Bt corn. These two key production inputs now account for over $0.40 in expenses for each bushel produced—between one-fifth and one-quarter of gross income. A little over a decade earlier, these expenses accounted for less than 10 percent of gross income. This marked shift in costs is one major reason why the seed and pesticide industry has, in general, prospered financially throughout the last three decades, while the balance sheet and profits of corn growers has substantially eroded. . . . The technology fee and other premiums charged for Bt corn has shifted to the seed-biotech industry a portion of the economic return farmers have traditionally received when investing in advanced corn genetics.[45]

Increased production or not, the market has not been as great as expected, in part because of the EU's resistance. So now Monsanto and the other transnationals are turning to GM wheat and hoping to create markets in China, Southeast Asia, and Latin America. Iowa State University's Robert Wisner, an ag economist, told the Montana legislature that "many European and Asian grain buyers will likely refuse to buy any spring or durum wheat from states that grow GM wheat." To emphasize his point, he added, "Every available indicator of foreign consumer demand points to a high risk of GM wheat rejection in export markets." Nevertheless, the *Agri News* article that reported Wisner's testimony concluded, "Monsanto has plans to introduce a genetically modified wheat into Montana, North Dakota and other states by 2005."[46]

What can a layperson make of all this? At least this: as farmers or consumers, we have to take with a grain of salt any claims that certain products will feed the world, or save the family farm, or increase production to undreamed-of levels, thus increasing farmers' profits. Skepticism should be the order of the day. Whether farmers or consumers, we should insist on seeing the data, and then we should insist on seeing the data gathered by those who checked the original data, before we buy or invest. If there are none, forget it. If the USDA and industry scientists do not follow the precautionary principle, we as citizens should, even—perhaps especially—if it hinders sales and reduces profits of corporations who have shown little interest in or regard for public health. Unfortunately, as transnational corporate contributions to research at our land-grant universities go up, the independence of the science they produce goes down. The folks who pay set the limits of the investigation. Nevertheless, though the financial and public relations resources are on the side of the great institutions, the final say, as the Europeans have discovered, really belongs to the consumer.

THE CUSHION

One important factor that the literature on contemporary agriculture rarely mentions is what the loss of family farms would mean to our society if we should suffer another economic depression like that of the 1930s. Some economists claim that could not happen now, but some farmers are old enough to remember when economists claimed it couldn't happen in 1928. Of course, between government regulations and sophisticated computers, we do have greater control over the stock market than we have had historically. Yet stock market shifts in Southeast Asia, droughts, drops in the Nikkei Index in Japan, and wars in Afghanistan or Iraq send shivers through our economy. The coming oil shock may set off an international economic disaster. Recent declines in stock market values and their effects on the market make many wary. Given the stock market fluctuations of the last two years, the loss of income for those who are retired, and the ever-increasing number of layoffs, not to mention the recent revelations of widespread corporate corruption, who really knows if the next NASDAQ nosedive will quit short of disaster, or rebound quickly enough to prevent it?

One of the great cushions America had to assuage its economic pain in the thirties and early forties was the number of people still on the farm. There may not have been much cash, but there were chickens in the coop, sheep and hogs in the pen, cows to milk, and still some "cashless" horses to provide labor and transportation. And there was subsistence. Most midwestern farmers supplemented the milk, beef, chickens, and pork they grew themselves by hunting and fishing. In Iowa, where I grew up, we hunted squirrels every fall and cottontails through the fall and winter. States farther west included medium-size game such as deer in the annual harvest. The prairie potholes and the great Mississippi and Missouri flyways supported substantial migrations of ducks and geese, and lakes and streams held crappies and bluegills, bass and walleyes, catfish and suckers, all of which were fit to eat. Though we loved to hunt, one of our rural values was that we never hunted just for sport. Everything we took hunting and fishing we ate, and whoever killed it cleaned it.

Every farm family gardened, and they canned vegetables and often meat for the winter as well. All of that became part of the family larder. Even in town, our family had a garden: rows of beans and peas, sweet corn and potatoes, as well as lettuce, radishes and onions, carrots and cucumbers. My mother was a registered nurse, but even working as many hours as possible for the needed cash, she canned four hundred quarts of vegetables and, when they were available, beef and pork as well. On farms like my grandfather's and my uncle's, there were usually enough chickens that some surplus eggs could be sold or bartered in town. Milk, meat, even exotic flowers such as gladiolas and peonies, raised for sheer beauty and for Memorial Day visits to the cemetery, were all sold or traded in the nearest community to provide a little cash for things we could not raise ourselves. Oats and corn were ground at the local elevator, and oatmeal and cornmeal mush were staple breakfast—and frequently supper—items. The mush was fried in bacon grease and covered with a bit of milk gravy and molasses or sorghum, all the produce of the family farm. That diversity in crops and livestock allowed a significant portion of the society to survive in a cash-short, severely depressed economy.

Since then, two things have happened. The first is that that cushion, the sheer size of which took much of the welfare burden off the government and the taxpayer, is now gone. People no longer have access to land. Many of those who still live on farms are now on farms so large and demanding that they don't have time to garden. The second thing is that most of us who have been off the farm a generation or more have forgotten how to farm and garden. We can learn again—there are excellent books on the market—but that knowledge that comes from the cumulative experience of generations of observation of how the land really works has been lost. It may never be regained, and if it is, it will take a long time, perhaps more generations, to do it.

THE FEDERAL POLICY STORY

Many of us once believed that the loss of small farms and the subsequent migration to urban centers were the inevitable results of social forces and the natural evolution of social history. Mark Ritchie, former director of the Institute for Agriculture and Trade Policy, believes that the loss of small farms and political capital among farmers is no mere accident of history or economics but the result of deliberate policy. In a monograph titled *The Loss of Our Family Farms: Inevitable Results or Conscious Policy?* published by the League of Rural Voters back in 1979, he traced the direction of farm policy between 1945 and 1974.

One of the most powerful nonfederal organizations to influence farm policy was the Committee on Economic Development (CED), a national group of two hundred industrialists and academics with enormous political influence, some of them already working on policy within the government. The purpose of the policy created in part by the committee during those years was to get farmers off the farm. According to a 1962 CED report, quoted by Ritchie, "the movement of people from agriculture has not been fast enough to take full advantage of the opportunities that improving farm technologies, thus increasing capital, create." The old farm parities, based on an entirely different principle from current subsidies, were seen as a liability because they helped keep small farms alive: "The support of prices has deterred the movement out of agriculture." One of the CED's goals was to get at least one-third of small farmers to give up. "If the farm labor force were to be, five years hence, no more than two-thirds as large as its present size of approximately 5.5 millions, the program would involve moving off the farm about two million of the present farm labor force." The report indicates that, in the committee's view, even more people should be forced off the farm to make up for the numbers who might enter farming during that time.[47]

The CED understood that getting people to give up their land would not be easy. According to Ritchie, Kenneth Boulding, a noted agricultural economist at the University of Michigan and a member of the CED's Research Advisory Board, wrote in 1974, "The only way I know to get toothpaste out of a tube is to squeeze, and the only way to get people out of agriculture is likewise to squeeze agriculture. . . . If the toothpaste is thick, you have to put real pressure on it. . . . If you can't get people out of agriculture easily, you are going to have to do farmers severe injustice." In the thinking of the CED's members, eliminating farm price supports would be a "signal to farmers" that they should get out of farming, and it would put 2 million people off the farm and into town, where they could become part of the labor force available to their industries. These people would most likely be in debt and desperate for jobs, though of course this possibility was not stated. That, in turn, would depress wages, making industry more profitable. It would also lower prices on agricultural products, increasing foreign trade, and would provide cheaper raw materials for domestic food and fiber processors. Those domestic food and fiber processors were, of course, at the table during the CED's deliberations, along with representatives of the major automobile industries, steel and fabricating industries, federal policymakers already within the administration, and land-grant university scholars. Farmers were not invited. The CED charmingly admitted that it did not know much about agriculture, but it did not hesitate to publish policy papers pressing for reductions in farm income aimed at removing farmers from the land. Was the CED successful in getting government to implement its policy?

Its members certainly believed so. Its 1974 report took credit for the movement of people off the farm and pointed out that the successful implementation of its policy recommendations made it possible.[48]

Ritchie believed that the real reason that major corporations and processors wanted people off the farm showed up in *An Adaptive Program for Agriculture*, the report issued by the CED in 1962. "Although the actual strategies and tactics of the CED are carefully spelled out, it's never hinted until the very end of this report what their underlying goals and motivations might be," wrote Ritchie. They were aiming at agricultural stability, and the CED report makes one of their goals explicit: "an atmosphere relatively free of the political pressures from farmers experienced in the past." It is easy to see why this was one of the committee's goals. The power bloc that farmers represented in the 1890s fueled the Populist movement. Decades later, there were enough people still on the land to provide the impetus for the Farmers Union, Farm Bureau, Non-Partisan League, and farm labor movements of the twenties and thirties. As Hinrichs remarked, they were known in Washington D.C. as the "farm block"; they "held the balance of power," forcing both Republicans and Democrats to hear their case. Those farm organizations not only held plenty of power in the thirties and forties, they had a program that was anathema to industry: they opposed corporate farming and the development of "state-owned bank and grain-trading enterprises in North Dakota." Ritchie adds, "The political strength of farmers was displayed in their ability to lobby for and win favorable farm policies. These policies helped strengthen their economic, thus political power, making it difficult for corporations to dominate agriculture as they were attempting to control other industries."[49] That was precisely the power the CED wanted squelched, and it has been—by the loss of families on the farm. Farmers simply don't have the votes they used to and now are so few in number that the U.S. Census Bureau no longer counts them as a profession in data collection.

This loss of political power, generated by deliberate federal policy and formulated under the influence of the CED, now hampers farmers' efforts to influence state and national legislation. The economic power of the merging and emerging transnational food corporations, and their massive infusions of campaign funds for congressional elections, means that their voices are heard in both state legislatures and the U.S. Congress. These corporations can afford to keep a lobbyist at the table for every committee hearing, floor debate, and mark-up session that creates legislation. They also can afford to have their attorneys sit in on every meeting to establish the regulations that actually determine the final character of legislation after it has been passed. Farmers and nonprofits cannot afford the expense of such counsel. Thus corporate attorneys exert a powerful influence on the law as it is finally promulgated and implemented, sometimes thwarting legislative intent entirely. Political power for farmers now would be useful in getting Congress to lean on the regulatory agencies that deal with antitrust laws, but the loss of farms has leeched away the political power of small farmers.

FIVE GREAT LOSSES FOR AMERICAN FARMERS

As we consider the implications of the farm crisis indicators outlined above, we might sum them up by pointing to five great losses that have occurred for American farmers over the last fifty years: loss of productivity, loss of efficiency, loss of access to markets, loss of independence, and loss of political power and influence.

The idea that we have lost productivity and efficiency flies in the face of all conventional wisdom, but the numbers put together by Peter Rosset support the contention (see "Twin Myths," above). The loss of access to markets follows the increase in contracting and the growing influence of transnational megacorporations. We no longer farm for the market; we farm to make the payment to the bank or to meet the contract with ConAgra, Tyson, Monsanto, or Cargill. Or "we farm to meet the conditions of the federal farm bill [another kind of contract] instead of the conditions on the farm," as one fourth-generation farmer put it in a community meeting sponsored by the Experiment in Rural Cooperation. Those same contractual arrangements may also lead to a loss of independence, "turning farmers into indentured servants," a phrase used by a man who began farming in 1937, speaking in a discussion about agriculture held in a local public library. Finally, the loss of the sheer number of farmers and their families living on the land means a loss in the authority and voting power of farmers. These losses are not the results of market forces or changing economics; they are the desired outcomes of federal policy. Those recommendations to the government came not from farmers but from manufacturers, academics employed by land-grant universities, and executives of the major corporations represented on the CED.

AN OLD STORY

The ties among good agriculture, good government, good business, and entrenched power—and our all-too-human impulses toward self-enhancement and, yes, greed—have been recognized for centuries. We began this chapter with the old Roman farmer and naturalist, Lucretius, of the first century. We'll close it with a story far older. Mencius, the Chinese sage of 2,500 years ago, said to the emperor, Hui of Liang,

> If you do not interfere with the busy seasons in the fields, then there will be more grain than the people can eat; if you do not allow nets with too fine mesh to be used in large ponds, then there will be more fish and turtles than they can eat; if hatchets and axes are permitted in the forests on the hills only in proper seasons, then there will be more timber than they can use. When the people have more grain, more fish and turtles than they can eat, and more timber than they can use, then in the support of their parents when alive and in the mourning of them when dead, they will be able to have no regrets over anything left undone. This is the first step along the kingly way.
>
> If the mulberry is planted in every homestead of five *mu* of land, then those who are fifty can wear silk; if chickens, pigs, and dogs do not miss their breeding season, then those who are seventy can eat meat; if each lot of a hundred *mu* is not deprived of labour during the busy seasons, then families with several mouths to feed will not go hungry. Exercise due care over the education provided by the village schools, and discipline the people by teaching them the duties proper to sons and younger brothers, and those whose heads have turned grey will not be carrying loads on the roads. When those who are seventy wear silk and eat meat and the masses are neither cold nor hungry, it is impossible for their prince not to be a true King.[50]

It is interesting to note the deliberate pursuit of the multifunctional character of agriculture so many centuries ago and the equally deliberate support of agrodiversity and good practice as it was understood at that time.

But the political consequences of Chinese government indifference, the greed of the ruling class, and the responsibility of government to support good practice are equally important to note. Mencius went on, still talking directly to the emperor, whom he apparently did not view as a "true emperor." There is no doubt that Mencius understood the implications and the danger of his next words, but he said them anyway—a courageous step that might have cost him his head.

> Now, when food meant for human beings is so plentiful as to be thrown to dogs and pigs, you fail to realize it is time for garnering, and when [people] drop dead from starvation by the wayside, you fail to realize that it is time for distribution. When people die, you simply say, "It is none of my doing. It is the fault of the harvest." In what way is that different from killing a man by running him through, while saying all the time, "It is none of my doing. It is the fault of the weapon." Stop putting the blame on the harvest and the whole Empire will come to you.[51]

Similarly, if members of Congress continue to say, "We're all for the family farm, but it's the market, it's the inevitable result of equally inevitable globalization, it's the way things are," while passing farm bills that penalize small farm diversity and conservation and protect affluent agribusinesses, absentee landlords, and transnational industries, then their action is no different from stabbing America's small farms and small towns to death and saying, "It's not my fault. It's the nature of the market." Since there is nothing natural or free about free markets, whose entire history and current development are fraught with backstage manipulation, and since there is nothing necessarily inevitable about globalization, our government is as complicit in the starvation of the poor as any ancient Chinese emperor.

But farmers' loss of political clout relative to that of agricultural industries also opens new possibilities. Collaboration among farmers, consumers who care about how their food is grown and processed, and environmentalists may yet create a political constituency that legislators and congresspeople will have to listen to. Those collaborations, including collective bargaining, though difficult and sometimes even painful to achieve, may lead to farmers' getting a bigger share of the agricultural dollar than they do now.

Late afternoon sun (Lake City, Minnesota)

Donna Christisen and pig (Plainview, Minnesota)

Field day at Art and Jean Thickes' (La Crescent, Minnesota)

Soil erosion after spring rain (Wabasha County, Minnesota)

(Right) Migrant worker and child (Owatonna, Minnesota)

(Below) Centro Campesino's Club Latino: Swinging at the piñata (Owatonna, Minnesota)

Pam Benike stirring the whey for cheddar (Elgin, Minnesota)

Migrant housing (Steele County, Minnesota)

Corn (Wabasha County, Minnesota)

Milking parlor, Dan Beard's farm (Decorah, Iowa)

Farm shoes: Sitting on the wagon, field day (Decorah, Iowa)

Checking the grass, field day (Decorah, Iowa)

Harvest (Goodhue County, Minnesota)

Ralph Lentz, winter feeding (Lake City, Minnesota)

Farmers' market (Lake City, Minnesota)

Yard light and full moon (Dodge County, Minnesota)

Part IV

It All Works Together, or It Doesn't Work at All

CHAPTER 10

Agriculture and
Community Culture

What are the ties between agriculture and community culture? What is
the relationship between small farms and small towns? Phil Abrahamson's family has been farming the highlands above the Root River for
generations. He told me about his purebred Angus herd and about his
great-grandfather, his grandfather, and his father, the last two of them
Angus men. We sat in the solid farmhouse, built at the end of the nineteenth century, and talked about his community and his church as well
as the farm business. I marveled at the way this small, frugal farm had
affected his industry, and at the fact that its progeny had been featured in
artificial insemination stud catalogs. Phil has one tractor fifty-seven
years old, currently being overhauled, and another over thirty. There are
some newer as well. Just as I was getting in the car to leave, Phil made
some of the ties between agriculture and community culture very
explicit: "I have to run into town this afternoon. We do a lot of our shopping locally. I shop the local hardware store in Lanesboro, and implement dealers there and in the little towns right around here. We've always
done it that way, been doing it that way since the farm first started. We
bank in town. We have these relationships that go way back. It all works
together. It's the way you help make things go."

It's no wonder that they say that if seven farms, operating in that
spirit, go under, a retail store in a nearby town has to close. The farm and
the town are parts of an integrated system. Unfortunately, not enough
folks see it that way; since I talked to Phil, the implement dealers have

disappeared, and the locally owned hardware store and the Ford dealership in Lanesboro have closed. Yet Phil continues to shop as close to home as possible.

IN A NESTED UNIVERSE

Everything exists inside a system larger than itself. There is an ecology for everything. In the world of sustainability, so much larger than the world of biology, ecology is not only a scientific concept but an appropriate metaphor for the myriad relationships of our human experience, and even for the ideas we hold. Thus sustainability is a concept that comprises many interrelated ideas, an ecosystem that includes many species of thought. One of its primary ideas is, simply, that everything is related. That idea is related to two more: that you cannot do any one thing, and that what we do in this place reverberates through and affects every place. And those ideas, in turn, are related to the idea that everything needs to be treated with respect. If I have respect for any part of creation, then self-interest might act as a brake on destructive behavior toward it, because everything is related to me. One outcome of this line of thought is this: If I am disrespectful toward anything, I reveal my disrespect for myself. If any one of those ideas becomes unhealthy through my failure to observe it, then the habitat for sustainability becomes unhealthy and unsustainable—and I myself become unsustainable sooner rather than later. Therefore, if we have any self-respect, we show it best through our respect for everything else.

This is true in every sphere of life, not just the environment. It applies to economics, a system in which the wealthy impoverish the poor and are, in turn, impoverished by their poverty. Lack of respect for the poor is one indication of our lack of respect for ourselves, an admission that we are not people of enough character to take care of our own. Yet a universe in which "we are all related," as the old Lakota song has it, is one in which rich and poor are relatives, and what kind of a family does not see to its relatives?

There is also an ecology of agriculture. Farmers are a species in an environmental, economic, social, psychological, and spiritual ecosystem that includes all the species from soil microorganisms, creatures small and large, and plants to neighbors, nearby towns, institutions like schools and churches and merchants, and large urban areas, as well as wilderness, geological formations, and light from distant stars. The water we depend on for agriculture and sustenance is related to it all too, never quite contained, even in the quiet sloughs and backwaters of our major streams. Where rain meets the welcoming hospitality of biomass, it seeps down into the soil through root systems, percolates through porous limestone, fills hidden aquifers or slides along impervious strata of shale, and eventually pools where our wells can reach it, or it bubbles, pours, and seeps out onto the surface again to become springs and streams that feed trout or pike and water our cattle and our children. It becomes the thread, like air, that unites us all and is our common heritage.

That buried shale we never see in southeast Minnesota, lying under towns extending a hundred miles from Preston to Zumbrota, has a name: it is Decorah shale. And those underground pools have names too; the upper carbonate aquifer is one. There are fancier names as well, like the St. Peter-Prairie du Chien-Jordan aquifer, though our hydrologists, perhaps more interested in the flow of water than in the flow of syllables, reduce the latter to SP-PDC-J. Even our language, so full of the names of farm creatures and tools, weeds and seeds, and the great grasses and grains, is part of an ecosystem that includes both the intimate and the far-reaching, taking in much, both on and beyond the farm, joining everything we name with words borrowed from American Indian languages, as well as German, Norwegian, Swedish, all pulled together in a single linguistic system that is very local— and also, if linguists like Noam Chomsky and Stephen Pinker are right, universal.

Everything is related in this system, and when it is healthy, everything is mutually supportive. When a component of that system is diminished —the farmer, for example (not the farms themselves; the amount of land being farmed in America remains pretty much constant), or microorganisms that keep the soil "alive," permeable and fertile— then the whole system is apt to fail. When a component of that system gets too large—when the city expands to overwhelm available agricultural land or when small farms are merged into larger and larger farms— then agriculture suffers, businesses close, small towns become ghost towns, and cities sprawl in a fashion that is not organic, but soon becomes uncontrolled and chaotic.

There is also a social ecology, in which a society's various components— neighborhoods, cities, languages, ethnic identities, mores, customs, traditions—all exist as ideas that compose the concept of society. Racism, sexism, and poverty work against a sustainable society because they treat with disrespect a social ecology in which everything is related. In the ecology of community, racism, poverty, and ethnic cleansing destroy the social ecosystem, rendering it monocultural in a cosmos where mono-anything is doomed to failure. They diminish and ultimately destroy us all, perpetrators and victims alike. As we saw in our conversations with migrant workers, that social ecology comes into sharp focus in agriculture.

There is an individual ecology too, comprising all the components, from cells that permit our physical bodies to flourish, to thoughts, to spiritual awareness, that grow together to make "me." But the genius of science and religion agree that an "I" never exists in isolation, only in relationship. Each of us is not all that different from the beat-up, aging Boeing 727 whose pilot told me, "That's not an airplane; it's 56,000 separate pieces of aluminum flying in formation." When my relationship with the smaller component parts of me, the bacteria or viruses or certain cells, gets out of whack, when cells go wild and grow beyond certain limits, then internal violence erupts and health declines. My self is attacking me. Similarly, when I run out of patience, understanding, and compassion in my relationship with the larger world, then I may damage the minds and hearts of others and myself. When anger grows too fierce to be contained, another kind of violence, physical violence, may erupt. All these are symptoms of a personal ecosystem out of balance, lacking in harmony. When pesticides, herbicides, and fertilizers of unknown effects are loosed in the soil and the atmosphere, agriculture may have a direct impact on my personal ecology, setting cells free to run amok. The farm system that starts with microorganisms in the soil extends throughout the individual ecology and the social ecology, which is why agriculture becomes the intersection where all those boulevards of environmental and social concerns meet, all parts of still larger systems.

In their introduction to *Confucianism and Ecology*, Mary Evelyn Tucker and John Berthrong say Confucius understood this. He believed that there were five concentric circles, nests, extending from immediate family through neighbors and friends and government to the natural world and the heavens to which we owe loyalty and compassion. In *Coming into Being*, William Irwin Thompson says the Buddha understood this; the five skandhas represent a kind of archaeology of consciousness. According to Thompson, we live in a "nested universe," and "part of the beauty and mystery of the universe comes from its nested quality." The pre-Socratic philosopher Empedocles declared, "All is one. . . . Everything has its share of scent and breath and intelligence. . . . Everything breathing in, breathing out."[1]

It is important, then, to look at agriculture within its context as a whole, dynamic system that includes the environment, the larger society, economics, and even intelligence, for farming has been from the beginning one of the most demanding intellectual exercises ever to employ the human mind. The connections here are not only biological, chemical, or physiological. They are also geographical. Farmers, wherever they live in the world, are all facing the same difficulties. This is in part because the connections are also economic and political. As globalization has ripped across the planet like a hurricane, all farmers have been caught up in a single economic, political, and corporate wind so powerful that, whether they are Ugandan, Mexican, or American, and whether they see themselves as local or national, they cannot stand against corporations that see themselves as transnational. That transnational system has so much political support at the moment that it sweeps all before it.

If the ties between poverty-stricken third world farmers and American or European farmers seem remote now, be warned: this free trade world cares no more for farmers with sophisticated equipment than it does for those who use digging sticks. The monarch butterfly and the snow goose know that Mexico and the United States and Canada are connected, all related, "all one," as the Lakota song goes on. In our species pride, believing that humans are the only creatures who can think, we do not want to acknowledge that monarchs and snow geese understand relationships that we do not even recognize. All these connections are also temporal; each connection has a history, and each present moment carries the past with it. Inescapable past and imperative present, impossible to separate, together shape and name our future.

When I speak of "community," then, I acknowledge that it is endless, stretching everywhere, far beyond the narrower definitions we artificially, for the sake of convenience, impose on the concept. Community, for many, means a small town. For others it is an inner-city neighborhood. Some speak of the community of ethnobiologists, the community of saints, the community of O scale train builders, the community of doll collectors. All such definitions are too small for the real community we live in, the larger community that is an absolute essential for life. Life does not come only from the 'hood or from like-minded colleagues, important to us as they are. It comes from the interplay of every element, microbe to star. I'll admit that this talk of cosmos and stars may seem insufferably lofty and vague. But we are not only rooted in our place on earth; we are citizens of a universe—call it cosmos (which can encompass several universes), if you will. Either way, it is easy to see the connection. If the earth is the only place we are rooted, why worry about a hole punched in the ozone layer? We worry about it because our place is not only on earth; we have a place in the cosmos as well. It is our place in the cosmos that makes our roots in the earth possible, for we are surrounded by the stars, and as dependent on them as on the earth. With photosynthesis, dependent on that light from the sun, we have oxygen to breathe. Without it, there is no life; with too much, sickness strikes. We are rooted in earth, yes, but we'd better heed the stars as well. Cut off their light, and we die before our time.

We are creatures of both the earth and the cosmos, the followers of Confucius insisted, with wisdom that sounds postmodern. Our science reinforces their insight every day. That gives us humans, who have developed so much power because of our brains, a huge role to play in how the earth and the cosmos develop their mutual future. We may, as some of Confucius's followers thought, help realize the heavenly on earth, or we may truly blow it and send evolution on a new course without us. The cosmos leaves such a choice almost entirely to us; it does not seem to care. Yet it has given us the capacity for respect as well as for thought, wisdom as well as knowledge, compassion as well as self-interest, reverence as well as greed. It has given us the capacity to love family, all the earth, and beyond. Those innate possibilities in each of us, whether latent or expressed, may indicate that the cosmos is inclined toward including humans in its future.

Rather than rendering the concept of community meaningless because it is too large and too vague, the cosmic view makes respect for every part mandatory, attention to detail rewarding, attention to the environment essential, and attention to society, including its economics and politics, both poignant and demanding. The cosmos is our home at least as much as the earth is, and our sense of community inescapably includes not only this earth but those farthest stars.

Three perspectives are at play in this chapter. The first stems from systems thinking as a way to look at agriculture and community. The second comes from within the system itself, from discussions with farmers, where the largest body of reliable information resides. The third comes from the influence of food and farm systems on larger systems—local communities, regions, even the United States and the transnational trade corporations, with all their trade tentacles that stretch around the globe. Whether they behave in sustainable ways or not, all those systems are related to a conceptual whole, the concept of sustainability. That notion drives this chapter on agriculture and community culture, for agriculture is not solely about raising crops or livestock. It has multiple functions in a larger system that begins in the tiniest cells that create and enrich soil and expands beyond the farm to include local communities and local institutions such as schools and commerce. It reaches beyond the region to include our urban megalopolises and the entire environment that surrounds us. Ultimately, our U.S. agricultural trade policy and our giant trading corporations reach out to include other countries and other continents, all of which, smaller and larger, surround the farm. This systems approach is so embedded in my thinking that I have to wonder how we can have any serious, comprehensive discussion of sustainable agriculture without mentioning community or culture. And how can we talk about urban sprawl and sustainable cities without talking about agriculture?

Whatever else the cosmos is about, in addition to stars, planets, and moons, light, heat, and energy, it is perhaps most about relationships, respiration, and reciprocity. The science of ecology helps us see many of those connections and understand that reciprocity. But one could also make a case that agriculture even more clearly outlines the complexity of the relationships, for agriculture, unlike most ecological science, not only includes the environment but also plunges us immediately into those social connections among farming, small towns and great cities, immigration, ethnicity, food production, and economic systems. Agriculture allows us to see the reciprocity that makes both farming and community possible. In agriculture, the ties between country and city, between individual farmers and community, between environmental concerns and social justice issues, have been clear for millennia. As far back into history as we can go, we find that one characteristic of successful agriculture has always been that difficult but essential balancing act between individualism and community. The farmer feeds the community; the community, in other ways, feeds the farmer. Neither survives long without the other.

The great reciprocity built into our lives by the cosmos requires our attention and participation. There is an inescapable price to pay for the uses we make of the earth and of one another: We have to give back what we take from the world. When we do not, the bills always come due. The bill for exploitation beyond our needs comes due much earlier than the bill that accrues when we live in closer harmony with the universe. We live primarily by the schedule and the rules of the cosmos, not those of the nation or the corporate world, not our own individual desires, nor the digital watches we wear on our wrists. The rules of nature surpass every law adopted by our legislatures. No matter how we squirm or wiggle or make adjustments, the latter are always secondary. It may take a while to reveal itself, but built into nature is a system of justice that is more powerful and rigorous than any that humans can devise. It is, alas, a retributive justice. If we damage nature, nature will always, in some perhaps unforeseeable fashion, damage us. There is no escape from this. Treat Ground Squirrel without respect, and Bear, his grandchild, will exact revenge. Deplete the ozone layer, and cancer will call us to account. Burn toxic materials, and rain will become an acid avenger; the trees will turn their backs on us. Nature may really believe in capital punishment. Damage any part of it enough, and it will take us right off the face of the earth. "Nature," says cattleman Mike Rupprecht, "always bats last."

But it is possible for us to be more open to what nature has to teach. What if, instead of imposing our view on nature and structuring it in light of our perceptions, we changed ourselves and our view, and aligned our lives and thoughts so that human life fit the patterns of nature? Then we would be free to work with the healthy elements of nature, like ozone, and not assist the elements that clearly have destructive characteristics when damaged or depleted. When we learn to cultivate the cosmic reciprocity between ourselves, our communities, our earth, and the stars—rather than seeking to evade it or ignore it—we may finally grow into an agriculture that will both nourish farmers and feed the world community.

Confucian scholars see a triad at work here. Our human role is to mediate between the cosmos (*tian*, "heaven," which has a connotation of "natural process," or, perhaps better, *tianli*, "heavenly principle," "the ultimate principle or norm for the good or perfect") and the earth, to realize the great principle of heaven right here on the ground we till, seeking to create sustenance (food; sustainability), beauty (flowers; harmony), and health (healing; medicine; the earth's natural balance and harmony, the most "natural process") for our families and our communities. Talk about balance! This will require some changes on our part. Our present system nourishes corporate wealth, impoverishes individual farmers, and leaves way too many in the world community hungry. Such an economy reveals its lack of balance. Its ubiquitous violence broadcasts its lack of harmony. No matter how lofty this talk about the stars may sound, we have some pragmatic, mathematical science on our side here. British astronomer Martin Rees, no Confucianist, also sees a triad, with humans poised with nearly exact mathematical precision between heaven and earth. "We are each made up of atoms numbering between 10^{28} and 10^{29} power. This 'human scale' is, in a numerical sense, poised midway between the masses of atoms and stars," explains Rees. "We straddle the cosmos and the microworld—intermediate between the Sun, at a billion metres in diameter, and a molecule at a billionth of a metre."[2]

Whether we believe that the farm crisis is a contemporary phenomenon or an ancient issue, it is clear that it is a crisis not only for farmers but also for our small towns, major cities, and the nation as a whole, and, if my reading of the circumstances faced by small farmers around the world is accurate, it is an international crisis. If we can point to any one, clear indicator of the relatedness of things, the impact of the trauma of farming on small towns, cities, and whole nations is it. That is one reason for seeing agriculture as a keystone. The indicators critical to the health of small-town and rural America, as well as food and farming, include social capital, availability of farmer support from local banks, level of need for ambulance and emergency personnel and vehicles, and local costs to cope with child abuse, spouse abuse, alcoholism, depression, and suicide. When I visited a small-town high school guidance counselor in 1999, he was in tears about a narrowly averted teenage suicide and about the increase in teenage drinking. "I know where the party's going to be on any given Saturday night, but I can't do much because the parents know about it too, but either don't care or even approve." This is the depressed ecosystem for tragedy. All these community indicators are also indicators of the ties between our agriculture and our community culture, and when they are out of balance, the root of that imbalance is stress.

Stress is always embedded in some relationship that is suffering. It is related to a fraying of connections—in rural America's case, the fraying of our connections with agriculture. The current stress also reflects a rip in the social fabric that grows directly from a tear in the fabric of the farm economy. Indeed, as was true in ancient Greece, the social and economic health of small towns is a direct measure of the social and economic health of the farms around them. Other community indicators, as the number of farms declines, are the increased pressure on schools and school budgets. As school enrollment drops, state supports diminish accordingly and tax revenues go down, increasing the pressure on an already full job market as the number of farm family members working town jobs to support the farm grows. Migration to urban centers increases along with hunger and food insecurity. As we will see, food security is not assured anywhere, even in the United States.

THE STRESS STORY

Numbers do not make the only, or the most telling, argument about anything other than mathematics and music. Stories may be as—or more—revealing of the truth. The real story of rural America is shown not only in the dismal statistics of the dismal science of economics but also in the stories farmers and rural communities tell about how the world works. The interdependence of small farms and local economies has been clear to the world at least since classical times, and in the United States since the "improvement movement" of the 1800s and Walter Goldschmidt's classic study of the San Joaquin Valley in the 1940s. One might think that after numerous repetitions we'd get the point, but federal farm policy over half a century apparently never got the message. What that historic litany from Lucretius to now makes clear is that, when the health of the farm economy declines, the economic health of small towns and all their institutions declines as well. Rural towns, their populations and markets, shrink. Churches, schools, banks, and businesses cut back or close.

These changes have a high cost in terms of community and individual stress. The experience of boomtowns, like Anchorage and Fairbanks during construction of the trans-Alaska oil pipeline, shows that social indicators like child abuse tend to increase exponentially as population and stress factors rise arithmetically. In the 1970s, when the Anchorage population nearly doubled with job-seekers, taxing the infrastructure and social workers' caseloads and increasing stress, the child abuse rate did not double but increased by seven times.[3]

There is another high cost in terms of social capital. When whole populations are uprooted, our intellectual and spiritual resources fade. People in high-stress communities lack the time, energy, and inclination (another sign of stress and depression) to volunteer for civic activities. Then emergency workers, for example, get caught in a classic bind, not unlike grain drawn into the movement of giant mill wheels. One wheel is declining city or county revenues; the other is increasing demand for services. One county in South Dakota cut back on emergency services because there was no money to hire and train new emergency medical workers or to maintain or replace emergency vehicles. Such circumstances generally occur just as the number of emergency calls for child abuse, spouse abuse, alcoholism, and depression are rising. Longtime EMTs say, "We need more personnel, not less," but the county commissioners say, "The resources are just not there." Both are right. The predicament is absolute and predictable.

The stress on family farming creates another kind of stress in town – on the job market. As farm family members look for work in town to supplement the falling farm income, they jam a market that is already crowded. They take up any slack that might have been filled by new families moving into the area. Further, the jobs that are available are most often lower-paying jobs in the service sector or high-risk jobs in food processing.

As farming gets tougher and less profitable, banks become more reluctant to make farm loans. Some farmers find other ways to support the farm, like a Dakota producer who confided over supper, "The local bank hasn't supported me for three seasons now. My seed company does my banking for me—through a bank in Iowa." This farm operator acknowledges that he does not like that system and fears being held in thrall to the seed company, but he also feels helpless because there is no local support for his operation. Bankers know how marginal once-profitable farms are now, and they know the exact financial status of their customers. They are wary of making another loan to a small operation. A South Dakota banker who is happy to make loans to large farmers told me, "My job is to tell the little guys to get out of business as fast as they can. I know guys who are making eight hundred dollars net on their farm for the year but feel they have some God-given right to farm; it's their heritage or something, so the wife has to work twelve or fourteen hours a day at minimum wage in town to support his habit." If this banker is reducing his own customer base, he doesn't seem to mind. His goal is simply to hang on to his bank till he retires. After that, he says, "I don't care. If I can't sell it, I'll close it." Meanwhile, he aims to stay afloat with "loans to the big guys who are buying out the little guys." In contrast, Dean Harrington, the Plainview, Minnesota, independent banker, is looking at the loss of dairy farms in his region, and his bank board members are thinking about the implications for their bank and trying to discover ways to slow, stop, or reverse the trend. Dean reports that his board "is starting to say, 'Well, wait a minute, why does anybody need us in the dairy industry if it's going in a direction that needs fewer farmers or goes to larger ones who aren't dependent on local suppliers like our bank, and what can we do about it?'"

In addition to the stress suffered directly by farmers, farmworkers, and their families, entire communities are affected by the shifts in demographics caused by the consolidation of food processors and the development of industrial-model food factories. Those stresses are revealed in the circumstances of food processing workers and the impact of the new industries on local communities.

WORKING FOR THE PROCESSOR, LIVING LIKE SLAVES

The consolidation of meatpacking and other agricultural processing industries has had an impact not only on those who work in the plants but on the communities where the plants are located. Those impacts have become increasingly clear as studies over the last twenty-five years have tracked the growth of processing plants. Many of them are located in small communities, which have had significant immigrant populations move in to work in these food factories. Experience with these industries clearly shows the ties between agriculture and the sustainability of our communities.

In *The Case Against Immigration: The Moral, Economic, Social, and Environmental Reasons for Reducing U.S. Immigration Back to Traditional Levels*, Roy Beck describes the sociology of immigration and the meatpacking industry in Iowa, tracing its early development and its impact on communities and labor. His work shows that Iowa launched a deliberate, aggressive campaign to attract Southeast Asian immigrants to work in Iowa's meatpacking plants. The state subsidized 50 percent of immigrants' wages as an incentive to bring the industry to Iowa and to encourage employers to use Southeast Asian labor. All of this was a generally approved method to create "economic development."[4]

Prior to 1981, Beck reports, Iowa workers had supplied the labor needs of the meatpacking industry. Till then, meatpackers paid middle-class wages to their workers, supplied health benefits, and still earned a profit. But when Iowa Beef Producers purchased a plant and renovated it for pork processing, the new hires did not come from among Iowa workers who had formerly worked in meat processing. Hundreds of experienced Iowans applied, but fewer than thirty were employed. When the plant opened, beginning wages were six dollars per hour, and no health benefits were offered to workers till they had been employed for six months. By 1999, the starting wage had risen to seven dollars per hour. The new plant had high expectations for production; pressure for speed outstripped the desire for safety, and injury rates climbed.[5]

High employee turnover meant that a steady flow of immigrant labor was necessary. The industry employs one production worker for every ten hogs that are slaughtered. Thus a plant that slaughters sixteen thousand to eighteen thousand hogs per day needs sixteen hundred to eighteen hundred workers. Most of them stand in line with several kinds of knives, making particular cuts on carcasses as they move by at the rate of one every three seconds.[6] Consuelo, the Centro Campesino organizer, described her experience to me.

> They treat workers very badly. They are like slaves. I did that in Iowa, at IBP, processing pork, and at a turkey processor's in Minnesota. You wear linked metal [like chain mail] to protect yourself. You work in a

very cold area, below freezing. You have to bundle up and wear plastic over everything. The line moves very fast. You stand there with a knife and make a cut or two every few seconds and you just do that. You can't make a mistake or miss a cut. There are many accidents and injuries. You get a fifteen-minute break in the morning. By the time you get your plastic gloves and the regular gloves off and get out of your work clothes, it is time to get back to work. There is no time for coffee. If you cannot make it to the bathroom during that break, there is no other time to go; you cannot leave the line the rest of the morning.

The new workforce in Iowa included three main groups of newcomers: Lao refugees, a small influx of Mennonites from Mexico, and more recently, a large and growing Mexican and Latino workforce. Neither the state nor the industry planned for the integration into the community of the minority workers they recruited to fill jobs. The tensions that quickly developed between immigrant workers and local people should have been easy to anticipate, but no one did. Now they are not easy to reduce or resolve. Improving the situation is now more complicated, for Iowa Beef Producers has become a subsidiary of Tyson Foods.[7]

Ironically, there is a direct tie between such increased employment and poverty. The traditional notion that economic development means creating jobs and bringing new industry to town is flawed if the jobs that are created pay poverty wages. Six or seven dollars an hour is less than the amount necessary for a family of three to live at the federal poverty threshold, an amount that the government sets so low that we should be ashamed of what we call living standards. Standards are what we aspire to. We set them as levels of achievement for our schoolchildren. But the living standard set by the government is a minimum, indicating how little we can make to get by. Further, wages across the processing plant are relatively flat; they do not increase much beyond starting pay for any job on the floor, so there is no incentive for employees to stay and no way for them to work their way up or improve themselves in the system. High turnover rates continue, and it is common for the company to issue two hundred IRS W-2 statements for every one hundred jobs. That turnover rate just happens to save the company money on health insurance premiums for the coverage that does not begin until an employee has worked for six months. New jobs did increase employment, but not for local residents. Because the pay was so low, the new jobs also increased poverty, and because they introduced unfamiliar cultural patterns, social stress mushroomed. There was a net loss of social capital. The concessions that the state makes on behalf of industry create poverty for the workforce and increase the profits of the already wealthy.[8]

Other social stresses wrack such an industrial system. They work both ways: locals feel the stress generated by strangers from other cultures; newcomers feel the stress generated by strange settings and strange ways. About 24 percent of the children in school in Storm Lake, Iowa, one of the towns that began processing through Iowa Beef Producers in 1982, needed bilingual instruction in school. Housing, waste disposal, and electric service were all strained by the influx of newcomers. The stresses on new workers who were trying to make their way in a foreign country, using a second language, and unfamiliar with the local customs resulted in a sense of alienation that made their dislocation worse. The Storm Lake experience has been repeated in other states, including Kansas, Utah, and California.[9]

Beck makes a powerful case throughout his book that immigration is the primary cause of the unemployment of white skilled workers and black labor leaders who once had a voice in improving conditions for union workers. But his analysis is flawed in three ways. First, immigration neither creates nor re-creates rural poverty; poverty wages do. Second, immigrant workers have education and skills enough to do their jobs, and do them well, but they have no incentive to gain more education or increase skills, since there are few higher-skilled jobs available in the industry. Third, immigration penalizes established workers with skills and education—including local Iowans, in this case—by hiring vulnerable immigrant workers at half the pay rate that most skilled and educated workers would have demanded and under conditions they would have refused.

A more important criticism can be made of Beck's attempt to use this information to make a case against immigration. Beck's data really make an argument against letting new industries increase social and economic stress in local communities and pay poverty wages that hurt workers and diminish social capital in small towns. Pointing out that those wages are very likely higher than the workers would have made at home is a disgracefully flimsy excuse. That approach does not create economic development. It creates economic and community erosion, confusion, and social decline. Why should any state impoverish itself and waste its social capital to subsidize industries that are already profiting from a larger piece of the agricultural pie than they really earn?

What, then, can states do when industry or commerce says, "We need to have some tax concessions made or land donated or roads built for us, or we'll go elsewhere"? Since such requests are the first sign that this industry will never be a responsible, contributing member of the community, sensible states and communities respond by saying, "We cannot do that. However, what we will do is require an infrastructure tax so we can recover the expenses we will incur for the added costs required of our schools, our water and sewage infrastructure, our social workers, and our loss of social capital." They might even add, "If you cannot afford to be a contributing citizen, we can't afford to have you in business." If such an industry goes elsewhere, the community can heave a big sigh of relief and keep its tax savings in the bank or spend them on more worthwhile projects. We should remind our Chamber of Commerce development committee that the purpose of economic development is not to create jobs but to create prosperity—for the community, not for an industry or two whose corporate profits go elsewhere.

We need to pay attention to such connections, for as Susan Combs, the Texas commissioner of agriculture, told the *Dallas Morning News* in 2000, "Four years of drought in Texas is not only a threat to farmers, but to the entire state economy.... If you see one segment of the economy in deep trouble, you are going to see it show up in others." Two out of five employed Texans still work in agriculture. A $4 billion loss that agriculture suffered in 1996 and 1998 "reverberated into an even greater $11 billion impact" on the state's economy, according to Combs.[10] She may have been even more right than she knew.

FEELING THE PINCH

As the number of farmers declines and the debt-to-asset ratios of those remaining increase, industries that support farming, whether large or small scale, also feel the pinch. In 1999, the *Mitchell (S.D.) Daily Republic*, a regional daily, explored the problems of implement dealers who have large inventories of "big iron" or "heavy metal," as some farmers call their combines and high cab tractors. One Case dealer said his sales were down only 10 percent from 1998 but added, "I've heard of dealers being down 30, 40, 50 percent." One small-town dealer reported that his sales were half of the previous year's and the bills were coming due for the inventory he had on hand. "I've got $800,000 for six combines sitting out there," he said, "and the interest is ticking every day." Another small-town dealer, a member of the Farm Equipment Association of Minnesota and South Dakota, reported that 1999 was the second bad year in a row for the industry. It was "the first time in 13 years that the numbers were down in all segments of the industry—new and used equipment and parts." The association reported that sales were off about 20 percent in 1998, and at least that much again in 1999.[11]

Normally, implement sales are pretty much tied to commodity prices. When corn and bean prices are up, farmers trade up; when they are down, "the first thing they cut back on is big-ticket items," said Dennis Van De Werff, president of the Farm Equipment Association. Nobody blames farmers for tightening their belts, and everyone understands when farmers are hurting. Mel Ptacek, executive director of the Farm Equipment Association, said that one response of manufacturers to the scene was to move more of the risk onto dealers by shortening the time they had to pay for inventory. Once, the payments could take up to eighteen months, and if equipment didn't sell one summer, it could be carried over the winter and might sell the following spring. Now, however, payment schedules had been reduced to four months, a very short—some would say shortsighted—turnaround time.[12]

For dealers, that inventory represents more than sales potential or debt to the manufacturer or the lending agency. Steve Marcus, general manager of Olsen Implement in Huron, South Dakota, told the *Daily Republic*, "Some of the corporate people don't understand that this is more than just a job. My retirement is on that lot, in this building—it's not in the bank." Those dealers who hoped that their personal future would pay off when they could sell equipment, or even sell the building on the assumption that real estate increases in value over time, may well find themselves disappointed. As the number of farmers declines, the number of implement sales declines, and as the community declines, so do the property values. The regulations imposed by manufacturers have also tended to proliferate, and some of them cost dealers money. The dealership rules, created by the manufacturers, about floor space, signage, and paint are increasingly strict, and failure to meet sales quotas may put the contract with the manufacturer at risk. Ptacek said, "This is worse than the 80's. If you don't get market share . . . your contract is in jeopardy."[13]

The difficulties of implement dealers have not changed in the last few years, and the situation is not likely to improve soon. *Agri News* reported that, in 2002, "sales of big tractors and combines dropped by 20 percent." The Association of Equipment Manufacturers estimated that in 2002, dealers in the United States sold 5,000 combines, compared with 6,400 in 2001. "That's a little bit more than one per dealer," the association reported. The group gave several reasons for the decline: the uncertain economy, the high cost of equipment, the drought that plagued much of the country's farm belt, and the decline in the number of farms. "If you had 10 farms owned by 10 people, they had 10 tractors and 10 combines," explained Mike Kraemer, spokesman for the North American Equipment Dealers Association. "One person now owns that, and a 10,000 acre farm doesn't need 10 tractors." The manufacturers do what they can to protect themselves. What they can do is shift risk to their dealers and press them to move inventory as quickly as possible. Their strategy replicates that of the food industrialists who contract with farmers and shift their risk to them. "It's in both the manufacturer's and the dealers' interests to get that equipment moved," said a spokesman for John Deere Corporation.[14] Of course it is; it always has been. But when prices are down and farmers' capacity and desire to buy diminish, dealers feel the pinch while farmers and corporations try to protect themselves.

When dealers and farmers feel the pinch, many futures seem uncertain, and the communities where they reside also begin to wilt like plants without water. In the economic drought, spending goes to the Wal-Marts and Targets of other communities or other counties. Dean Carlson, a farmer from Kennedy, Minnesota, writes in the *St. Paul (Minn.) Pioneer Press*'s special series "Harvest of Risk," "Services and conveniences we once enjoyed have disappeared along with the population." Carlson spells out the impact of the loss of small farms on rural society: "As populations decline, schools consolidate, resulting in longer bus rides for students. With fewer dollars to work with, school districts must trim their teaching staffs and reduce class offerings." But that is not all. "Many communities are losing their health care, restaurant and entertainment facilities. Longer trips must now be made to obtain those services as main streets are filled with empty buildings." Every aspect of the community's culture, including recreation and religion, is affected: "Recreational facilities such as golf courses and swimming pools have more difficulty remaining open. Churches find it difficult to pay clergy a decent wage, forcing them to either merge or share speakers."[15] Carlson accurately summarizes what those of us who have served on small-town school boards, town councils, and church boards have observed. And it is not new.

THE IMPROVERS' STORY

The exodus from our towns around the time of the War of 1812 led to a whole movement to revise the way we farm. Known as the improvement movement, it became a force in New England and spread to the South and toward the frontier through the 1820s, 1830s, and 1840s. In *Larding the Lean Earth: Soil and Society in Nineteenth-Century America*, Steven Stoll traces the improvement movement from its beginning to the time when its effort blurred into the fledgling U.S. conservation movement. What George Perkins Marsh and other "improvers" understood even before the movement began, Stoll writes, was that, "no matter the mode or the cause, where topsoil is removed or degraded, agriculture is impossible and a settled people cannot persist. . . . Most of all they recognized a link between an enduring agriculture and an enduring society in the long-settled places and picked up an old word to name their efforts: improvement. Improvement makes anything better by raising it from a rude to a more refined state or condition, but in American usage it specifically applied to the condition of land."[16]

Improvement in the nineteenth century grew from two concerns, both of which sound utterly contemporary. One was concern for the collapse of small towns, the result of the loss of farms and population, which was, in turn, the result of the loss of productive land. The other was concern for the land itself. In these nineteenth-century farmers' view, agricultural land was deteriorating because of bad practice. The improvers understood that agriculture is inherently destructive. It takes nutrition and fertility out of the soil to produce crops, feed, and livestock. If agriculture does not replace what it takes out, it finally undercuts the possibility of both farms and community. In the years Stoll covers, the movement away from farms was the result not of a rigged economic system that favored the giant corporation or industrial-scale farmer but of the availability of open land to the west. When the land was no longer productive, farmers moved on, taking "free land." As farm population declined, communities were left desolate. Numerous observers attested to the fact that the land had been devastated. In 1819, writes Stoll, George W. Jeffreys called his contemporary agriculture "a land-killing system" and accurately foresaw that it would "ultimately issue in want, misery, and depopulation." Stoll explains, "Farmers never simply altered their private property . . . they tampered with environments and became implicated in the rest of nature." He continues, "The deterioration of soil is the inescapable injury of agriculture to the environment, so its severity is a sign of the fealty or failings of any people who husband land."[17]

What the improvers took up to maintain their farms and their communities was a system known to the ancient Romans: the careful cultivation and replenishment of soil using manure, preferably from livestock that were herded carefully to leave their manure scattered throughout their fields where it was most needed. Those farmers of the early nineteenth century frequently quoted Cato's admonition: "Make it an aim to have a big manure pile. Preserve the manure carefully." They would also have understood Seneca's forlorn disparagement of surveyors "who can tell me precisely how many acres I have, but not how much is enough." The improvers knew they had some things in common with both the ancient Greeks and Romans. Cato and Seneca were warriors, lawyers, senators, statesmen, and farmers, deeply involved in the life of their country. So were Washington, Adams, and numerous other signers of the Declaration of Independence. America's farmers in those years were also legislators and citizen-soldiers who left their plows in the field in a minute to respond to the call to arms.[18]

Crop rotation and animal husbandry were key elements of what was known in the 1820s by a variety of names: field-grass husbandry, alternate husbandry, convertible husbandry, up-and-down husbandry. What's in a name? Rotational grazer Ralph Lentz, who taught agriculture in high school for years, might say, "A lot!" He notes that before he finished teaching, "animal husbandry" changed to " animal science." The

192

other agriculture course titles changed in the same way. "The implications of that change are just huge!" says Ralph. Stoll's descriptions of the old articles in journals and newspapers of the day, with suggested rotations and ideas for winter pasturing, keeping 50 percent of the land in pasture or hay, have a very contemporary ring. A rancher in Montana once told me that his father, who homesteaded the place in the 1890s, told him, "Take half and leave half, and you'll always have plenty of grass." And he did. In addition to crop rotation, manure, and animal husbandry, confinement was an issue then as it is now. Monoculture and commodity crops were issues then (though they were not called commodity crops) as they are now. One exasperated New England farmer urged a new kind of agriculture altogether, to replace the one they had been taught by preceding generations who thought that "the more corn they could plant, and the more wheat they could sow, the more profits they derived." Those earlier farmers who followed poor practice "lived in the illusion of limitless gain," notes Stoll, a warning that contemporary legislators, farmers, and the general public might well heed.[19]

Once again, in our own agriculture we are plowing ground that has been plowed before, rehashing arguments articulated centuries ago, learning to understand what others understood long ago. How is it that what is clear to the science and practice of one generation is ignored or forgotten by the next? We seem to need to relearn old lessons over and over again. Ralph's comment on the difference between animal husbandry and animal science offers a clue to this. Have we replaced agrarian values with commercial values? Such a shift could have implications that "are just huge!"

HUNGER STORIES

Unfortunately, all of these farm and community stresses have an impact on children, either indirectly, as their parents feel the pinch, or directly, as hunger stalks families on the margins. Now, as the nation gets wealthier, the hungry get hungrier. A 1999 report from the Government Accounting Office concluded, "There is a growing gap between the number of children living in poverty—an important indicator of children's need for assistance—and the number of children receiving food stamp assistance." That gap has increased every year since. In 2001, the Food Research and Action Center reported, "Approximately four million

American children under age 12 go hungry and 9.6 million more are at risk of hunger according to estimates based on the results of the most comprehensive study ever done on childhood hunger in the United States—the Community Childhood Hunger Identification Project." The survey included more than five thousand families living on incomes below 185 percent of the federal poverty level, and the council applied that to the best available national data. Note the level it set for its inquiry: 185 percent of the poverty line, a level that is beyond people who work for minimum wage for a family of three or four. The survey results showed that as many as 13.6 million children under twelve—29 percent of the U.S. total—"live in families that must cope with hunger or the risk of hunger during some part of one or more months of the previous year." A 2003 report by the Food Research and Action Center found that "38.2 million people lived in households experiencing food insecurity compared to 33.2 million people in 2001 and 31 million people in 1999." In 2004, the center reported that "11.9% of all U.S. households were 'food insecure' because of lack of re-sources," up from 11.2 percent the previous year.[20]

Other surveys reinforce these conclusions. The USDA reported that, in 1998, 10.5 million households (10.2 percent of the U.S. total) were "food insecure," meaning that they did not have access to enough food to meet their basic needs. Of the total number of food insecure individuals, 40 percent were children. This represented an increase of 6.4 million adults and 3.7 million children between 1997 and 1998—years of fabulous growth in the stock market. A 2003 report from the USDA showed 20 million adults and 13 million children living in families that suffer from hunger or live on the edge of hunger. Yet, overall, poverty declined 11.3 percent between 1998 and 2000. Nevertheless, as the recession gathered steam in the final months of 2001, December food stamp participation increased by 246,070 over November, the ninth straight month of in-creases, "representing a growth of 1.8 million persons between February, 2001, and December of that same year." Figures released in June 2002 revealed 1 million new U.S. families dealing with food insecurity, won-dering how to feed their children. The Women, Infants, and Children program serves low-income pregnant women, new mothers, infants, and children at nutritional risk. In 2000, the average monthly participation in the program was 7 million. As the economy declined in 2001, partici-pation increased, reaching 7.5 million.[21]

The U.S. Conference of Mayors has also documented a steady increase in the demand for emergency food assistance since 1983. In 1997, as the nation's prosperity increased under the false promise that a rising tide lifts all boats, requests for assistance rose by an average of 16 percent from the previous year, though the number of families with children increased only 13 percent during that same time. In 2004, the conference reported that 96 percent of major U.S. cities saw increases in requests for food stamps and hunger assistance such as food banks. The increases averaged 14 percent. The organization's report of 2005 showed similar numbers. Catholic Charities USA reported that more than 5.6 million persons used their emergency food services in 1996 (another banner year for stocks), food banks served more than 2.7 million, and soup kitchens, a reminder of the lines during the Great Depression, served another 1 million.[22]

To me, there are three striking things about these reports: First, on average, 38 percent of the adults qualifying for food assistance were employed full-time. Second, hunger in America was increasing during a time (1983–2001) defined by increases in our gross national product, a flourishing economy, and a readily acknowledged—yet relentlessly ignored—rapidly expanding gap between the well-off and the poor. The more recent data show the requests for assistance growing continually, even as the economy has "recovered" from its shocks early in the new century. It appears that there really is a trickle-down effect in national economics; what trickles down, however, is the lowest end of the economy: its circumstances trickle from difficult to desperate. Eventually, the poor fall off the dock as the tide rises, drowning those who have been going under all their lives. As John B. Cobb describes it in *Sustaining the Common Good,* "A rising tide, we are repeatedly told, raises all ships, and the rising of the tide is to be measured by GNP." But Cobb isn't buying it. "The truth is that despite or because of the rising tide, many ships sink, and . . . these policies [expending natural resources and energy on the wrong priorities] will only hasten the destruction of the remaining ships in the inevitable storm."[23] As the nation's wealth increases, increased incomes never reach those who experience the threat of hunger. It is clear that, as long as the minimum wage stays below poverty level, employment programs will not solve the problems associated with poverty and its concomitant hunger. Third, this increase in hunger coincides with an extraordinary growth of yield (measured in single crops) in agriculture. This growth fuels a continuing but unsubstantiated advertising claim by large producers that we feed the world. It is also linked to the shift to monoculture land use and the rise of a global economy. We cannot bear to admit that our vaunted farm technology will never feed the world because our major grain traders will not sell to the hungry because the hungry cannot afford to pay. The disconnect lies between our agricultural skills and high yields and our capacity to feed our people.

In early 2002, pundits reported that the recent recession was over. It wasn't. The claim was repeated in September 2003. It wasn't then either. In 2006, the government is still telling us that the economy is on the upswing. The corporate economy appears to be. But the personal economy continues to fall. What the pronouncements make clear is that economists at the highest levels of government and in academe do not have a clue about how to evaluate the wealth or the health of a national economy. They appear wholly unaware that a wealthy economy may not be healthy at all.

SECURITY, INSECURITY, AND FOOD

The concepts of food security and food insecurity came into play during an earlier food crisis that gained worldwide attention in 1972–1974. Newspapers were filled with haunting pictures of emaciated children in Bangladesh and Biafra. People around the world wanted to help, and many did. In the midst of that sadness and concern, the UN Food and Agriculture Organization sponsored a series of conferences, beginning in 1974, that came to see food security as having two aspects: on the one hand, food security means that there is food available locally, nationally, internationally, and on the other hand, it means that individuals and nations have access to those food supplies. The Overseas Development Institute tracks growth in delegates' understanding of food security. In 1983, according to the institute, "FAO expanded its concept to include a third prong, securing access by vulnerable people to available supplies. Attention should be balanced between the demand and supply side of the food security equation." As for food insecurity, there are two kinds, explains Majda Bne Saad of the Centre for Development Studies at University College Dublin. "Transitory food insecurity is a temporary decline in a household's access to enough food. Chronic food insecurity is a continuously inadequate diet caused by the inability to acquire food." The latter type is caused primarily by poverty; the former may be caused by drought, flood, war, widespread disease that prevents the tending of crops, or agricultural dumping of surpluses that put small, local farmers out of work. In some cases, as in the recent crisis in Zimbabwe, governments have access to food but, through sheer ineptness, corruption, or lack of infrastructure, fail to deliver it to their people.[24]

NO ROOM FOR THE UNEXPECTED

There is a parallel, of course, between economic security and food security. They are not perfectly contiguous, for there are times of natural and social disasters when it does not matter how much money one has; there is simply no food available for purchase, and both rich and poor go hungry. But in most of the world, most of the time, the poor go hungry. For all of us, when the margin in our bank account dwindles, the household budget get a little crazy. If anything unexpected goes wrong—the car's engine fails, the washer or dryer breaks for the final time because the manufacturer no longer makes parts—we begin to juggle expenses. Who can we stall this month so we can take care of the rest of the bills? It takes little imagination for a middle-class person with some credit card debt to wonder how long his or her family might stay afloat if the loss of a job left them without an income. "Two months and I'd be on the street," a friend told me a few years back. Being "on the street" means being hungry, searching for food in the trash bins behind restaurants and grocery stores.

We run a similar risk when food in the world's pantry runs low. When the unexpected happens—a drought here, an infestation there—food supplies go a little crazy too. For all the wealth of our markets, the world has not achieved economic security for the majority of its population, and for all of U.S. farming's commodity production successes of recent decades, the world has not yet come close to achieving food security. One thing we might learn from this is that increasing U.S. commodity production, for which there is a continual cry from the agribusiness industries, will not touch hunger in poor countries or reach poor people in our own nation.

Further, increased production of commodities may not be a realistic option for us or for the world. In 1999, world grain production fell nearly 2 percent, while the population increased by some 78 million. A severe drought in Russia and a slighter one in India undercut the gains made in the United States and in the European Union. Meat production has also been slowing down the last few years, and the per capita supply from the world's fisheries showed a 2.4 percent decline in 1998. Saad writes that, for the last twenty-five years, food production has "successfully kept ahead of population growth." Some believe that we can continue to do that for another twenty-five years, "but against this reality, hunger still persists in many parts of the world," she notes. Some "800 million are chronically malnourished. . . . Millions more become blind, retarded or suffer other disabilities that impair functioning because of lack of vitamins and minerals. Moreover, hunger and poverty are the root of much political turmoil and armed conflict, and of a growing tide of refugees and migrants."[25]

All these statistics lead the Worldwatch Institute to conclude that world food security will not be achieved soon. We may always have the poor with us, which means that we will always have the hungry as well. But why? The institute's concern is based on several factors, also reflected in the statistics. It reports that "a substantial portion of the world's current grain harvest is based on the unsustainable use of resources, primarily land and water." It cites the impacts of such practice in Kazakhstan, which, thanks to "big iron" and chemicals, produced and exported more grain between 1980 and 1998 than Australia did. Since then, however, half of Kazakhstan's grain lands have been lost to wind erosion.[26]

Another factor that contributes to food shortages down the road is the overpumping of groundwater. Unfortunately, in the major food-producing countries that are heavily dependent on irrigation—China and India—and in North Africa and the Middle East, water tables are falling. The Worldwatch Institute describes a report by a joint Sino-Japanese research team indicating that water tables are falling almost everywhere in China that the land is flat. Under the North China Plain, which produces 40 percent of China's grain harvest, water tables are falling by an estimated five feet a year. Throw in poor conservation practices with crops, combine them with drought, and the result is severe dust storms that outclass our own Dust Bowl of the 1930s, choking off and closing down cities as large as Beijing and as far away as Seoul. The Worldwatch Institute also points out that, in India, the drawdown of underground water for the nation as a whole is double the rate of recharge, and in the United States, aquifers upon which we depend for irrigation agriculture are also being drained more quickly than they can be replenished. The Ogallala aquifer, which supplies irrigation to agriculture in parts of eight western states, is now estimated to have a life expectancy of twenty to forty years, although it may not last even that long. Under the pressure of drought conditions, the groundwater in western Kansas fell five feet in the summer of 2002.[27] One might think that drought years would trigger conservation. Instead, we just pump harder. "The frog," the old Aztecs said, "does not drink up the pond in which he lives." Is it possible that we humans might someday acquire the wisdom of frogs?

In 2000, the Worldwatch Institute concluded, "The reality is that hunger is the product of human decisions—especially decisions about how a society is organized. Whether people have a decent livelihood, what status is accorded to women, and whether governments are accountable to their people—these have far more impact on who eats and who does not than a country's agricultural endowment does. . . . Poverty —rather than food shortages— is frequently the underlying cause of hunger." Their report also indicates that "poverty and hunger . . . result from a range of misguided government policies, from inequitable distribution of land and other resources to poor management of foreign debt." The uproar over the World Trade Organization meeting in Seattle and subsequent meetings is another reflection of the impact of poor policy on hunger and poverty, according to the Worldwatch Institute: "The growing emphasis on free trade in agricultural products brings additional nutritional vulnerabilities. Current trade arrangements, such as the Agreement on Agriculture under the World Trade Organization, permit the industrial farmers of Europe and North America to sell subsidized grain, oils, and other commodity surpluses cheaply in developing nations, undercutting local farmers and forcing many off the land."[28]

Perhaps even more telling in terms of food security for the world is that both wheat and rice supplies are down. World stocks of corn are up, but a rapidly increasing portion is grown for ethanol and other fuels. Globally, 95 percent of soybeans are now fed to animals, which do provide meat or milk. Wheat carryover stocks were at seventy-eight days in

1999, the third lowest on record—higher only than those in 1996 and 1997. Wheat and corn stocks need to be higher than rice supplies, the Worldwatch Institute points out, because they are rain-dependent. Rice, always irrigated, rarely fluctuates in production more than 2 percent, but year-to-year swings of wheat and corn may reach 10 percent. The implication of this for farmers and for consumers is that, "when carryover stocks drop below 60 days of consumption [exactly the amount reported in 2005], prices become highly volatile and can easily double from one year to the next." It is necessary, then, to hold world carryover stocks at approximately seventy days to maintain price stability in world markets and to cushion the effects of a poor harvest. The Worldwatch Institute makes clear that, although recent declines in grain production have hit mostly small countries like Kazakhstan and Saudi Arabia, larger countries will soon have to wrestle with the same problems. That means that American families, too many of whom already struggle to put food on the table, will also face food shortages in years to come.[29]

The good news is that grain production was up in 2005 as a result of fair weather and an increase in acreage planted around the globe. Meat production also increased in 2004, by 2 percent. The bad news is that hunger also increased during that time. About 852,000 people "go hungry each day, about 18 million more than in the mid-1990s." Further, "hunger now kills more than 5 million children each year—roughly one child each five seconds."[30]

The fallout from these international community pressures ramifies through every aspect of our own communities as well. We take America's food security for granted, believing that we are the world's greatest food suppliers. It has been a given in our thinking, reliable as a mathematical equation. Yet some agricultural economists now predict that by 2007, the United States will become a net food importer—an indicator normally associated with the least-developed countries.[31] Indeed, the last few months of 2004 showed net food imports for the first time since the mid-twentieth century.

CHAPTER 11

Farming in Developing Countries

One striking feature of comments from participants in an eighty-nation agriculture "e-conference" in 1999 was the broad agreement in many nations about what is happening to small farming, the environment, farmers' access to markets, and the circumstances of farmworkers and food processors. The issues are similar around the globe. The economic, social, and psychological symptoms of change, and perhaps even the causes of change, are pretty clear on every continent.

The exchange of e-mails in this conference drew a wide diversity of people: farmers, agriculture experts, university faculty members, and policymakers. What was surprising, though perhaps it should not have been, was the worldwide awareness of the multifunctionality of small farms. There was a clear recognition among these international observers of the value of small farms and their connection to social and environmental health, a concept featured only minimally in the agricultural policies of the United States. There was also widespread agreement in every quarter about the nature of the issues, and their causes, that farmers must deal with. Participants believed that, in nearly every case, globalization separates farmers from their markets, reduces family farm income, and finally separates them from their land. The rise of corporate control of farming, "mining the earth" instead of husbanding it, as several participants put it, leads to more and more contractual arrangements that strip farmers of their freedom. It creates a new form of indentured servitude. Participants also expressed widespread concern for world food security.

The environmental degradation caused by extensive chemical inputs led to some of the saddest comments from small farmers around the world. One of the most touching came in an e-mail that began pleasantly enough: "We, the farmers of Sri Lanka, believe that we speak for all of our brothers and sisters the world over when we identify ourselves as a community who are integrally tied to the success of ensuring global food security." What followed was devastating.

> We have watched for many years as the progression of experts, scientists, and development agents passed through our communities with some or another facet of the modern scientific world. We confess that at the start we were unsophisticated in matters of the outside world and welcomed this input. We followed advice and we planted as we were instructed. The result was a loss of the varieties of seeds that we carried with us through history, often spanning three or more millennia. The result was the complete dependence on high input crops that robbed us of crop independence. In addition we farmers, producers of food, respected for our ability to feed populations, were turned into the poisoners of the land and living things, including fellow human beings. The result in Sri Lanka is that we suffer from social and cultural dislocation and suffer from the highest pesticide-related death toll on the planet.

The e-mail concluded, "Was this the legacy that you the agricultural scientists wanted to bring us? We think not. We think that you had good motives and intentions, but left things in the hands of narrowly educated, insensitive people."

Reg Preston, working with the UN Food and Agriculture Organization in Vietnam, seconded and then extended the comment of the Sri Lankan farmers with his e-mail description of the current dilemma: "The developed countries continue to mine their natural resources while most poor farmers in the developing countries actually practice integrated farming in the face of huge pressure from multinationals to promote monoculture for their own short-term ends, lack of technical and financial support from their own governments, most of whose advisors have been trained in 'agricultural mining technology' in the industrial world, and markets distorted by 'dumping' from the industrial countries." E. R. Orskov of the Rowett Research Institute in Aberdeen, UK, lent support to both the comments made above. "If multinational companies were rewarded for their contribution to poverty alleviation rather than maximizing profits for their shareholders we would do a lot better! Most of the high tech solutions have contributed more to poverty creation than alleviation. What have the bankrupted intensive animal production systems in South East Asia, supported by Western feed, Western animals, and Western medicine done for poverty alleviation? They have left a trail of poverty."

For decades, as the United States engaged in a cold war with the Soviet Union, we saw a world conflict between East and West. Much of the rest of the world had a different vision: that of a world divided between a wealthy North and an exploited South. Miguel Holle of Centro Internacional de la Papa in Lima, Peru, suggested one reason the richer North might be lagging behind the poorer peasant farmers of the South in understanding the issues facing farmers around the world: "Our most common thinking is mono-functional, and the reality of small or poor or traditional or marginal farmers is multifunctional with quite a complexity."

An Australian farmer, identified only as Danderto, saw the current international issue as follows: "Emphasis on production agriculture has led to a lose/lose situation in both the countries of the North and South. In the North we have vast tracts of land being degraded and not producing any profit. In the South we have land being converted to 'modern' cash cropping and causing loss of employment, community cohesiveness, and cultural and environmental degradation." If one doubts Danderto's charge that we in the Northern Hemisphere are following a production agriculture that does not produce any profit, one has only to remember Illinois's production costs for raising corn and soybeans in 2005 or the most recent numbers about farm income and non-farm income. (See chapter 9.) Danderto's analysis is on target, not only for Australia but for America as well. Danderto does see a possible solution: "Perhaps one way of turning this around is to acknowledge that land is more than a food and fibre factory, so the technologies chosen should seek to improve the land's ability to serve many functions." That seems easier to say than to accomplish. Doing so would require replacing some of our most powerful myths with accurate information. Acknowledging some of the myths that drive us, like the myth that larger is more productive and more efficient, is a first step, as Vance Haugen told us early on.

What is clear and indelible is that small farmers in many countries are struggling. The losses farmers have suffered over the past fifty years cross both national and climatic boundaries, and they are similar in nearly every case. American farmers are neither isolated nor insulated from this system's apparent pitfalls.

CHAPTER 12

The WTO, NAFTA, CAFTA, and the FTAA

Thanks to the publicity surrounding the 1999 World Trade Organization (WTO) meeting in Seattle and subsequent meetings, most U.S. citizens know about the WTO. Because so much of the news focused on isolated incidents of property damage, there is considerable wonder among our own farmers and other citizens at the opposition, hostility, and outright anger expressed toward the WTO. In spite of the media reports, however, the real threats posed to democracy, farmers, laboring people (including white-collar laborers), human health, and the environment by our free trade negotiations are becoming better known.

Many Americans also know there is a North American Free Trade Agreement (NAFTA), but fewer know any details or real outcomes of it. Fewer still, I suspect, know about the Free Trade Area of the Americas (FTAA). The latter was promulgated as the means to extend free trade agreements beyond Mexico, Canada, and the United States to include more than thirty other countries in Latin America. It would have extended the economic boundaries of free trade, "liberalizing" it beyond the agreements already gained by the WTO and NAFTA. Farmers and their nations are already implicated in the WTO and NAFTA and now will be deeply affected by the development of an agreement among Central American countries, the Dominican Republic, and the United States, generally referred to as CAFTA, that narrowly passed our Congress.

The names of the International Monetary Fund (IMF) and the World Bank are also well known, though what they do and how their policies work are less well known and understood. Yet many of their policies, programs, and lending actions radically alter the nature of agriculture for the countries involved, without farmers' or agriculturalists' participation in the negotiations. Those who care about our local farms and about social justice need to know the possibilities inherent in these trade agreements and the activities of the IMF and World Bank. Understanding their impact on the lives of our citizens, especially farming citizens, and our democracy is critical. Though the WTO, NAFTA, FTAA, World Bank, and IMF are not all the same, decisions and actions taken by any one trigger decisions and actions by the others. Their purposes, actions, and impacts are not the same either, but they are interrelated.

The WTO has not conducted any studies about the impact of free trade thus far. A recent study by UN University's World Institute for Development Economics Research, however, looked at the results of IMF and World Bank involvement with forty-seven of the poorest countries receiving the most aid. That report revealed that they are poorer today than they were when the World Bank was established in 1944. In its own study in 2003, the IMF reported that countries that follow IMF requirements for loans often suffer a "collapse in growth rates" and also face "significant financial crises." The open currency markets encouraged by IMF actually "amplify the effects of various shocks." Summary statements in the report reveal the disconnect between economic theory and practice: "While financial globalization can, in theory, help to promote economic growth through various channels, there is as yet no robust empirical evidence that this causal relationship is quantitatively very important." There doesn't seem to be much less-than-robust evidence to support the contention either, but there is a great deal to refute the list of benefits claimed for free trade in agriculture and other affected sectors. The section concludes, "One of the theoretical benefits of financial globalization, other than to enhance growth, is to allow developing countries to better manage macroeconomic volatility, especially by reducing consumption volatility relative to output volatility." Since food security trumps economic development for most of the world's people and is one expression of genuine economic development, the volatility that the IMF hopes to curb has a direct effect on food, food security, and hunger around the world. How beneficial has its policy been? Its own report continues, "The evidence suggests that, instead, countries that are in the early stages of financial integration have been exposed to significant risks in terms of higher volatility of both output and consumption."[1]

Both the IMF and the World Bank have endured serious public criticism by social justice activists in the United States and abroad for many years. Yet they have not much changed their practices to reduce the misery they have created around the world. In terms of hunger and trade in agriculture, we are still steaming downriver into the darkness, without lights to probe the fog, and the buoys have gone silent. To our peril, we are ignoring the warning bells of those studies that have been done in many areas of the globe.

The negotiating documents of the WTO, NAFTA, and FTAA all have sections called Agreements on Agriculture. They have been a primary stumbling block in the negotiations. In the recent CAFTA negotiating documents, there is no Agreement on Agriculture; the sections that affect agriculture are scattered throughout, perhaps in an effort to make the real effects of the agreement more difficult to discover and weigh. Throughout the history of the WTO and NAFTA, the advantages somehow have not accrued to farmers, or developing nations, or the hungry —all of whom should benefit from free trade, according to the theory. Neither do the advantages accrue to the nations that negotiate the agreements, for signatories, including the United States, sacrifice important elements of their own sovereign authority and integrity. The agreements forbid nations to establish their own agricultural policies, for example, or pursue their own goals, or protect their own citizens. Not only must impediments to free trade, such as tariffs and export subsidies, be eliminated, but national, state, and even local laws must be changed to fit the new agreements. These specifications are true of CAFTA as well, and they will be true of the FTAA negotiations that will take place in some form soon. Before it could sign on to NAFTA, Mexico, for example, had to amend its national constitution to eliminate land protections for peasant farmers.[2] Further, "non-tariff trade barriers" have to go, so that those nations that agree to the WTO, NAFTA, CAFTA, and FTAA regulations also agree that they will not protect their farmers or their crops, regional or national businesses, their workers, their environment, or even human health, from the encroachments of any trader.

So who does benefit from free trade? Certainly, transnational corporations, and the politicians who receive campaign contributions from those corporations—not only those in the U.S. Congress and the White House, but those in farm states' state legislatures as well. These agreements do not make trade free; they set free the transnational corporations that can now trade without regulation, control, or accountability to the public. As the Friends Committee on Unity with Nature tactfully put it in 2003, "In truth, these agreements have primarily promoted the productivity and profitability of large corporations by reducing legal constraints on their activities."[3] When there are complaints about trade, it is nations, not transnational traders, that are penalized with economic sanctions or fines, thus creating a burden on the nation's taxpayers if the agreements are not met. Further, the agents that decide such penalties are not national or state courts but tribunals of free trade experts that do their work in secret. Appeals cannot be made to any nation's court system but must be addressed to the tribunal that made the original decision.

NAFTA studies have focused primarily on industry and jobs rather than agriculture, but nongovernmental organizations (NGOs) have made several analyses of NAFTA's impact on farming and have also looked into the impact of free trade on agriculture under the WTO, IMF, and World Bank. These studies reveal a wide disparity between theory and practice, a disconnect between the reasons given for agreements and results on the ground. The results show that farmers in the United States and around the world get squeezed so that farm prices drop, consumer prices escalate, and agribusiness profits soar.[4] Meanwhile, the U.S. Congress is prohibited by the trade agreements from taking steps to regulate U.S. agricultural trade. As we have seen, the text of those agreements makes clear that trade agreements take precedence over the U.S. Constitution and any other federal, state, or local laws. To whom do citizens who want to change the system turn? Article 1, section 8 of the Constitution grants members of Congress the power "to regulate commerce with foreign nations, and among the several states, and with the Indian tribes." According to article 1, section 7, initiatives to increase revenue must begin in the House, though the Senate is free to "propose or concur with amendments as with other bills." But, as we will see below, each free trade agreement renders Congress powerless to change the treaties it has signed. Democracy is disabled. In signing each treaty, Congress surrenders its power to "regulate commerce with foreign nations" and turns over some of its authority to initiate increases in revenue to the secret trade tribunals that can impose penalties that inevitably require raising revenues, whether Congress wishes to or not.

These trade agreements are based not on economic theory but on sales pitches. Theories may be supported or disproved by experiment or experience, but no matter what outcomes the WTO and NAFTA produce, the theories do not change. They proceed as accepted fact rather than hypotheses. Such theories do not have the substance of myth, but they have myths' persistence and believers. Since the results rarely seem to match the promises, proponents of the theories rarely examine the outcomes; even after the evidence is clear that the theories are wrong, they keep promoting them as mantras for personal and corporate gain. Even when hunger is growing around the world, steps that could feed the world's impoverished hungry are ignored, undermined, or prohibited by free trade agreements.

DISSENTERS AND PROPHETS

In the debates over world trade and NAFTA, there were a few dissenters early on, but their caveats were ignored. According to Sophia Murphy, director of International Trade Studies for the Institute for Agriculture and Trade Policy (IATP), the UN Food and Agriculture Organization held that the free trade Agreements on Agriculture "would accelerate the already-established trend of developing-country dependency on food imports," raising concerns about food security. That prediction turned out to be accurate, but the information was buried under the sea of pretentious claims favoring free trade in agricultural goods. Murphy also describes the forecast of Kevin Watkins of the Catholic Institute for International Relations that, under NAFTA, "over-production would not be curtailed, and that under-priced imports to developing countries would increase under the AoA." Watkins also believed that cutting price support mechanisms would not reduce production in the developed countries.[5] He turned out to be a prophet, as we'll see later, in a report from Mexico.

Public Citizen's Global Trade Watch, in its report on the first seven years of NAFTA, reminded us, "The National Farmers Union was concerned that NAFTA would undermine sustainable agriculture in all three countries"—Mexico, Canada, and the United States—and make it easy for them to reduce or eliminate the farm support systems in place to protect family farms. The National Farmers Union, like the Food and Agriculture Organization and Watkins, was dead on. Public Citizen subsequently praised Minnesota's IATP because it "pointed out that specific provisions in NAFTA encouraged concentration in the food processing industry and the expansion of factory farms and agribusiness in all three NAFTA countries." IATP's view was also prophetic, and that agency has monitored developments since, maintaining a page on its Web site that provides up-to-date information on the status of current negotiations in the WTO, NAFTA, CAFTA, and other industrial and agricultural trade agreements.[6]

The results of WTO and NAFTA thus far make all those dissenting agencies look as if they could foretell the future. What they really did, of course, was study the negotiating documents, think very carefully about what they found, and provide more accurate analyses than did corporations and federal governments, all of which had reasons for distorting their analyses or keeping them secret. One might wish that the U.S. Congress had seen fit to study the documents as carefully. As we will see, labor and consumers had good reason to fear unrestrained agricultural trade, and the environment has suffered new onslaughts as well.

DOING THE NUMBERS ON NAFTA

NAFTA has perhaps had a more direct impact on us than has the WTO, though the continuing rancorous debate over genetically modified crops indicates that the WTO also has a powerful influence. Trade theory in NAFTA predicted essentially the same five outcomes for all three agreeing nations. In the United States, it was sold to us with the following claims:

1. Production for export in both industry and agriculture would create U.S. jobs, at least two hundred thousand overall and fifty thousand in agriculture alone, to gear up for export production.

2. In developed countries, there would be reduced subsidy payments to farmers, which would lead to reduced commodity prices.

3. Lower prices would reduce output, because commodities without subsidies would not be profitable.

4. Farm incomes would increase as a result of the demand for agricultural exports.

5. Consumers would pay lower retail prices as a result of lower world food prices.[7]

In any assessment of NAFTA, there are landmarks along the way. Just prior to the fifth anniversary of NAFTA, Robert Weissman, editor of the *Multinational Monitor*, commented that it had been a "five-year party for multinational corporations.... Unfortunately, the corporate CEOs have been dancing on the heads of the rest of us."[8] Now, after more than eleven years of NAFTA, it appears none of the predictions or promises listed above have been borne out.

1. After just five years, "200,000 verifiable specific U.S. jobs [were] lost to NAFTA," Russell Mokhiber and Robert Weissman report in *Corporate Predators*. Robert E. Scott writes that, since NAFTA was signed, "the rise in the U.S. trade deficit with Canada and Mexico through 2002 has caused the displacement of production that supported 879,280 U.S. jobs. Most of those lost jobs were high-wage positions in manufacturing industries." But the number of jobs lost rather than created is only one reflection of the impact of NAFTA on workers. Scott continues, "NAFTA has also contributed to rising income inequality, suppressed real wages for production workers, weakened workers' collective bargaining powers and ability to organize unions, and reduced fringe benefits."[9]

2. Instead of reducing agriculture subsidy payments in the developed countries—which insist through the IMF and World Bank that developing countries must reduce theirs—subsidy payments to U.S. and European commodity growers have continued, and even increased, in all three countries since NAFTA's 1994 implementation.[10]

3. Rather than reducing agricultural production, Mexico, Canada, and the United States have held production steady or increased it. "The U.S. has fewer farmers, larger farms, as much land in production as before, and similar, even increasing levels of production," Murphy reports. Her argument is supported by data that extend from 1987 (pre-NAFTA) through 1999. During those years, hog producers dropped off as if they'd been imbibing atrazine on Saturday nights—a 50 percent reduction. Farmer-owned dairies swirled down the drain

at a 40 percent loss. Wheat growers disappeared like chaff in the wind, falling by 30 percent. At a rate six times greater than during the pre-NAFTA era, Murphy reports, thirty-three thousand American small farms have disappeared during the NAFTA years, according to our own USDA. Yet commodity production has held up or increased in all three countries.[11]

4. Farmgate income in all three countries has gone down during the NAFTA years. Annual total farm income prior to NAFTA's implementation was approximately $59 billion. In 2001 it was down to $41 billion, a decline of about 30 percent in seven years. Between 1995 and 2000, the bushel price U.S. farmers received for corn dropped 33 percent. Wheat fell 42 percent during the period, soybeans 34 percent, and rice 42 percent. In all three countries, government and grain traders hyped the value of increased exports to farmers and policymakers as a sure way to increase profits for farmers. To NAFTA's credit, Canadian agricultural exports grew between 1993 and 1999 by 6 billion Canadian dollars—almost double the previous rates. Still, "in 1998, the average profitability of Canadian farms, measured by return on equity, was only 0.3 percent, but the return on equity for the cereal companies buying and using Canada's commodities such as Kellogg's, Quaker Oats, and General Mills was 56%, 156%, and 222% respectively." And farmers' net incomes decreased by 19 percent during those same years.[12]

5. Consumer prices did not decline as farm incomes fell; rather, they rose higher than before. In its 1999 study *Compare the Share Revisited: The Family Farm in Question*, the Centre for Rural Studies and Enrichment, a Canadian NGO, reported that, "when pork packing companies in Canada pushed wages down by 40 percent, they also gained from a decline in pork prices to farmers." The report added, "They became more profitable than ever." Though exacerbated by NAFTA, this process has been going on for a long time. In the 1970s, for example, Canadian farmers received $0.50 to $0.75 per pound for their hogs, and the price of a pork chop at the store was between $2.10 and $2.50 per pound. In recent years, the Centre for Rural Studies and Enrichment's report showed, pig prices fell to $0.23 per pound, but the retail price of pork chops rose to $5.00 per pound. Grains fared no better. While the price farmers received for the 450 grams of wheat used in a box of crackers went from $0.07 to $0.08, the retail cost of a

450-gram box of crackers made from that wheat went from $0.77 to $1.94 (all figures Canadian dollars). According to Public Citizen, here in the United States, the Consumer Price Index shows that U.S. consumer prices also increased—almost 20 percent during the first seven years of NAFTA—despite the declines in farm income that abound in the documents cited in these pages. As the cost of commodity goods has gone down, consumer prices have gone up, and transnational grain trader and food producer corporate profits have increased almost geometrically.[13]

These outcomes made the dissenters look prophetic indeed, maybe even clairvoyant. Have the economic theorists and free trade supporters changed their views? Not at all. They make the same promises over and over again as they defend each new agreement.

TRANSNATIONAL CORPORATIONS VERSUS MEXCANUS

In this lopsided game, no one wins but a handful of CEOs and a few investors. NAFTA was advertised by the U.S. trade representative as "a win-win-win" for all three countries. Despite the results of NAFTA, the CAFTA agreements were described by the same office as "a win-win for everyone." Not quite. As Murphy points out, even the investor group that really did win is not as large as one might expect, since seven families control the world's five largest grain companies.[14] The rest of us pay to watch as peasant farmers are led up the great stone steps of the free trade temple to be sacrificed.

The free trade agreements have had additional outcomes that, though they were anticipated by the Food and Agriculture Organization, the IATP, and other NGOs, are nevertheless startling in their reality. As documented by Public Citizen's Global Trade Watch, they demonstrate further the lopsided score.

- As farmer incomes went down 4 percent between 1993 and 2000, the transnational traders' profits rose. "ConAgra's profits grew 189% from $143 million to $413 million; and Archer Daniels Midland's profits nearly tripled between 1993 and 2000 from $110 million to $301 million." Other transnationals did equally well.
- "Agribusinesses have been able to create new export platforms

which play farmers from the U.S., Mexico and Canada against one another in a fight for survival as prices paid to producers are steadily pushed down."

- U.S. fruit and vegetable growers were severely undercut by produce from Mexico. Some of the Mexican plantations "are owned and operated by transnational agribusiness concerns which, by relocating to Mexico, are able to use pesticides banned in the U.S., exploit $3.60-a-day Mexican farm labor and avoid U.S. regulations regarding safe and sanitary working conditions."
- Taking advantage of their new freedoms from constraint, the grocers and transnational agribusinesses began to merge. For example, Smithfield Foods merged with Murphy Family Farms, Cargill bought Monsanto, and Tyson Foods took over Iowa Beef Producers.

In addition, the U.S. International Trade Commission data cited by Public Citizen indicate that agricultural trade balances are shifting dramatically.

- In 1995, the cattle and beef sectors had a $21 million trade surplus. By 1999, it had become a $152 million deficit.
- Between 1995 and 1999, the U.S. poultry trade surplus fell 14 percent.
- The frozen fruit trade had a $9 million surplus in 1995 and a $37 million deficit in 1999. Prepared and preserved U.S. fruits, already in trouble, plunged from a trade deficit of $236 million in 1995 to a $396 million deficit in 1999. The fruit and vegetable juice industry watched an $18 million surplus plummet to a $48 million deficit.[15]

HOW AGRICULTURAL "FREE" TRADE REALLY WORKS

If U.S. and EU farmers, and those in other developed countries, are struggling, farmers in developing countries are facing even more difficulties because they have less access to credit, less access to information, poorer transportation infrastructure, and less well developed markets, all of which means they have even less market power than farmers in the United States have. Their plight is exacerbated by U.S. and EU transnational traders' dumping surplus subsidized corn on their markets when their own corn subsidies have been reduced or terminated by WTO or NAFTA rules or by the demands of the IMF and World Bank.

In Mexico

In a study titled *The Environmental and Social Impacts of Economic Liberalization on Corn Production in Mexico*, Alejandro Nadal describes some of the promises and the results of NAFTA in Mexico. Since corn was developed in "Indian Mexico" five thousand years ago, it has evolved to become not only a staple for the Americas but "one of the three staple food crops in the world," Nadal points out. "Forty-one racial complexes and thousands of corn varieties are recognized in Mexico, forming a rich reservoir of genes for coping with drought and other adverse environmental conditions." Approximately 60 percent of the agricultural land in Mexico included in NAFTA was cultivated in corn, and corn production accounted for 40 percent of the Mexican people working in agriculture. Yet, Nadal reports, "NAFTA negotiators failed to recognize the important role played by this part of Mexico's corn sector in maintaining the country's genetic diversity, and see these (inefficient) producers merely as a drain on the economy. As a result, it was considered desirable for this group of producers to move from corn production to cultivation of other crops, or to non-agricultural sectors of the economy."[16] No one asked, or apparently cared, whether the farmers wanted to quit raising corn or move off their land to nonagricultural sectors—as if there were jobs available in town. NAFTA has thus become a tool for a brutal social Darwinism.

Nadal describes the promises made by NAFTA to Mexico and its farmers. The first was that NAFTA would allow a fifteen-year transition period for Mexico's corn prices to come into line with international prices. To accomplish this, the Mexican tariff and import permit system would be changed to a "tariff-rate quota" system that would gradually be phased out over that time. During that period, the tariff-free amount of corn allowed into the country would grow at 3 percent per year (approximately 2.5 million tons, according to Nadal), and ad valorem taxes would be reduced to zero by 2008. A second promise was that a series of "adjustment-assistance policies" would be put in place to protect Mexican corn growers from economic disaster. These would range from "direct income support mechanisms to credit, infrastructure investments, and agricultural research and development." Nadal describes how that came out: "A key finding of this study is that the planned fifteen-year transition period was actually compressed to roughly thirty months. Between January 1994 and August 1996 domestic corn prices fell by 48%, thereby converging with the international market some 12 years earlier than provided for under NAFTA, and forcing Mexican corn producers into a rapid adjustment." This abrupt end to the adjustment process had approximately the same shock effect as suddenly eliminating farm bill benefits to U.S. corn and bean growers would. Further, it "took place against a background of declining state support for agriculture," says Nadal. Rather than keeping the adjustment assistance policies in place, "the policy instruments supposed to assist transition rapidly lost most of their effectiveness. As a result of inflation, PROCAMPO, the mechanism for direct income support, lost 40% of its value in real terms. In addition, credit provision declined to extremely low levels and public investment in infrastructure was severely curtailed. Finally, the price regulation agency, CONASUPO, which was to have been gradually phased out, was completely dismantled in late 1998."[17]

Pre-agreement computer models, based on the trade theories, showed that Mexican corn output would decrease gradually as domestic prices fell, also gradually, so that farmers and farmworkers, their land, and their capital could be moved to "more productive activities," as Nadal paraphrases NAFTA. Further, there would be an environmental benefit, since the models assumed that "marginal land vulnerable to erosion would be left fallow." From 1994 to 1998, corn prices did fall as a result of imports of cheap corn, mostly from the United States, but "at a much faster rate than anticipated." The drop in corn production that was supposed to follow did not occur. Instead, production remained at historically high levels, and the acres under production actually increased, "implying a drop in average yield per hectare and suggesting that more marginal land was being cultivated under increasingly stressful conditions." What went wrong? In Nadal's view, "The fundamental flaw in the official pre-NAFTA studies was the assumption that only one variable, market price, would determine changes in the behavior of Mexican corn producers. However, producers' decisions are made in the light of many other factors." Though Mexican corn prices went down, the prices of beans, cotton, barley, rice, sorghum, wheat, and soybeans still remained lower than the price of corn, "so that corn remained relatively more profitable." Further, technological changes meant that more corn could be grown on less favorable land—if the farmer could afford the inputs. Those who could afford to do so purchased land, including marginal land, bought the pesticides and herbicides, and simply grew more corn, making up for lower prices with more production.[18]

Some of the outcomes Nadal points up are even more discouraging than the displacement of many farmers and farmworkers. The result of increased production through increased inputs is that more chemical fertilizers and pesticides are being used than ever before, water is being used at higher rates, aquifers are being drained faster than they can replenish themselves, "and salinization and accumulation of chemical residues are widespread," Nadal observes. Further, he notes, there is more compaction from more heavy machines. Farmers are using transgenetic corn, which pollutes the traditional seeds that account for a major portion of the diversity storage cupboard for the world. And, because there has been a decline in hired labor, "there is also likely to be a reduction in capacity to maintain soil conservation structures and practices," many of them traditional efforts such as "terracing, hedging, groundcover crops, mulching, minimum tillage, ridge planting and alley cropping."[19]

The trade theories also assumed that there was no need to worry about, or even consider, what might happen to subsistence farmers; the assumption behind that assumption was that subsistence farmers do not sell their produce. Yet subsistence farmers in Mexico, like U.S. farmers during the Great Depression and poor people the world over, make what economists call "petty sales" of their corn to earn enough cash to pay household bills and to meet emergency needs. Unfortunately, because they operate on such tiny margins, peasant farmers often have to sell as soon as the crops come in, when prices are lowest because of abundant supply. Then they must repurchase corn to restore their family larders in late winter or spring, when supplies are down and prices are high. This is a perfect description of the way connections, especially those we do not recognize, lead from one outcome to another. We are reworking a system we do not understand, creating an economic ecosystem that leads to tragedy. One result has been migration. Most of the migrants move to the cities of Mexico, straining their infrastructure and resources —a process that began long before NAFTA. Others cross the border to work in the United States. According to Nadal, "social institutions that play a key role in resource management tend to deteriorate under the combined pressure of poverty and migration."[20] He does not mention the wrath that migration stirs among many Americans, some of whom are corn growers eager to export their corn to Mexico and are thus indirectly responsible for setting the migratory process in motion.

Nadal concludes his study by saying, "The inclusion of corn in NAFTA is a clear example of how trade liberalization planned with a short-term view can have long-term negative impacts."[21] How seldom we know what we are doing, especially to cultures other than our own, despite our modern science and learning, our sophisticated technology, and our higher education. And Nadal is not the only one who reports such negative outcomes. Emilio Lopez Gamez is a professor of agronomy and one of the founders of The Countryside Can't Take It Anymore, a coalition of Mexican farmers protesting globalization. Speaking in Minneapolis on July 16, 2003, Gamez and Jose Luis Alcocer de Leon, another farmer and founder of the coalition, reported on Mexico's experience with NAFTA. They told essentially the same story as Nadal, but they put it in a slightly more sinister context.

To sign on to NAFTA, Mexico had to revise its constitution's chapter 27, which protected the land rights of peasant farmers against foreclosure, sale, and seizure. Those protections were stripped from the constitution, and now that land is up for sale. Transnationals, following a different tack than the one they take in the United States, are buying up large amounts of Mexican land made available by this change in the constitution and by the mass exodus of farmers. The transnationals' purpose is to develop a broad agricultural enterprise, growing many crops in a fully integrated system that eliminates local farmers and is essentially regulation free. Gamez and Alcocer de Leon believe that the outcomes reported by Nadal and others are the deliberate results of a hidden agenda. What concerns them most are the hidden objectives of the NAFTA Agreement on Agriculture. First, creating poverty: The desire was to push one-half of all Mexican farmers off the land. "This has been accomplished," said Gamez. Second, increasing imports: "This too has been accomplished." Mexico now imports twelve times as much corn as it did before NAFTA. Third, destruction of the rural economy: NAFTA has dramatically increased the gap between the rich and the poor, between those who benefit from free trade and those who do not. "Ninety percent of the total producers are small," and more than 40 million Mexicans live in extreme poverty, yet "out of 6 million producers, only 20,000 benefit from NAFTA," Gamez reported. Fourth, increased migration: The numbers cited by Nadal and other monitors of Mexican agriculture support Gamez and Alcocer de Leon's contention.[22]

One of the promises of NAFTA was that, through free trade in agriculture, Mexico's economy might look more like that of a developed country—the United States, for example. This "economic convergence" was seen as desirable. Indeed, the economic growth that the World Bank attributes to NAFTA does make Mexico's economic recovery quite like the U.S. recovery in at least two regards: the wealthy get wealthier, and the poor get poorer.

In these NAFTA cases, one hears the echoes of farmers in Sri Lanka: the international economists who developed the free trade scenario were likely "too narrowly educated and insensitive." Either they did not have any understanding of social psychology, anthropology, or history, or else they advanced their theories despite their knowledge, grasping for another corporate opportunity. Perhaps the education of our economists should broaden their perspectives to include the human implications of the numbers and equations they rightly love.

In Other Countries

How "free" trade really works in agriculture is further revealed in other stories from around the globe. *Trade and Hunger: An Overview of Case Studies on the Impact of Trade Liberalisation on Food Security,* a collection of case studies by NGOs around the world, examines the impact of free trade and structural adjustment programs (SAPs)—the name given to the "conditionalities" imposed on developing countries by the World Bank and IMF. (Joseph E. Stiglitz, former World Bank economist and Nobel Laureate, explains in *Globalization and Its Discontents* that "'conditionality' refers to more forceful conditions, ones that often turn the loan into a policy tool.")[23]

In Tanzania, *Trade and Hunger* reports, "the overall impact on food security of the liberalisation of agricultural trade is profoundly negative. Farmer incomes are declining, and, at the same time, school and medical fees have been reintroduced under the SAP. Farmers have to part with some of the little money they earn, and have less to meet farming costs and to buy food in times of shortage. Food insecurity has thus increased."

In Kenya, where women make up 70–80 percent of agriculture's labor force and produce three-quarters of the food, the case study showed that, "as a result of the country's SAP and the liberalisation of agricultural trade, many women cannot afford adequate chemicals and fertilizers, and farm output has declined. Liberalisation has led to an increase in food imports into the country and caused food dumping (cheap surplus food from the North) in local markets, hitting the country's own farmers. Liberalisation has also led to an increase in the prices of farm inputs, putting them beyond the reach of most small farmers." The report from Thailand includes the following:

> Presently Thailand is reeling from the EU's restructuring of the Generalised System of Preferences. . . . Rural suicide is up and mental health problems are climbing as evidenced by consultation statistics of regional health services. Rural crime is up and other social indicators are starting to blink as well. . . . It is hopefully obvious at this point that the mess isn't simple: devastating weather patterns, massive unemployment, foreign food shortages putting pressures on domestic supplies, a need to earn foreign exchange to bail out an unbelievably irresponsible private sector (both domestic debtors and foreign creditors), [and] external pressures to increase trade liberalisation.[24]

In his essay "Sell the Lexus, Burn the Olive Tree," Greg Palast reports that the average number of conditionalities on such loans is 114—a bit like having a two-thousand-pound albatross around one's neck. In Tanzania's case, it took about fifteen years for the per capita gross domestic product to drop from $319 to $210, while poverty rose to include 51 percent of the total population and literacy declined, Palast tells us. Food security decreased as incomes went down. *UN Wire*, the United Nations Foundation's daily news update, reported that, in 2000–2001, one-third of Tanzanian children between the ages of five and seventeen were forced to join the workforce to help their families make ends meet. In all, 4.8 million children were working, 4.3 million of them in rural areas in the informal economy.[25]

One reason farmers in Tanzania and elsewhere have less to spend on food is that they want their children to have an education. They are so marginal, economically, that their dilemma is this: if I pay the school attendance fees, my children face hunger; if I buy my children food, they go without schooling. What kind of choice is that? IMF believes that's not a choice it has imposed with its conditionalities; it's a decision that people make on their own. The IMF tied a package of 157 conditionalities to Tanzania's loan, most with similar results. Stories like these lead Eunice Kazembe, Malawi's ambassador to Taiwan, to lament, "As long as we keep talking about the issue of liberalisation, structural adjustment and now globalisation largely in terms of GNP, economic growth, competitiveness and such technical terms, we will continue to gloss over the real issues and the dehumanisation and indignity that supposedly well meaning initiatives of the World Bank, IMF, WTO and such institutions can promote." Kazembe's comments become a litany of times past contrasted with present circumstances in the small-farm-dependent villages of her country. As in the United States, there is a direct tie between the loss of small farms and the loss of food security and schools.

> Times were when, in the villages I know, there was not just enough to eat, but of adequate variety to ensure healthy growth for children and physical stamina for the adults. Not anymore. Children are hungry and listless most of the time, their mental and physical potential sabotaged and limited from childhood. Adults, physically weakened, are unable to concentrate their minds and to work long hours as they used to. Times were when schools had books and writing materials, teachers had motivation because they earned enough to live on and had respect in the local community. Now most children go to school, yes, but with nothing to write on, nothing to read. . . . Time was when what the villager produced from the land had value.[26]

These are important voices for us to hear, since many of those agricultural trade agreements have been initiated by U.S. corporations and promoted vigorously by the U.S. government in international negotiations, and by the IMF and World Bank, which are essentially our country's instruments for enforcing our foreign policy. These stories give us a glimpse of our own American farm future if it is left in the hands of transnational corporations and our own government. After fifty years of such experiences, is there any wonder so many people around the world resent us and fear our influence? Stiglitz concludes, "While conditionality did engender resentment, it did not succeed in engendering development. Studies at the World Bank and elsewhere showed not just that conditionality did not *ensure* that money was well spent and that countries would grow faster but that there was little evidence it worked at all" (Stiglitz's emphasis).[27]

TRIPPED UP BY TRIPS?

Small farmers in other areas of the world also face an intellectual property obstacle. Corporate scientists tap the indigenous knowledge of plants, learned over many generations by farmers in, for example, India, Southeast Asia, or the jungles of Latin America, and then patent that knowledge as their own intellectual property. These patents on trade-related intellectual property (TRIP) then prohibit local farmers from continuing to use the plants, even for their own traditional, local, medicinal purposes. The patents often prohibit farmers from saving or using seeds from the plants they have grown or harvested for generations. Instead, they have to purchase new seed every year from the seed companies that hold the patents.[28]

Exclusive claims to intellectual property complicate the system in other ways. Our land-grant universities, originally friends to the farmer, often conduct the research that creates intellectual property and thus have more recently been better friends to chemical companies, farm equipment manufacturers, and agribusiness. As the percentage of state support declines, says Jerry Caulder, CEO of Mycogen, "the farmer is going to become more and more at the mercy of the few who own intellectual properties. Again, it goes back to the shortsightedness of funding basic research in such a parsimonious fashion. The universities are becoming branches of whoever they can get a contract from." Major agricultural corporations have become a primary source of such contracts. As a result, large components of the ag departments at our land-grant universities now work for the industrial sector of the agricultural economy that makes its profit at the expense of small farmers. They assist in turning Jefferson's "independent yeomen" into serfs. In the process, neither the universities nor the corporations mention the loss of academic freedom imposed by these economic constraints, or how they might affect the free inquiry upon which science depends.[29] The land-grant University of Minnesota, invented to serve our state's people, has expressed its Department of Agriculture's commitment to become a leader in the development of genetically modified crops and the artificial agriculture that benefits transnational corporations. It is no accident that the newest crop research building on the University of Minnesota campus bears the name of Cargill.

Sustainable farming still gets scant attention, save from a few courageous faculty members willing to buck the trend and work directly with Minnesota farmers. "I've been kind of disappointed in the university," says dairy farmer Bonnie Haugen's friend Judy, perhaps thinking about fifty years of the university agriculture department's education for bigger and bigger farms, higher production, commodities, and chemicals, and support for multinational industry while neglecting sustainability. Farmers remember the original role of land-grant universities—to serve regional agriculture—and lament the surrender of that mission.

THE DYNAMICS OF DECLINE

How does all of this happen? It begins with U.S. government farm subsidies, which do not go to the average small, diversified farmer following sustainable practices but to large-scale commodity producers of corn and soybeans. As the small farmers go down, their land is purchased by larger farmers or by agribusiness corporations, and production remains the same—or may even increase in the short term, as the result of more destructive industrial farm practices. Murphy cites a 1998 survey that showed net farm income at $8,616 (excluding subsidies), while government subsidies averaged $30,000. As Murphy summarizes it, "Farmers have become a flow-through channel for government payments." Well, some farmers have—those who raise industrial-scale corn and soybeans—but the rest have not. Of U.S. farmers, 60 percent "receive no subsidy at all."[30]

According to Murphy, "almost all of the US$22.7 billion of direct government payments" go to only eight crops, primarily corn and beans. And most of the funds go to a small group. "Nationally, over 60 percent of the (farm) payments will go to 10 percent of the recipients," who are principally nonfarmers. They are landowners who rent, "rather than risk losing money in production of their own," says Murphy. Some are older farmers who were unable to keep going, physically or financially. For them, the rent check, underwritten by the commodity payments, is a means to supplement retirement savings or Social Security payments—and to stay in touch with the land. Since "payments are made to land-owners, and not necessarily to the person farming the land," landholders also benefit from federal programs. Murphy cites a 2000 report indicating that Iowa owner-operated farms constituted only 30.8 percent of the total land farmed in that state. Other benefactors of taxpayer largesse are "grain and processing companies, who buy their inputs at less than cost of production prices," and agribusiness corporations—often those same processing companies—"that dominate input supplies."[31]

Further, free trade theory held that grain surpluses would not be a problem because lower world prices would reduce supply. Who argued—and is still arguing—that we need lower grain prices to compete in world markets? The transnational grain sales industries that do the trading. They can afford lower prices because their volume is so great: they can survive on a slim percentage of a cent's profit, whereas smaller competitors and farmers go under at that rate. "The reduced floor prices set by the U.S. government reduced the number of farmers without reducing production," Murphy reports. Costs to agribusiness thus "were reduced, and the volume available to trade or process has been stable and even increased."[32] The corporations win either way.

Another result of the pressure to lower prices, exacerbated by the concentration of the ruling corporations in agriculture, is that the price of inputs (such as pesticides, herbicides, gasoline, and veterinary bills) has increased. Murphy echoes soils scientist Gyles Randall when she notes, "Farmers have next to no choice from which company they buy equipment, seed, pesticide or fertilizer. Nor do they have a choice of buyers for their produce or of haulers for their grain." The buyers and haulers are the same folks who sell farmers their seeds, pesticides, and fertilizers and own the trucks and ships that carry the produce to processing and markets. "Since 1948," Murphy continues, "the use of off-farm inputs such as inorganic fertilizer and fuel has increased 84 percent, the use of capital has increased 33 percent, while labor as an input has dropped by 70 percent." She concludes, "All this has diminished the value of agriculture to the local economy and reduced the independence of farmers as businesses."[33]

The relative diminishment of producers' agricultural income means that corn and soybean farmers are now "a net drain on local economies: they do not earn enough profit to pay taxes, they need at least one other job off-farm to survive, and their farm purchases are from companies whose production base is out of state, and possibly outside the country altogether," Murphy explains. Further, little of the corn production goes to healthy food. Much of it goes to the high fructose corn syrup that is hidden in so many foods that it has been linked to epidemic obesity, and into the production of ethanol. The new demand for synthetic fuels will require more soybeans, along with high-cost inputs that will reduce farm profits further. There are links solid as those in a new chain here: these crops require toxic pesticides, nutrients, and herbicides; they also reduce ground cover and increase erosion, which increases stream pollution that expands the toxic bloom that kills marine life in the Gulf of Mexico.[34]

To increase production in order to make ends meet, farmers turn to new technologies, "such as genetically modified crops, some of which reduce the need for labor," or sign "contracts with corporations that guarantee a price (even if that price remains below the cost of production)." The result is "reduce[d] economic autonomy of U.S. farmers," according to Murphy's report—and greatly increased profits for transnational corporations. The final results have been noted above: exports increased, farm income decreased, food imports grew, and consumer food prices rose in the United States and around the globe.[35]

THE STORY OF THOSE PROTESTERS IN THE STREETS

We can be sure the U.S. negotiator in trade discussions has pressed for considerations as close as possible to those contained in the FTAA. Because the playing field is nowhere near level, the United States gets its way, putting serious pressures on smaller nations if it has to. But, because the talks are secret, most of us do not know exactly what the agreement holds us to. Nevertheless, a growing movement, gathering force as it collects and disseminates information, is working to develop and maintain some control over the corporate license to exploit the hemisphere. U.S. citizens are involved, and representatives of many other nations are also making strong efforts. In 2002, 50 million Brazilians voted for Luiz Inácio Lula da Silva for president after he ran on an anti-FTAA platform.

Lula himself speaks pretty plainly about his goals: "Agrarian reform is also fundamental if the Brazilian economy is to be rebuilt. And it will play a crucial role in making the country fully democratic."[36]

The unrealized promises and sad outcomes of the WTO and NAFTA outlined above are better known to the rest of the world than they are to us in the United States. Elsewhere, they arc analyzed and argued in the media and are part of the public discourse. They have informed the developing world's view of how free trade works. The talks in Cancún, where the WTO agreements were to be settled in September 2003, collapsed, and negotiators went home early, thanks in large part to international protests, as well as to internal tensions over agriculture and food security. Similarly, at the FTAA negotiations in Miami that fall, disagreements over agriculture and food security broke the meeting up early. Getting acquainted with these stories about agriculture helps us think about how world trade might benefit all citizens instead of a few, and how it could make life more tolerable for peasant farmers and agribusiness farm families rather than increase the wealth of a few corporations.

Chinua Achebe's novel *Things Fall Apart* describes what happens when tribal peoples try to fit into a different society with different values. What happened in Cancún and Miami fits that pattern. The developing world's workers and farmers, coming from an entirely different culture from that of the negotiators in the hotel, were able to make their voices heard. They were very articulate, and they formed a coalition broad enough and strong enough to stand up to entrenched international corporate power. The stand-off was between rich and poor, yes, but it was also about agriculture and the role of farmers, not just transnational corporations in world trade. Food, food security, and farm income (as opposed to corporate profit) were key issues.

When the talks collapsed in Cancún and Miami, our trade representative was outraged that the developing countries did not want to forge ahead into a disaster so transnational corporations could reap the benefits. The *Economist* excoriated the poor who wanted to survive as naysayers to progress. The media tended to portray the protesters in Cancún and Miami as wild-eyed radicals, leftists, or unsophisticated peasants who did not understand what was happening and could not recognize their own self-interest. American viewers may have thought of the street protesters as hooligans. But the protestors in the streets were the real conservatives in Cancún and Miami, as in Seattle and elsewhere. The really radical voices came from the fine suits gathered in the swank auditoriums—the voices raised against national sovereignty and in favor of altering the U.S. Constitution, diminishing civil rights, revoking local control, and increasing constraints on individual private property to gain freedom for corporate property. The protesters saw the expensive suits of the trade representatives as symbols of bullying power, and they refused to be bullied. The police in cities where WTO and FTAA discussions were held were later found to have used unnecessary and excessive force to "control" what were, in reality, largely peaceful demonstrations.[37]

For now, CAFTA holds center stage, but we may be sure that the FTAA is eagerly waiting in the wings. Under some new guise, it will rise, phoenix-like, from the ashes. "FTAA has become the template for the other agreements," Dennis Olson, director of the IATP's Trade and Agriculture Project, told me. During the maneuverings in Congress before the vote, some commentators described CAFTA as a "stalking horse" for the FTAA, a description that supporters of CAFTA denied vigorously. But there was good reason to sing those FTAA resurrection blues. The final goal outlined in the preamble to the Dominican Republic–Central America–United States Free Trade Agreement (still generally referred to as CAFTA) is to "contribute to hemispheric integration and provide an impetus toward establishing the *Free Trade Area of the Americas*."[38] More of us should therefore seek a better understanding of why protesters from so many countries jammed the streets in Miami, just as they had for earlier WTO gatherings, from Seattle to Genoa and Cancún. We must look carefully at what the FTAA negotiating documents actually say. They indicate the normative ideal that transnational corporations seek and the U.S. government supports.

STRAIGHT FROM THE HORSE'S MOUTH

I have talked with farmers, extension agents, and businesspeople in southeast Minnesota, including our U.S. senator's agriculture representative, and have yet to find one who has actually read the negotiating documents of the FTAA. Some of them do not believe that the concerns I express over these agreements could possibly be real. They do not believe—perhaps do not want to believe—that our federal government would give up its sovereignty or that it would protect transnational corporations at the expense of the health of its own citizens.

The FTAA documents, however, reveal what a merciless process we are engaged in. Once an agreement is made, it cannot be reconsidered. Implacably, the process moves on, leaving all second thoughts in the dust, as if we could never toss and turn, fret about things in the night, and wake up with changed minds. Or come upon new information, learn from the experience of others, and thus alter our perceptions so that we might wish to alter our behavior or our agreements. Any allowable protections for producers and consumers have to be built in before negotiations end. It is possible to withdraw from the agreement later, but only after compensating member nations for their loss of potential profit or suffering retaliation, including sanctions. All that inexorability provides a kind of rigor that will one day rise up and bite us with poison fangs, as we will see later in "The Two-Way Street of Exports."

In the FTAA draft agreement issued in November 2002, general article 4 is key. Every section of this article, below, applies to agriculture as well as other kinds of trade.

4.1. Each party is fully responsible for the observance of all provisions of the FTAA Agreement, and shall take such reasonable measures as may be available to it *to ensure such observance by regional and local governments and authorities within its territory.*

4.2. The Parties shall ensure that their laws, regulations, and administrative procedures are consistent with the obligations of the Agreement. The rights and obligations under this Agreement are the same for all the Parties, whether Federal or unitary states, *including the different levels and branches of government, unless otherwise provided in the Agreement.*

4.3. This Agreement *shall co-exist with bilateral and subregional agreements,* and does not adversely affect the rights and obligations that one or more Parties may have under such agreements, *to the extent that such rights and obligations imply a greater degree of integration than provided hereunder.*

4.4. The Parties confirm the rights and obligations in force among them under the WTO agreement. *In the event of conflict between the provisions of the WTO Agreement and the provisions of this Agreement, the provisions of this Agreement shall prevail to the extent of the conflict.*[39] [my emphasis]

Thus the FTAA would require that not only national laws but state and local laws change to reinforce FTAA agreements. Nothing a nation might do to protect its farmers, workers, the environment, human health, or small businesses could take precedence over the FTAA. The nation that did not bring its laws into conformity would face disciplinary action, including enormous fines and sanctions, by the FTAA. At present, the WTO Agreement on Agriculture takes precedence over any of our own national laws. Because the FTAA would offer even more concessions to free trade than the WTO, if it conflicted with WTO agreements, FTAA would take precedence over both the WTO and our own national, state, and local laws.

Sovereignty thus passes from our nation's legislative, executive, and judicial branches to the WTO participants, and from the WTO to CAFTA, the FTAA, or other regional agreements. Since transnational corporations have such powerful influence on the agreements, and NGOs are not permitted to participate or observe, sovereign authority is, in a sense, transferred from the nation to the transnational corporation. Once the agreement is approved, the freedom of corporations to exploit any resource and profit from any practice is protected, and the desires of citizens and nation-states are rendered irrelevant. A secret panel of trade experts resolves any disagreements that arise; there is no recourse to national courts. In place of the Constitution, we have the Agreements on Agriculture, among other agreements. Since the U.S. Constitution holds that the regulation of commerce is reserved for Congress, FTAA clearly undermines our Constitution, and the constitutions of other nations, all of which must become subordinate to FTAA.

A look at the FTAA negotiating documents related to agriculture confirms the fears raised above. Article 15 of the chapter on agriculture begins to explain where the process is heading:

- Article 15.1 notes that the negotiators recognize that "domestic support measures can be of crucial importance to their agricultural sectors, but *may also have severe distorting effects on the production and trade of agricultural products*" (my emphasis). Since anything that "distorts" production or trade is forbidden, except in cases of extreme emergency, domestic support measures must go.

229

- Article 15.2 holds the parties to "continue to work toward an agreement in the WTO negotiations on agriculture to substantially reduce and more tightly discipline trade-distorting domestic support," a constraint that thus far has been applied to developing nations but ignored by the United States and the EU. Because the United States has ignored its agreement, it is currently being sued through the WTO for its failures to reduce farm subsidies. If the WTO agrees with the petitioners, it will cost the country millions, which U.S. taxpayers will be obligated to pay.
- Article 15.3 specifies what FTAA hopes to achieve through the negotiations, including, in part, the following:
  (a) "the elimination or maximum possible reduction of production and trade distorting domestic supports"
  (b) "an overall limit on the amount of domestic support of all types"
  (c) "a review . . . to ensure that such support does not distort production and trade"
- Article 15.5 defines domestic support: "Domestic support means any policy or measure that affects decisions to produce, applied by a Party, *to sustain the prices of agricultural products, increase the revenues of farmers, and/or improve production and/or marketing conditions*" (my emphasis).[40]

What would be limited here is our nation's, our states', and our local governments' right to develop policies that might pay farmers a fair price for their agricultural products, increase farm income, or improve a farmer's production or marketing conditions. Our farmers, en masse, would thus be limited by the FTAA to stagnant or falling incomes, falling production, and low market value for their goods, while transnational corporations would be free to increase profits at an exponential rate.

The total of all possible domestic support for local agriculture is known as the "aggregate measurement of support." According to the negotiating documents, this "means the annual level of support, expressed in monetary terms, provided for an agricultural product in favor of the producers of agricultural products, or of non-product-specific support provided in favor of agricultural producers in general, other than support provided under programs that may qualify as exempt . . . under the provisions as established in this Article."[41] The translation is simple: There will be little or no domestic support, such as our U.S. farm bill, to enable U.S. farmers to improve their revenue, production, or marketing conditions, or to protect their fields from chemicals and erosion. Neither will there be a way to protect U.S. consumers from food imports that have been subjected to fertilizers, pesticides, and herbicides, including known and suspected cancer-inducing chemicals. If U.S. health and environmental regulations are more stringent than the FTAA's, ours will have to go. To provide such domestic regulations on behalf of producers would diminish transnational corporate freedom and control of corporate traders.

The declaration that came from the 2002 meeting in Quito, Ecuador, of the ministers of trade who represent the countries of this hemisphere spells out some other objectives of the FTAA. Declaration 2 states, in part, "We reiterate that the negotiation of the FTAA will take into account the broad social and economic agenda contained in the Miami, Santiago and Quebec City Declarations and Plans of Action with a view to contributing to raising living standards, increasing employment, improving the working conditions of all people in the Americas, improving the levels of health and education and better protecting the environment." Declaration 7 states, "We reiterate that one of our general objectives is to strive to make our trade liberalization and environmental policies mutually supportive, taking into account work undertaken by the World Trade Organization . . . and to promote sustainable development in the Hemisphere." Declaration 8 takes that objective a step further: "We . . . recognize the importance of strengthening throughout the Hemisphere, national actions and cooperation in order to ensure that the benefits of trade liberalization, the protection of the environment, and human health are mutually supportive." So far, so good—these objectives appear positive, even noble. But the real meaning of making policies mutually supportive, expressed in declarations 7 and 8, is ensuring that all policies regarding the environment, human health, and labor, are subordinate to and mutually supportive of corporate trade. Finally, declaration 11 reveals the outcome that is really sought in the negotiations for FTAA: "We reject the use of labor or environmental standards for protectionist purposes"[42]—that was, for any purpose that interferes with transnational corporations' capacity to utilize low-wage labor and exploit natural resources. One has to assume that "human health" was not included in that last declaration because of the public relations problems that might ensue.

This is where all our U.S. free trade negotiations, under whatever name, are headed. We have seen the future, and it is the FTAA. The scope will shift, the number or size of the countries participating will change, but the purposes and principles—and the sales pitch—will not. Indeed, it was soon after CAFTA was approved that our government began to work on the agreements of the FTAA, with President Bush making his moves at a summit of mostly unreceptive Latin American countries.

The folks who run agribusiness in Brazil believe that they are ready to compete with the United States under the trade agreements on agriculture—even if we do not actually implement the negotiated agreements by eliminating tariffs and subsidies for agriculture. (As a nation, we have broken our promises to other nations so many times that we are no longer expected to keep our word.) Indeed, in 2002, Brazilian farmers produced four times as many soybeans as U.S. farmers and, by some accounts, of better quality. However, as Fatima V. Mello writes in *The Post-Cancún Debate*, under Lula, NGOs and social movements were for the first time not only allowed in the Brazilian delegation but encouraged to participate. Their views of the issues are diverse. One group, Mello says, "contends that the WTO rules only benefit the large global corporations and financial capital, keen on freeing themselves from any type of regulations established by national legislations. The WTO agreements, in this view, do not serve to regulate the corporations' activity, but rather to limit the capacity of governments and to prohibit them from regulating the corporations." Other NGOs and social movements hold that "the WTO's problem is that its rules are distorted and that double standards are applied which especially penalize the poor countries. They contend that international trade should be governed by multilateral rules, in the absence of which, countries with less power are left at the mercy of the more powerful."[43]

THE ONLY THING WORSE THAN THE WTO?

After the reversals of the WTO in Cancún, America's trade representative at the time, Robert B. Zoellick, divided the world into "can-do" countries and "won't-do" countries. His new plan was to reach agreements by negotiating with individual developing countries one-on-one or in small regional groups.[44] This method was clearly in the American trade interest, for what developing country, standing alone, could withstand our blandishments and threats? Sometimes the sophisticated public relations rhetoric is simply the velvet glove over the mailed fist. You want medicine for AIDS? Then you must buy our grain. You want grain? Then you must refuse to sign the international agreement to support the International Criminal Court. We are not above such bullying tactics; we have already used them powerfully in Africa and elsewhere. We will get our way under those negotiating circumstances. If we do not, others will be made to suffer.

The only thing worse than the WTO and FTAA may be these one-on-one talks and the smaller, regional trade negotiations, like CAFTA, that apparently are the result of the collapse of trade talks in Cancún and Miami as well as of the resolve of Trade Representative Zoellick and the persistence of his successor, Rob Portman. There is a certain irony here. In the WTO and FTAA negotiations, we have been working on a large scale, seeking to bend many nations to our will in secret meetings. Now we are learning again that small is beautiful. Everything about American and international agriculture and business in the past fifty years has been about size: get big or get out. The WTO was seen as a means to organize the whole world along American principles in one fell swoop. One size fits all; if it works for this nation, it must work for all nations. Fortunately for many people and for democracy, it didn't work to bully many nations at the same time. There were too many; they were too different; it was all too big. So now we are back to taking small, one-on-one negotiating steps with individual poor nations. And small, America is discovering, has its charms, because it is apt to be successful, a hope that may be fulfilled by CAFTA. Perhaps at a federal level, we will now discover what many farmers in the United States already know, and what Abraham Lincoln warned us against in an earlier chapter: big efforts can be very hard to sustain; they will wither on the vine without the irrigation provided by taxpayer subsidies. Small does work. But if the incremental impact of small trade agreements means achieving our will at the expense of the developing nations' food security, and our own, that will be more than ironic. It will be tragic.

HOW THE DECK GETS STACKED

Murphy's report for IATP sets out five "peculiarities" of the current market system for farm products. These peculiarities mean that both small farmers and agribusiness commodity producers are playing with a stacked deck when competing with transnational corporations.

- Transnational grain companies are not always interested in base prices. For grain traders, Murphy points out, "profit is a percentage of the sale . . . but high volume of sales at lower prices is also profitable. In fact, because the grain companies have a significant interest in keeping the barges, rail cars and ships they own busy, higher volume may at times be more important to the companies' profits than high prices. . . . High grain prices make it more expensive to feed hogs and cattle and to make tortilla flour, all of which bear on the companies' profits."

- Transnational corporations have an information advantage "in keeping competitors out of the market. Their global reach and presence gives them access to information that, even in the age of the Internet, is otherwise scarce and difficult to obtain."

- Commodity trade is very costly, and transnational corporations "have access to enormous amounts of capital."

- "Because they own banking, processing and shipping businesses, transnational agribusinesses respond to different economic pressures than do farmers."

- Finally, "the wealth and size of the transnational business[es] also make them politically powerful."[45]

The corporations contribute not only to politicians in this country but to those of other nations as well, Murphy points out. In the United States and perhaps elsewhere, the transnationals' corporate executives move easily in and out of government and the private sector. In their public sector capacity, they have opportunities to slant legislation to suit their private industry interests. For example, Murphy notes that Dan Amstutz, a Cargill vice president, moved into government work at the U.S. trade representative's office. It was he who drafted the original text of the WTO Agreement on Agriculture—and then later returned to the grain trade.[46] Since Murphy wrote her report, Amstutz has become a government employee again, now responsible for the agriculture of Iraq as it seeks to recover from the devastation of war. It should be no surprise that Iraqi agriculture is heading toward privatization. It will be placed in the hands of our U.S. corporations, which will "help" those farmers by shifting to high-production export crops. Experience tells us that will lead to domination by U.S. transnational corporations bent on private gain and will further impoverish local farmers. Experience also makes clear that food insecurity will increase when such policies are implemented. "Public service" has become a euphemism for furthering corporate interests at the expense of the public, a smiling mask for conflicts of interest.

And, sometimes, the way international trade works is not international at all, except as it is disguised by data. Murphy cites an example: what shows as increased trade under NAFTA, she writes, "is in part the result of intra-firm exchange—from Cargill's grain division to Cargill-owned maize mills of Mexico City or to Cargill beef feed lots in Alberta."[47] It is not always Mexico, Argentina, or Brazil we are competing against. It is sometimes Cargill in Brazil, growing cheaper crops that can help Cargill in the United States drive commodity prices down in the United States. More and more, the WTO seems like a monster descendent of the mid-twentieth-century Committee on Economic Development.

One reason for the struggle that small farmers and the rural poor are facing, whether in the United States or overseas, in least developed countries, is that the agricultural market is not like industrial markets, even though we use terms like "industrial agriculture" and "agribusiness," and farmers like Ron Scherbring are right to think of farming as a business. But the farmers are not the traders with the most influence, no matter how big individual agribusiness producers may be; transnational corporations are.

THE TWO-WAY STREET OF EXPORTS

The first part of the export myth is that there is an ever-expanding international market available to U.S. agriculture. China provides a case in point. The IATP traced predictions and outcomes for export to China. In 2001, Mark Muller, a senior associate at IATP, reported a prediction that farmers would soon see "a $2 billion increase in agriculture exports from recent trade agreements with China." But Muller explained why "farmers don't jump up and down—they've heard it all before." More recently, Muller pointed out that exports were not supposed to be a myth. "The technology and expertise of U.S. farmers was going to overwhelm the competitors, drive them out of business, and subsequently increase demand for our exports and eventually raise prices," Muller noted. "Unfortunately," he continued, "Brazil, Argentina, China and other countries didn't read the script. Brazil is opening up a region of newly cropped land larger than our Corn Belt. And this year, it is believed that Brazil and Argentina's combined soybean production will exceed that of the U.S."[48]

Muller's prediction was understated in two ways: the time it would take and the amount of soybeans that would be produced. As we learned above, in 2002, Brazil raised four times the soybeans that U.S. producers grew, and they were judged by transnational grain traders to be of superior quality. Without traditional protections and commodity subsidies, American farmers are as much under threat as Mexican peasants, and the transnationals will give them exactly as much consideration. It is not Brazil or Argentina that is raising the soybeans, it is Cargill in Brazil and Argentina, and it is Monsanto in those countries that is providing the Roundup Ready seed. These corporations apparently encourage foreign competition with their U.S. customers to drive American farmgate prices and farm incomes down to increase their profits. In light of the real outcomes of NAFTA and the WTO's Agreement on Agriculture, it is also clear that exports are a two-way street and that free trade creates a superhighway that will surely impede our capacity to trade our own agricultural products here as well as overseas.

The story of farmers' desire for exports (or, rather, corporations' desire for exports, sold to farmers by politicians and public relations firms) puts me in mind of a village I know in Alaska. Like most villages, it was isolated from town because there was no road to the nearest city. Air travel was expensive and, especially in bad weather, more dangerous than automobile travel. Villagers thought it would be good if they had more convenient, cheaper access to supplies and medical care. There was a feasible route for a modest highway. They appealed to the state government for one, and the state finally built it. The first fall the road was open gave villagers reason to wish they had thought harder. Though supplies were handier and medical care easier to reach, city dwellers also used the highway. They descended on the village environs to pick berries, hunt moose, and catch salmon and pike—in far greater numbers than there were villagers. Folks from town threatened to pick them clean in a few years, while village hunters and gatherers who depended on subsistence for their livelihood had to range farther and farther afield to maintain their food security. Evidence that exports are a two-way street is as old as the great clipper ship trading companies of nineteenth-century England or Holland. Perhaps Phoenician sailors knew it even before Greek and Roman traders ruled the Mediterranean. It is also as current as the morning news.

WHAT DOES THIS HAVE TO DO WITH US?

All this has to do with access to healthy food, consumer prices, and tax rates, just for starters. I can't estimate the impact that shipments of Brazilian soybeans have on the price of American soybeans, but they clearly don't help American farmers market their crops. *Agri News* of April 10, 2003, for example, worried that "cash soybean prices from September onward have considerable downward price risk from an estimated 460 million bushel increase in Brazilian production this spring."

The profits of some transnationals may be increased in the next few years as the U.S. Army Corps of Engineers looks at investing billions of taxpayer dollars to replace the dams along the Mississippi. That is a project currently being sold as a benefit to Midwest farmers, whose grain—already owned by Cargill and Archer Daniels Midland at loading dock time—is hauled downriver in Cargill and Archer Daniels Midland barges. What the improvements will also accomplish is allowing Brazilian grain to come upstream in barges more easily to compete with Midwest commodities. Lobbyists' calls for improvement of the Mississippi River locks and dams system are accompanied by Cargill's pressure on Brazil to develop the great Amazon River system. The plan is to make it easier to ship corn and beans from Brazil to world markets, including the United States, and have taxpayers cover the expense. Cargill opened a new soybean terminal at Santorem, on the Amazon, in April 2003. That same month, Daryll E. Ray, director of the Agriculture Policy Analysis Center at the University of Tennessee, estimated that Brazil would open 420 million acres of cropland in the near future. How soon will depend "on improvements to transportation infrastructure," says Ray, "including roads, railroads, river barges and seaports."[49] If Cargill is successful, it will have its private corporate transportation issues solved with U.S. and Brazilian citizens' tax dollars. The result will be that Brazilian and U.S. farmers will be in immediate competition for sales to Cargill, commodity prices will fall in both countries, and Cargill will profit from the cheap purchases and the transportation infrastructure changes paid for by taxpayer dollars. Meanwhile, as the trade agreements mandate, producers' income will fall. U.S. farmers' support of such overhauls of the Mississippi seems about as wise as those Alaskan villagers' calls for a road to town. Fortunately, the outcome of this debate is still in doubt, though pressure to rebuild dams on the Mississippi River continues unabated.

And would that we had to worry only about Brazil. *Manufacturing News* reported that, in 2000, China shipped 3 million pounds of fresh garlic to the United States. In 2001, it shipped 16 million pounds, and in 2002, 54 million pounds, cutting our own farmers' production by more than 30 percent. An additional 12 million pounds of garlic was illegally "transshipped" from China to the United States that same year. It came in through Thailand, Vietnam, the Philippines, and Pakistan, according to Jon Vessey, a garlic producer in Coalinga, California, giving testimony before the U.S. Congress. Vessey said that the Justice Department has refused to look into such shipments.[50]

Apple growers in the United States face the same onslaught by imports. Between 1995 and 1998, concentrated apple juice from China increased more than 2,000 percent, *Manufacturing News* reported. During that period, "the price of Chinese concentrate fell 53 percent from $7.65 per gallon to $3.57 per gallon." As of July 2003, "60 percent of apple juice consumed in the United States is from foreign concentrate." The *Manufacturing News* quotes Philip Glaize, a former chair of the U.S. Apple Association, as saying, "We are producing crops that have no market value not because of overproduction by me or my fellow growers but because China is dumping its overproduction on the U.S. market"—just as U.S. grain traders and farmers have dumped their cheap corn into Mexico and elsewhere.[51]

One of the links between farmers in other locations and our own local farmers in the United States surfaces in *Trade and Hunger*'s report from Thailand about what has happened under increased liberalization and restructuring. While U.S. farms look vastly different from Thai farms, there are some commonalities in practice. The report indicates that, "generally, policy has been trending towards increasing acreage under high-return crops. . . . The June [1998] policy paper of the MAAC [Ministry of Agriculture and Agricultural Cooperatives] suggested crops to replace rice including fast-growing trees and cotton."[52] We do not have to think twice to recognize the parallel with U.S. farm bill policy: Grow high-production corn, beans, and cotton, and you will be rewarded. The more produced, the greater the reward. Like Thailand, the United States has moved to high production, much of it for export, and toward an increased use of "marginal" lands. In Thailand, those marginal lands had been used to grow rice for the family; in the United States, they provided vital habitat for a host of species and helped maintain a healthy ecosystem.

My purpose here is not to contrast the vastly different farming practices one would find in the United States, Mexico, Thailand, and Tanzania. Nor is it to argue for a return to subsistence farming—although that system has survived for millennia longer than our present system has, in part because it not only offers diversity on the farm but guarantees diversity for other species as well. Our farms around the world are comparable in this regard: international agreements on agriculture mean that we are all caught up in the same political and economic process. That process is at work in the policies of those who want to empower transnational corporations, opening up the world to trade liberalization while it seeks to engulf and disempower the rest of us, farmers and consumers, in every nation. Differences in how we farm are now far less significant than the monoeconomics of globalization, and the stress will be the same for farmers whether they are using a digging stick or a John Deere.

If the numbers from the USDA that indicate that we are losing farms six times faster under NAFTA than we did before NAFTA are even half right, those policies are having the same effect on us that they are on peasant subsistence farmers elsewhere. Our "big iron" tractors and our computerized systems won't protect us any more than their hoes will protect them. Neither will high-production outputs, nor, if the numbers are any indication, will free trade and exports. U.S. government farm bill subsidies won't protect us either, for they will end soon. Farmers around the globe are now participants in a system that generously subsidizes transnational corporations, increasing welfare for the wealthy while it bitterly resents subsidies to individuals and families.

One of my purposes here is to argue that, if our own farmers are to survive, the present system must alter its course, away from the oligarchy of mammoth corporations controlling the lives of individual farmers. We farmers, like commercial fishermen, are notoriously independent, at least in our own minds. I've done some of both, and I know that pride. Why, then, do we surrender our options so quickly and easily to the public relations blandishments of Archer Daniels Midland or Monsanto? Though there has always been international trade, many cultures around the world have survived very nicely for twenty thousand years without costly free trade agreements, and we can survive without them now. Why, then, do we pursue the chimera of high production when what we need is a wider margin between expenses and income?

Increased profit, not increased production, ought to be our goal. Yet many of our farmers seem hooked on high production. It gives them bragging rights. "Got two hundred bushels off that lower cornfield this year." As if bragging rights were more important than money in the bank. Increased production may be a red herring. To get it, we eliminate fence-rows that provide habitat for all kinds of animals and birds and act as a buffer for wind-blown soil and snow. We plant snug against the edge of the county road's right-of-way to raise corn that costs us three dollars to produce and has a market value of two dollars. Who needs it? We already produce more corn than we can use for human food, so we invent uses it was never intended for: Feed it to animals that are supposed to eat grass, so their stomachs are strained and we have to shoot them full of chemicals to ward off disease. Put it in our automobiles for fuel. Dump it in foreign lands so we can get rid of our surplus and drive their farmers off the land. Turn it into sugar and insert it into food like a surreptitious drug so we find ourselves addicted, obese, diabetic, and prone to heart disease, and wondering how that happened.

So why subsidize more corn? If profit is the goal, increased production may be the best means to it for industry. It may not be the best means to it for farmers. For an independent farmer, the way to beat the system is clearly to go for profit enough to keep the farm in the family. That profit margin is as apt to come from reduced inputs, as Vance Haugen explained, as it is from high production—if that high production requires high-cost inputs.

In "An Essay on Farm Income," Richard Levins, the University of

Minnesota ag economist, holds that one way, perhaps the only way, to increase income for farmers is for farmers to stop competing with each other ("I got better production than you did") and to work together as a collective bargaining unit. In that united front, farmers can have some influence on supply and a greater effect on the price that processors will pay. But that collective bargaining must be very selective. Rather than a general strike, refusing to provide certain foods or crops, farmers might, for example, threaten to withhold wheat from Cargill for a whole season. Cargill, in that case, might pay more for wheat. Bill McMillin, the dairy farmer from Kellogg, Minnesota, also makes a case for collective bargaining. He writes, "In the last twenty years we have watched as the food processing industry and the grain merchandising industry have become more concentrated. It is now at the point where three or four big players control 70–80 percent of the industry. In just the past few years we have seen the same changes occurring in the food retail industry." McMillin cites ag economist C. Robert Taylor, who testified before Congress in 1999 that "the rate of return on equity for retail food chains and manufacturers was 17 percent during the 1990's. The corresponding figure for farm income was 2.3 percent." McMillin points out that farmers are not competing with multinational companies. Rather, "we provide them with a raw material, which they *need* in order to profit" (McMillin's emphasis). He goes on to outline a plan for milk producers to negotiate with a food processor, "to bargain with their milk." The purpose of the negotiation is not only to focus collective bargaining, as Levins suggests, but also "to stabilize prices at a level profitable to most dairy farmers," avoiding the "wildly fluctuating prices . . . making it very difficult for the producer to do any financial planning." Farmers are notoriously independent and hard to organize, but collective bargaining may be an appropriate next step toward turning around a farming system that is hard on farmers.[53]

We cannot forget about production, but the idea that increased production is the key to increased profit bears closer examination. It wouldn't hurt to adopt a healthy skepticism about what we are told by our land-grant universities, our politicians, and our corporations. I talked with one dairy farmer who cut back on milking from three times daily to two, a step that conflicts with most institutional advice. One result of that appears to be decreased stress on his animals. Another appears to be a higher pregnancy rate. The system is so new that he is not ready to make reducing production *the* cause of increased pregnancy rates, but he is clearly interested in seeing how this might work out.

What other steps can we take to deal with the large forces arrayed against farming? Whether we farm or not, we can first of all inform ourselves, read the documents that are changing our lives, and stay in touch with our state and national legislators, letting them know that we are aware of some of the implications of trade agreements. If we farm, we can get our act together and bargain collectively with the Cargills of the world. We can refuse contracts. We can refuse to buy inputs. We can change our farm practices from the high-tech, chemically dependent, maximum production transnational systems and get back to more diversified systems that rely on more natural and sustainable techniques. It is issues like these, not farm size, that will determine the viability of farming in America and provide healthy food security to our people.

As Mike Rupprecht and others told us earlier, we consumers have an important role to play as well. We can shop locally, a healthy alternative whose benefits we will explore in the next chapter. We can also insist to our legislators that producers provide full and accurate labels on the food that we have to buy in stores. Those labels should include ingredients and indicate country of origin or, if from the United States, where in our country it was grown and processed so we know how far it has been transported. Corporations will protest—they have already marshaled their lobbyists and public relations companies to defeat such labels, declaring them far too costly (although they would change labels in a heartbeat if a new slogan or logo might increase sales). In the 2002 farm bill, Congress included a provision requiring country of origin information on food labels by September 30, 2004. The industry response was extremely critical, and Congress backed off, agreeing to delay implementation for two years while it considered alternatives, including making the labels voluntary.[54] Yet if U.S. corporate food processors cannot afford to meet the simple challenge of labels, then how are they different from small farmers in Mexico who cannot compete with them for grain prices because of onerous regulation by trade agreements to which they were not party, or IMF conditionalities they never had a say in? The transnational stance toward those farmers has been to say too bad, and then go right ahead to destroy their livelihoods. If that is the world that transnationals wanted to create and want to live in, how can they complain when they feel the pinch, or even go broke, because of safety regulations designed to protect our health?

We can also ask our congressional representatives whose campaigns are supported by contributions from those agricultural transnationals when they will bring us the representative government our founding fathers had in mind—that is, one that represents the interests of citizens before the interests of world commerce. There is great faith among commercial enterprises that those interests always coincide; there is little evidence available to support that faith.

When the economic language of commerce gets embedded in our political rhetoric, then wealth and power come together in our legislative halls, both state and national, and create havoc. Poverty, hunger, and the loss of arable land, potable water, and natural resources around the world increase as corporate policy becomes national policy. A man asked Confucius, "What would you do first, if you could run the country?" "Rectify the names," Confucius said promptly. "The first task of government is to call things by their right names." As a culture, we haven't come a long way; we've forgotten a lot we should have remembered.

Where do we find hope when the culture and Congress grant economic power not only their envy but their respect? And where lies hope when the government surrenders its authority and our national sovereignty to economic powerhouses? For me, hope comes from what I have learned talking with farmers in this region. Not all, but many in our own community of farmers know a better way: "Maybe we have to work together as one, and work to solve our problems and think of this as our world, this is all we have, and we have to start working to save it," said Bill McMillin at the conclusion of our conversation. I have talked with others who hold similar views. Larry Gates's comment, one I keep going back to, still hovers, inescapable above it all: "The question isn't whether or not it is a good system, but whether or not we make it a good system."

Part V

Alternative Visions, Hopeful Futures

CHAPTER 13

Healthy Food, Healthy Economics

As nutritionists have expanded our knowledge about the importance of diet to health and longevity, consumers' awareness of what they eat has grown as well. There has been a shift in interest from fatty foods to leaner meats, and to vegetables no one used to like, such as broccoli, because they are thought to reduce the risk of heart disease and cancer. The shift among consumers has increased as they have become more aware of small producers' uphill struggle against food business mergers and vertical integration that mean only a handful of huge corporations controls our food supplies. Many consumers today are willing to pay a bit more for an organic label on food, and throughout the country there has been a surprising growth in regional markets, community-supported agriculture, co-op stores, and local farmers' markets. The desire has grown for fresh, seasonal foods, rather than tasteless food frozen for shipment or juiceless fruit bred for thick, bruise-resistant skins and transported two thousand miles.

With the increase in consumer awareness of healthy food, healthy agriculture, and healthy cooking, a new way around the onslaught of the global economy has opened for small producers. While hog prices fell to record lows ($0.08 to $0.10 per pound in 1998, and down again in early 2003), several farmers in southeast Minnesota and Wisconsin sold directly to consumers and got $3.50 or more per pound for their product —enough, they said, that they could still turn a profit.

Restaurants, too, are noticing the new attitude. Several are taking

advantage of produce from local growers. Local meat and dairy producers supply their meats, including local buffalo and elk, and milk and
cheeses. Menus are deliberately planned to take advantage of seasonal
supplies, and some menus list their local producers' names and addresses and encourage customers to shop locally for their vegetables, meats,
and dairy goods. A restaurant in Boulder, Colorado, does this, and it also
advertises Niman Ranch meats on its menu board. I happily ate my
morning sausage there, pretending it had come straight from Dennis
Rabe's farm.

With funding from the Experiment in Rural Cooperation, several
regional agencies, such as the Omega Cooperative and Community Design Center, and more than thirty food producers of vegetables, various
meats, dairy products, and flowers now collaborate on food issues, looking at total food systems in a region of about seventy miles' radius around
their core city, Rochester, Minnesota. They created a food producers network that, after some initial wariness, is becoming a mutually supportive
marketing enterprise. The number of local producers in the region that
are trying to market directly keeps growing.

Kamyar Enshayan, a professor at Northern Iowa University in Cedar
Falls, has been developing "buying local" programs from a slightly different perspective. At the 2003 meeting of the Southeast Minnesota Sustainable Farming Association, he explained that he works with local
institutions, hospitals, school districts, and universities, encouraging
them to shift from buying highly processed foods, shipped for miles in
semis, to buying local. He points out the economic and health advantages of buying from farmers nearby and provides names, lists of available produce and meats, and contact numbers of food managers in local
institutions. In short, he does whatever is necessary to make local purchases easy. Kamyar and his students began with certain assumptions
and some important homework. They discovered that "the forces that
are systematically destroying agriculture in America are entirely domestic, and land grant universities have helped enable this." Kamyar believes
that the massive industrial food systems we have developed are more
dangerous than commonly realized. During his talk, he pointed out a
phenomenon most of us never noticed: "At the same time a mysterious
sniper was terrorizing the East Coast, twenty Americans were killed by
bad meat, fifty-four others were sickened, and two more suffered mis-

carriages." He continued, "The only difference was that the one was widely publicized and an extensive search made for the sniper. Only a few knew about the bad meat, and no arrests were made of meat producers." Kamyar calls the current meat production system of highly confined chickens, turkeys, pigs, and cattle "value subtracted agriculture." "And this is not just about chickens or pigs," he said, "but about two different futures for agriculture, and for a healthy society." Kamyar and his students also learned that folks in their area spend $2,769,691 per year on drinking and eating. They couldn't help but wonder, "What if we could keep 10 percent of those food dollars at home?"[1]

A few weeks later, I drove to Cedar Falls, Iowa, to continue a conversation with Kamyar that had begun after his talk. Kamyar's project to increase local food purchases and short-circuit the too often deadly mass systems has garnered some notable successes. The gains have been gradual, and he is modest about them, but the economic impact on the community is beginning to show. A local taco shop, "not a high-end restaurant," Kamyar says with a smile, "now buys 65 percent of its total food purchases locally, and 100 percent of its beef, pork, chicken, cheese, tomatoes, and black beans. The amount that circulates in the local economy from this one place was $185,579 in 2002." Other institutions have brought local purchases up to more than half a million dollars in just four years. The impact on local farmers' incomes is significant, and the healthy food served to customers is tastier and more satisfying. "But the biggest point," Kamyar notes, "is not just those institutions' contribution to the community's economy, but the relationship they have developed with the community."

There are some concerns that accompany this changing national attitude toward healthy food. One is that some farmers, primarily in the West, are already growing organic produce using unsustainable farming practices—growing large monoculture organic crops, for example—to control a market. Industrial organic farming can be damaging to small producers, and ultimately to the environment, as industrial commodity growers and agribusiness giants move quickly to control organic foods, dilute the certification standards to favor their produce, and take over the organic market, again usurping the place of more sustainable organic farmers. A case in point: Section 771 of the Omnibus Appropriations Act of 2002 contained a provision, slipped in without notice at the last minute by Representative Nathan Deal of Georgia, that would have allowed agribusiness producers to feed their animals conventional (non-organic) feed that had been subjected to chemicals and genetic modifications and yet claim their meat to be organic. Such a setback to hard-won organic certification regulations could have had no motivation except to pander to the transnational corporations whose interests would have been served by the provision. The action led one local organic farmer to comment, "Quick little devils! Have to keep an eye on those political critters twice a second." Fortunately, Senator Patrick Leahy of Vermont went to work to repeal section 771. He marshaled support from fifty-one Senate co-sponsors, and with both the House and the Senate confirming, the measure was repealed on April 12, 2003.[2]

Still, in just the last few years, it seems the pecking order among farmers has shifted. Only seven or eight years ago, sustainable farmers and consumers saw organic farmers as the top of the line. The order of prestige then was organic-sustainable-conventional-agribusiness-commodity farmers. Now the order appears to be sustainable-organic-conventional-agribusines-commodity farmers. Organic is fine—crucial, even—but to reap the benefits that consumers look for, and that we need in order to create a sustainable culture, it has to use sustainable means to achieve its certifiable organic ends.

The shift in consumer interest, and the desperation of farmers looking for ways to stay afloat and in business by changing their markets, has meant that many producers are catering to the new consumer interests in food and food production. One result of that attention is the greater variety of foods available now to the public. There is not only a wide array of organic and sustainable produce available, but new meat products as well. "Healthy" brats made of wild rice, buffalo, leaner pork, and beef have made an appearance and garnered a share of the market. True, this activity still tends to be small and local or regional. Chain stores like Wal-Mart are paying some attention to the trend, but in farm country at least, their interest is seen to be mostly sham for public relations purposes. One local farmer noted that when a new Wal-Mart opened in his town, a couple of bushel baskets of produce were artfully displayed in the entrance with a sign: "We support our local farmers." That was both the beginning and the end. A week later the baskets were gone, and no local farmer knew of any Wal-Mart purchases in the region. But some big chains, especially in England, are beginning to include healthy, organic foods in their grocery sections, trying to suggest that they care about local producers and healthy consumers.

The big chain stores may become genuinely interested in buying local if fuel costs continue to rise, since the long-haul operations that bring produce from Mexico or California cost them (and us) dearly. The Leopold Center for Sustainable Agriculture at Iowa State University studied what it cost to transport produce items consumed by Iowans using three different transportation systems, each of which was in use somewhere in the state. The first was a conventional system that "represented an integrated retail/wholesale buying system where national sources supply Iowa with produce using large semitrailer trucks." The food is transported 800 to 1,500 miles, from as far away as Mexico and California. The second was an Iowa-based regional system, in which a cooperative network of Iowa farmers supplies produce to retailers using both large semis and midsize trucks. The third was a local system, with farmers marketing via community-supported agriculture and farmers' markets or selling to local institutions such as schools, hospitals, conference centers, and restaurants. Transportation in this system uses the light trucks common on small farms. The analysis revealed that the conventional system "used 4 to 17 times more fuel" than the regional and local systems and "released 5 to 17 times more CO_2 from the burning of this fuel than the Iowa-based regional and local systems." If Iowa farmers supplied 10 percent of the twenty-eight produce items used in the study, the regional and local systems would save between 280,000 and 346,000 gallons of fuel—an amount equivalent to annual diesel fuel use on 108 Iowa farms. The authors of the study pointed out that these were small numbers and that local producers would be in competitive pricing with conventional systems only at high fuel prices. Of course, as fuel costs rise, savings will be sought in all three systems, but the short hop is always better in the long term. With high fuel costs, local looks not only better and healthier but inevitable and cheaper, as the long haulers' transport costs eventually drive them out of business.[3]

There are other virtues in farmers' markets in addition to fuel savings and healthy food. One of them is the community building that occurs. In southeast Minnesota, the first two weeks of farmers' market season are a time of universal pleasure, as farmers greet returning customers, ask how the family is, and help them select what they need among the early produce, meats, bread, garden flowers, and handcrafted gift items. Shoppers also talk to each other and become at least nodding acquaintances as the season moves on and the faces become familiar. By midsummer, broader conversations develop among three or four customers and a couple of farmers, all standing together, arms crossed, talking. Common concerns are raised about the state of the world, and nuances of policy or the local news are mulled over and discussed. Brian Halweil, the ag expert at Worldwatch Institute, says that sociologists report ten times more conversations between customers at a farmers' market than between customers at a supermarket.[4]

As the summer develops, thoughtful farmers exhibit a real diversity of vegetable offerings. They have several species each of tomatoes, squash, carrots, cauliflower, broccoli, beets, and eggplant. Cucumbers of several kinds are offered for salads or pickles. Onions and leeks, two or three or more species of potatoes, and green and yellow string beans lie alongside snap peas. The coolers in our little farmers' market in Wabasha, Minnesota, sport fresh eggs, bison, pork, beef, chicken, and turkey. A variety of regular and goat cheeses, Colby, the various cheddars, blue, and cottage all beckon for purchase. Signs describe the various cuts and choices: half chickens, quarter-dressed; wild rice brats. One sign—"Lamb chops, leg of lamb, or lamb kabob"—reminds me of the old sheepherder's joke to his visitor: "We've got three kinds of meat for dinner tonight. We've got lamb, we've got mutton, and we've got sheep. Take your pick." The diversity of vegetables and meats adds an important dimension to the value of the farmers' open market, a community-supported agriculture system. It renews biodiversity in the environment, restrains erosion, sequesters carbon, and has none of the monocrop's pesticides, herbicides, or antibiotics that threaten the long-term soil health of American agriculture or result in runoff that kills or contaminates fish near and far.

The farmers' market also brings an economic benefit. The Experiment in Rural Cooperation asked Ken Meter and Jon Rosales of Crossroads Resource Center in Minneapolis to look at where food dollars go in southeast Minnesota. I think we were all shocked at the numbers. Folks in this region—and it must be similar to many others—spend more than $500 million buying food. Another $300 million goes into the conventional food system described by Iowa State University, in which transportation is a huge cost and food is mostly shipped from elsewhere and sold in supermarkets. A total of $800 million thus goes skipping away to enrich folks outside the region.[5] That's enough money to make even Wal-Mart take notice.

The farmers' market in Rochester is much larger than those in Red Wing and Wabasha, where I often shop. One farmer who sells her produce in both markets looked around Wabasha this summer with a judicious mien and said, "You really can get anything here that you can get in Rochester." Yet Rochester's square awnings, larger space along the river, and cosmopolitan population draw a host of folks who walk the market, visiting and buying what they need. There, I have to say, the shoppers are at least more colorful: Somali women in their traditional garb mingle with Hmong and Vietnamese, black, Arabian, East Indian, American Indian, and European. Add live music to the kaleidoscopic movement, and there is a gentle pleasure in circulating with the crowd.

Lake City's market is held in the evening, right along the shore of the Mississippi River. The setting is bucolic, the live music is country and Irish, and the coffee, pastries, and homemade ice cream on sale and the food samples available mean that people sit down and visit over tables, listen to the music, get the news or the gossip of the day, and watch the great river slide by in blue or slate gray grandeur. The scene is richly reminiscent of a painting by Pieter Brueghel—perhaps *The Wedding Feast*. Only the clothes are different.

The economy, then, becomes a description of relationships, not just of money circulating or accumulating. At our farmers' markets, the relationships, unlike those at the supermarket, are local. The money changing hands here stays here, traveling in a small universe. It goes into the pockets of local people who grow what we eat and make what they sell or trade. It helps support bakers, local musicians, visual artists, and craftspeople, as well as farm families. And, yes, the local gas station attendant too, for even though the distance is short, it still takes a truck to haul that produce to market. And the farmers, musicians, artists, and craftspeople put the money in their local bank, pay their local property taxes, support their local schools, and buy other items from their local shops and neighbors.

The work is hard, no question, but when the money stays in the region, it eventually supports schools, churches, the drugstore, the bank, the pizza parlor, the hairdresser, and various community services. As

Phil Abrahamson says, "It all works together. It's the way you help make things go." This comment rings true on a Thursday afternoon in Wabasha; on a warm Saturday morning in the buzz, hum, music, and shuffle of Rochester; in the Thursday evening swing of Lake City. There is a complexity in this that is engaging rather than discouraging, and a sense of community that I don't think I'm exaggerating. Complexity and community both have inherent difficulties, but what's worthwhile that doesn't? They also have an innate, inescapable beauty. These are hard to achieve, but here they are, laid out around us like a festival, a swinging liturgy of economics and the common good.

Philip Snyder, eighty years old and going strong, still growing potatoes and much else, relishes his trips to the Rochester market. "Now I have corn, beets, onions, tomatoes, sugar snap peas, watermelons, cantaloupes, squash, pumpkins, cucumbers for slicing. And dill, parsley, coriander, cilantro. . . . The Vietnamese, they like it too. They're good gardeners too. I always sell out of parsley—take a big wash pan, heaped up. It'll be all gone. My potatoes too. People like my potatoes. I have plums and apples. . . . But I like potatoes. I like to see them in full bloom, you know, all those white blossoms. . . . I take as good care of them as I know how, so they'll be good to me."

What if, thanks to alert consumers who understand how you help make things go, local farmers could capture 10 or 15 or 20 percent of that $800 million that goes out of state in our conventional system, and see that money circulating throughout the region? One satisfying way to throw a little sand in the gears of the corporate, conglomerate, all-the-food-tastes-the-same-and-none-of-it-is-healthy approach of high-powered food production and processors in this country is to shop the local farmers' market, eat healthy and seasonally, enjoy your neighbors and yourself, and re-create the local economy in the process.

Is it important to you and the economy to do that? Richard Levins, the university ag economist, told me that Wal-Mart now makes up 2.2 percent of the entire gross domestic product of the United States. Eight years ago, it didn't have groceries. Now it is the largest grocer in the world. Poverty wages, strong-arm competition, and blind indifference to community are the hallmarks of such growth. The conventional system could use some sand in its gears, less gas in its tank. Farmers' markets, on the other hand . . . Talk about rewards! They are about time and money well spent.

HEALTHY SOIL, HEALTHY FOOD, HEALTHY HUMANS

Throughout the Midwest and elsewhere, a growing number of farmers are aiming for good health, like Lonny Dietz, who says, "I know we get a healthier crop from healthier soil." Art Thicke, a dairy farmer in southeast Minnesota, recently said to a group of farmers who came to look over his winter pasture system, "We're interested in healthy soil, healthy animals, healthy food, healthy people." It seems to be working. In 2002, he spent less than seventy dollars on veterinary bills and did not have to help a single cow deliver her calf. The winter of 2003, he had no veterinary bills. His cows never get mastitis, and when I talked to him about this in 2004, he did not plan to vaccinate at all. Ralph Lentz, another grazer, raises beef and has not had a single vet bill this year either. Of course, both Ralph and Art are still quick to sing the praises of veterinarians. "You never know when things might happen and you need one," Ralph says. "We've been lucky." But they help the luck along with their good practice. They do not have to buy chemicals to increase production, kill bugs, or stimulate growth. Neither claims to be organic, but the land itself helps them control pests and provides ample nourishment. They and others work toward profitability by "farming smart," staying afloat and profitable by reducing inputs.

There is growing evidence, and increasing public awareness, that the food that comes from large corporate processors has a detrimental effect on human health. The Institute for Agriculture and Trade Policy and the Sierra Club reported the results of a poultry project that tested, in an independent laboratory, whole chickens and ground turkey bought from supermarkets in Iowa and Minnesota. The chickens were tested for "three bacterial strains—*Salmonella*, *Campylobacter*, and *Enterococci*—and for resistance to a number of antibiotics." The project found that "overall nearly 18 percent of the fresh whole chickens purchased were contaminated with *Salmonella*, the leading cause of food borne illness in the U.S." The ground turkey showed even greater contamination with salmonella, at 45 percent.[6]

Further, as processed food has taken a larger and larger share of the market, portion sizes have increased in both fast-food and regular restaurants. Obesity and diabetes have increased in the last twenty-five years and have become major health risks. The media filled pages and air waves in 2004 with news that more than half the U.S. population is overweight—the largest population of overweight adults in the world. Part of that obesity comes from processed foods that are high in sugar and saturated fat. The increased consumption of "hidden" sugars is a primary reason given for the increase in obesity in a 1999 report from the Minnesota Institute for Sustainable Agriculture. According to an internal memorandum on the report, "Most prepared foods available to consumers today—everything from breakfast cereals to juices, salad dressing, pasta and ham—contain high-fructose corn syrup. In addition soft drink consumption has skyrocketed worldwide, raising health concerns about sugar and weight gain as well as the decline in the amount of calcium consumed, particularly among young women who have substituted soft drinks for milk." Soft drink manufacturers, among other food processors, rely on the availability of corn syrup for their products, and this drives a significant portion of our commodity agriculture. "It leads farmers to plant a single crop to meet the demand," says the memorandum. "This emphasis on corn-fed meat and high-fructose corn syrup leads to a reinforcement of demand for the high input monoculture systems that have low economic, environmental, and social resilience." Those same food characteristics have apparently been instrumental in increasing the incidence of heart disease and cancer.[7]

The current focus on the extensive planting of soybeans and corn in the Midwest is linked, via cheap animal feed supplies, to dietary changes around the world that are having a profound effect on human health. Recent changes in the American diet reflect changes in our cropping systems. We no longer produce healthy food for a market, but commodities to fulfill a contract. Meat and milk are important elements of the American diet, and livestock now consume more than 60 percent of the corn and 48 percent of the soybeans grown in the United States, according to Minnesota Public Radio. The USDA's National Organic Program shows even higher percentages: 72 percent of corn and 60 percent of soybeans.[8] These percentages have been high for so long they are by now almost folklore, but the rapid expansion of ethanol and biofuels is surely changing them. The chemicals leeched into water supplies from eroding commodity fields and the dispersal of genetically modified organisms have unmeasured and unknown consequences for human health and plant diversity over the long haul. Yet the newest farm bill still widely favors industrial agriculture and commodity crops while largely excluding from participation vegetable growers and those who wish to follow sound conservation practices. Whatever their rhetoric about the farm bill's protecting family farms, congressional supporters of the bill essentially say to many farmers, "You want to take care of the soil, do it on your time. Stay poor, and out of the hair of our corporate contributors."

Still, the Conservation Reserve Program (CRP), implemented in 1985, has shown significant gains for conservation. According to a report by the nonpartisan Congressional Research Service,

> FSA [Farm Services Administration] estimates that, compared with 1982 erosion rates, CRP has reduced erosion by over 440 million tons per year on the 34 million acres enrolled in the program. Other conservation benefits NRCS [Natural Resources Conservation Service] has documented on these lands include the sequestration of over 16 million metric tons of carbon annually; over 3.2 million acres of wildlife habitat established; and a reduction of the application of nitrogen (by 681,000 tons) and phosphorus (by 104,000 tons). Also, participants have planted about 2.7 million acres of trees, making it the largest federal tree-planting program in history.

One argument against the program was that retiring so much acreage from active cultivation would reduce the number of farms and farmers. The Congressional Research Service report quoted a USDA Economic Research Service finding that "population trends were largely unaffected by high CRP enrollment. It also found that high CRP enrollment was associated with some job loss in rural areas between 1986 and 1992—the years CRP was first underway—but that this was not true during the 1990's." There is speculation that, as the federal budget deficit grows, revenues get smaller, and more trade agreements prohibit income increases for farmers, all farm supports will get cut, and the conservation measures will be the first to go, a prediction that seems borne out by the federal budget cuts in 2005. The most recent farm bill does contain a new conservation program, the Conservation Reserve Enhancement Program, but two Midwest farming seasons passed with no support for it or for those who would like to raise healthy food and protect resources. Only farms in some watersheds scattered across twenty-five states were funded in 2004 and early 2005.[9]

These facts do not seem to discourage the sustainable farmers and organic farmers who are trying to do something different. "It's not the sustainable guys," says one, "who are sitting around complaining about how tough farming is. Being with them, like at the annual organic conference coming up, is a real high. I always come back from that pumped up and inspired."

Food may be the primary arena where we humans sort out our values. In an essay on the five precepts of the Buddhist tradition, poet Gary Snyder writes, "Our life in regard to food is the first question of economics and ecology. Our food is the field in which we daily explore our 'harming' of the world and how we deal with it." The first precept is ahimsa, which Snyder translates as "non-harming," or "no unnecessary harm." In the context of his essay, he is considering the taking of life, especially animal life. But his view is much more inclusive than concern for animals; he is concerned for the health of the land and the dependence of our farm system on fossil fuels and chemicals. "Any kind of gathering or gardening calls for compassion, purity, and respect on all sides: as much mindfulness is asked of the vegetarian as is of the hunter."[10]

MEANWHILE, BACK ON THE FARM

When farmers don't prosper, local communities don't prosper. The beauty of the farmers' market is that it is one more opportunity for farmers and the community to sustain themselves, perhaps even prosper. The loss of families on the farm, as we have seen, expands, ramifies, and becomes many losses, each with its own deleterious effect on the society as a whole. Keeping farmers on the land works the other way. As a bumper sticker on a pickup in the Wabasha farmers' market put it, "Everyone does better when EVERYONE does better."

Aside from articles in regional papers, there is a certain quietness about the movement of farm families to town or to larger urban centers. In a June 1999 interview with staff members of the Institute for Agriculture and Trade Policy, Jerry Kruger, a farmer in the Red River valley, says the exodus of farm populations in that area has been "real quiet." He remarks on the changes and uncertainties that many farmers live with now. "There are probably three layers" of exodus and uncertainty, Kruger says: "Those who have left or are leaving, those that are hurting, and the well-established layer that will weather the prices. That last stratum keeps moving. You may feel like one of the secure ones and all of a sudden, by golly, you're pretty close." The ramifications of that exodus inevitably reveal themselves first in the nearest town.

Kruger can recite a whole litany of impacts of the population decline on local institutions, all based on personal experience and observation. In his county, for example, the schools have "merged three times. Warren, Alverado, Oslo. And Argyle, to the north of us . . . Pretty soon it will probably be Warren-Alverado-Oslo-Argyle. Kennedy-Hallock combined a couple of years ago, and the circle just keeps getting bigger." The local health system in his town has transitioned from being a primary caregiver to a secondary caregiver, but "health care is very important to the community. Most of the community is retired and needs a health center," says Kruger. In his community as well as in others, many retired older people move reluctantly to the nearest large urban area to have quicker access to health care. Commerce, too, has declined in Kruger's town: "We've got one grocery store in Warren. There were two. The second one just closed up. . . . Farm parts, equipment dealers . . . We used to have several dealers in Warren. I have mostly Case-H and have to go to Grafton —that's fifty miles. It takes an hour and a half to go, bring the stuff and come back. . . . There was a Case dealer and an International dealer in Warren, and they combined. . . . We still have a John Deere dealer. We used to have several car dealers. There's no car dealer in Warren."

Kruger notes that among the effects of the ongoing crisis is a loss of farmer self-sufficiency. "Less than 1 percent" of local farmers produce some food for the household, he reports. "None of them have time. I love to garden. When I retire I hope to garden." He adds that his wife, Ginny, "loves plants. She does the flowers and the stuff around the house. She enjoys gardening." Kruger tells a poignant story that shows just how tough farming has been for many. He estimates that he needs twenty-one operating days to get his crop in. "They don't need to be consecutive. But I need three weeks." In 1999, the spring was wetter than normal. They had only five operating days. "I had her [Ginny] drive the tractor for me yesterday for the first time ever. (She drives the truck and stuff.) She was crying when I came home last night, saying, 'I ruined your corn.' And I said, 'Oh, it can't get any worse.'"

Kruger and his wife, like many of their neighbors and farmers throughout the country, exhibit a tremendous strength of spirit, com-

mitment, and grit as they work toward a very shaky future. But there is no way to create a sustainable culture without sustainable farming systems. Years ago, an artisan interviewed by the Canadian Broadcasting Corporation said, "Craftsmanship is getting the details right, and not taking any shortcuts." Sustainable agriculture is about getting the details right; industrial agriculture is about finding shortcuts. As we have seen, the issue is a concern not solely for farmers. Food security for us all is an essential characteristic of a sustainable culture. Our national and international food security is not assured. This, too, works both ways, as Wisconsin farmer Prescott Bergh told us earlier. "There's no point in talking about sustainable agriculture if you don't have a sustainable culture to back it up."

The whole system suffers when farmers suffer. We cannot let all our farmers disappear from the scene or let our small towns drift into abandoned storefronts and then oblivion, for one of the things we have learned in recent years is that our great urban centers also have an organic nature and an optimum size. When too many people move from the country to the city, the city sprawls in a formless mass that threatens the surrounding environment, whether it be farmland, wilderness, open space, or clean air or water. The sheer number of people also threatens social cohesion and true community. Safety then becomes an issue, whether we seek safety from crime or simply from the crush of folks too different from ourselves. The walls go up around some communities, and the gates slam shut. But "gated community" is an oxymoron. A more accurate name is "ghetto." In the economics of nature, there is no survival for single, isolated species. That rule of nature—and of agriculture —is a rule of society as well. Homogeneity leads to the extinction of community just as it leads to the extinction of a species. Who is there to commune with if everyone is the same? I'd rather argue at the top of my lungs with my friend Ralph Lentz than be locked up at night with a hundred folks who think just like I do.

Exodus from the farms has been going on for a long time, but if perceptions are correct, it will accelerate rather than diminish for a while yet, and we will all be the poorer for it. And hungry, too, for the current strains on agribusiness make clear that, despite its claims and great accomplishments, monoculture corporate agriculture does not feed the world over the long haul; it is not a reliable supplier of food, or of food security, for America or the world. The farmers you have heard from thus far are working against the outcomes Jerry Kruger describes above. Fortunately, there are alternative visions of how the future might open up for agriculture and for the rest of us.

CHAPTER 14

Alternatives for Agriculture and the Whole Culture

The trade agreements of the WTO, NAFTA, CAFTA, and the FTAA, with their narrow vision and dismal results, are not the only way to think about trade and the future. Some alternative visions come from new and (on the surface at least) unlikely amalgamations of nongovernmental organizations that are willing to take thoughtful stands for what they believe. That is one hopeful sign; another is that agriculture is the focus of trade concern, the central hub of the wheel of our rapidly whirling globalization. Without a sustainable agriculture, there will never be a sustainable and sustaining environment, and there will never be hope for social justice. Ending racism won't feed the world. Getting our notions about gender straightened out won't feed the world. Though ending poverty would be a great start, even that won't feed the world. But getting our thoughts together about the importance of agriculture and food may offer a clue not only about how to feed the world but also about how to take care of the environment, overcome poverty, end racism and sexism, and get a handle on other social issues.

Both the numbers and the anecdotal information about trade as it is presently configured indicate that it impoverishes more people than it enriches. Those it enriches are made so by the sweat of those it impoverishes. Misery is added to sweat labor, and farmers who lose their land lose their food security along with everyone else. Sandy Dietz figured all that out—"There's something wrong with a system where to keep our level of life we have to have poverty nations," she told me—so why can't the IMF? The system is long broke, and it needs fixing. The solutions projected by the World Bank, the IMF, and WTO, NAFTA, CAFTA, and the FTAA (what Jim Hightower calls the Forced Trade Agreement of the Americas) only deepen the problems. What other ideas for agriculture are out there?

ALTERNATIVES FOR THE AMERICAS

Compared to the WTO and FTAA rhetoric, which obscures the real outcomes of highly developed exploitation of local peoples, local landscapes, and local cultures for expropriation by transnational corporations for executive (priority one) and shareholder (priority two) profit, "Alternatives for the Americas" is a breath of creative fresh air. Expert researchers and other specialists from eight nations created the first draft of the document at the request of a coalition of national organizations from Chile, Canada, the United States, Brazil, and Mexico in preparation for the 1998 Peoples' Summit of the Americas in Santiago, Chile. There, in the Forum on Social and Economic Alternatives, more than two hundred people met to discuss and revise it. The second draft, though still a working paper, is a document more knowledgeable, realistic, sophisticated, and visionary than those drafted by transnational corporations and national trade representatives for the WTO meeting in Cancún and the FTAA meeting in Miami.

- Rather than promises, "Alternatives for the Americas" begins with a set of general principles that establishes a wholly different set of priorities than those set by other free trade documents and poses a radically different worldview.
- Rather than the obfuscation of "trade speak," the language in this document is clear and direct, and the thinking is people-oriented rather than corporate-oriented: "Trade and investment should not be ends in themselves," the summary of general principles begins, "but rather instruments for achieving just and sustainable development."
- Rather than being developed in secrecy and exclusion, the document holds as a general principle that "citizens must have the right to participate in the formulation, implementation and evaluation of hemispheric social and economic policies."
- Rather than corporate hegemony that destroys human rights and national sovereignty in favor of corporate welfare, the document

declares, "Central goals of these policies should be to promote economic sovereignty, social welfare, and reduced inequality at all levels."

The document tackles all the critical issues confronting both developing and developed countries in the Western Hemisphere. Its list of critical issues includes human rights, the environment, labor, immigration, the role of the state, investment, finance, intellectual property, sustainable energy development, agriculture, market access, and enforcement and dispute resolution. All these issues are analyzed; goals for each section, and. strategies to reach those goals, are established that align both with the general principles and with that section's specific guiding principles.[1]

The section on agriculture identifies and responds to a number of current threats to society and the environment. The consequences of continued liberalization of trade through the FTAA "include the acceleration of migration from rural to urban areas, and the growth of poverty zones and increased marginalization, both within cities and within rural regions, creating more pressure on local governments for basic services." Noting events in Latin America, the drafters say, "Large corporations are pressing for the sale of agricultural land to be converted into forestry plantations." This may seem to be a fairly neutral development, but it too has consequences: "a decrease in agricultural employment" and the loss of "basic agricultural capital." Moreover, industrial forestry makes "the hemisphere's food security increasingly dependent on volatile international market prices." What these threats mean for policy development is spelled out as follows:

- "Agriculture should be given special treatment in trade and investment liberalization agreements, rather than being considered an economic sector like any other."
- "Governments should also recognize that small-scale farming requires special policies concerning land conservation, appropriate technology (including biotechnology), agricultural research, credit and subsidies."
- "Therefore, any trade liberalization agreement for agriculture must include concrete measures for the upward harmonization of financial assistance for agriculture, with the eventual goal of spending similar amounts expressed as a percentage of GDP."

- "Laws and regulations designed to guarantee sanitary and phyto-sanitary standards to ensure high quality produce and protection for consumers and the environment should be arrived at through wide consultation with citizens."[2]

At present, one of the glaring deficiencies of all the free trade negotiations lies in the absence of citizen participation, the establishment of secrecy that requires police protection, and the exclusion of nongovernmental organizations, regardless of their pertinent expertise. "Alternatives for the Americas" addresses the exclusionary nature of the trade negotiations:

> Countries should work to strengthen the organization of [each country's] rural sector to ensure that this population is duly represented, both in its relations with the state and with the market. For example, small-scale farmers and their organizations, who have been previously excluded, should be allowed to play an active role in trade negotiations. This ongoing process of modernization of the rural sector must take into consideration the most vulnerable sectors of the society, and safeguards should be adopted to protect cultural minorities and social groups that do not have the means to adequately and efficiently integrate into the market.

The integrative power of agriculture is different from the power of integrated industries. The document puts it plainly: "Agriculture is a sector which fulfils a series of essential functions for the stability and security of nations: to preserve the cultural richness and multi-ethnicity of societies, to preserve biodiversity, to generate employment and sustainability (as much in agriculture as in related economic activities), to maintain the population of rural areas, to ensure basic food security and to contribute to a sustainable development with more economic, social and political stability."[3] One might only add that "basic food security" means having access to healthy food, for access to unhealthy food offers no security at all. Public health is thus included among the other values this approach to agriculture upholds. Though "Alternatives for the Americas" does not use the phrase "integrative power of agriculture" to refer to these culture-transforming possibilities, it seems an appropriate phrase to describe agriculture's highest role in any culture, the keystone of new agrarian values in the ecology of culture that holds it all in place.

The result of the power of integrated transnational food industries, on the other hand, is monopolistic control of our most basic human needs—food and water. There is a difference, apparently unrecognized by national trade representatives, even in democratic countries, between saying that small producers will find it hard to compete in global markets and saying that small producers have no right to compete in global markets. It is no wonder, then, that "Alternatives for the Americas" is so refreshing and challenging. It offers an alternative vision of the future that is inclusive of many values beyond corporate profit.

Perhaps most exciting and hopeful of all is that the document's principles, goals, and strategies appear to be all of a piece. Unlike the WTO and FTAA agreements, in which the prologue of promises is directly undercut by the goals and strategies that follow, "Alternatives for the Americas" shows continuity, with no surprises, no deceit. The general principles are upheld by the guiding principles, and both determine the goals and the actions to be taken. That the authors seem to believe in the principles they espouse gives the document a sense of genuineness and authenticity. In the current context of international tension and increasingly overt hostility and violence directed at the United States and other developed nations, "Alternatives for the Americas" offers a way to reduce antagonism and elevate the prospects for peace and justice. It exhibits the possibility inherent in agriculture itself to transform the whole culture into one that supports healthy humans, healthy economics, healthy soil, and healthy foods.

"Alternatives for the Americas" is not the only international view of how agriculture might change to feed and serve the world, including farmers. Another hopeful and comprehensive view of how things might work out, the National Family Farm Coalition's "Farm Aid: A Declaration for a New Direction for American Agriculture and Agricultural Trade," was signed by twenty-seven nongovernmental organizations in September 2003. It identifies the crossroads that we are facing and, like "Alternatives for the Americas," sets out principles on which a new direction must build. The crossroads it describes is this: "The World Trade Organization's Ministerial in Cancún, Mexico, this month helps to decide who will plant crops and who will be uprooted, and in many cases who will eat and who will starve in the global free trade of food. This is a time to affirm that agricultural policy must support human rights and livelihoods, not overrun or destroy them." The declaration holds that "environmentally and economically sustainable agriculture is central to each nation's ability to feed its citizens today and for generations to come," and it elaborates the principles that "international trade agreements must be designed to defend and support." The signatories to the declaration include five major church organizations—the Presbyterian Church USA, United Church of Christ's Justice and Witness Ministries, United Methodist Church's General Board of Church and Society, National Catholic Rural Life Conference, and National Council of Churches of Christ—as well as Farm Aid, the AFL-CIO, American Corn Growers Association, Defenders of Wildlife, National Campaign for Sustainable Agriculture, United Steelworkers of America, and Soybean Producers of America.[4]

For this reader at least, "Alternatives for the Americas" and "Farm Aid: A Declaration for a New Direction for American Agriculture and Agricultural Trade" represent an authentic concern for the fate of our ever-unfolding world. Together, the two documents can provide the leaven that our lumpy blob of global dough needs in order to rise to become the bread that feeds the world.

THE ALTERNATIVE VISION OF WES JACKSON

Individuals, too, are thinking hard about alternatives, taking into account the past, the forces at work in the present, and the prospects for a positive future. One of those is the visionary director of the Land Institute, just outside Salina, Kansas. Wes Jackson's view of the future may seem bleak to some; the image of an Old Testament prophet comes to mind among those who hear him speak. His vision is not popular with agricultural industries because he insists on a long view and won't promise quick results or immediate profits. His approach is also radically different in that he is out not to kill weeds or genetically modify other plants but to figure out how to make everything fit in its proper place in a natural system. But it seems hopeful to me for exactly those reasons, and also because it is based on one of the most honest and insightful assessments of the culture we are likely to find. It also offers an alternative vision of how the future of agriculture might unfold in a positive and hopeful fashion.

If life is but a series of choices, Wes Jackson might ask us to choose between big bluestem and corn. He would point to the root system of a hearty big bluestem plant, extending a thick, rich mass of tendrils eight feet or more down into the prairie earth. It requires nothing but sunlight and rain; "it restoreth our soil," he might say, "just as the Good Shepherd restoreth our souls." Big bluestem holds the soil and prevents erosion, creates habitat for creatures aboveground and a haven for microorganisms below, and sequesters the nitrogen that is essential to photosynthesis. Corn has a root system that is puny in comparison, leaves the soil without groundcover for months at a time, and requires not only huge amounts of gas and oil but also, now, chemical inputs to control pests, kill invasive weeds, and provide the nutrients that the corn and its chemicals have taken out of once-fertile soil. Those chemical fertilizers also kill the microorganisms that many believe are an indicator of healthy soil. Since corn makes erosion come easy, the lightest rains flood our city water supplies with chemicals and send nitrogen in such quantities to the ocean that "dead zones" are created where marine life cannot survive. Grown as a monocrop, as it most often is, corn is also susceptible to infestations and weather fluctuations. With genetic modifications, it infects other, pure species of plants, diminishing biodiversity. If early tests prove right, it also kills pollinator insects essential to plant life. But we can't eat big bluestem, you say. True, but in Wes Jackson's vision, it can be part of a diverse prairie mix that includes perennial grains and legumes that will withstand drought, infestations, and floods—and feed us too. And that is what the Land Institute is working on, gradually creating perennial grains that can be included in a mix of prairie plants that has the endurance of perennials and the nutrition values of our finest annuals. Part of the secret lies in creating polycultures that will have the same resilience that naturally diverse prairies have.

As he shows me through the test plots and we head out into the fields, Wes explains that the Land Institute's purpose is to work with nature rather than against it, as we so often do, trying to warp nature to our ends without reference to what nature itself might want. The means to work with nature lie right at hand, in the natural structures of the prairie before we altered so much of it, contributing to its poor health and costing us so much topsoil. "So there are three questions we have to ask ourselves," says Wes: "What was here? What will Nature require of us here? And what will Nature help us do here?"

The first question requires that we observe as closely as possible what natural prairie remains, trying to figure out how it works, why it works, and what its virtues and liabilities are. It also requires studying the earliest written accounts and observations of it and learning what we can of American Indian relationships to the prairie. One of the searching questions that Jackson raises in his book *Becoming Native to This Place* is "Were the natives more sophisticated at providing their living than we are?" The question comes up, Jackson writes, because, "on the basis of evidence from dwellings and villages that have been excavated, and that from written narratives by those who accompanied Coronado or came into the region later, archaeologists have estimated that 'within the confines of Rice County [Kansas] there were well over 25,000 people,' or about thirty-five natives per square mile. In 1927, Rice County had just under 15,000 people. In 1980, 11,800. In 1988, 10,800. In 1990, 10,400. Why this huge decline in the numbers of people?" Jackson asks, and our minds go back over our own prairie hometowns with so many empty storefronts. "We know that nearly all of the young people who want to stay or who leave but want to return would bring the population to well over 25,000. Why can't they stay or afford to return?"[5]

The second question, What does Nature require of us here? is, as I understand it, a cautionary one. The issue is not to see how much of our intervention prairie soil can tolerate and still produce but to discover how we can work with the constraints laid on us by this place. How do we stay within the boundaries imposed by an already healthy operating system? Another, ancient form of the question, is, How do we live here and do no harm?—the question that comes to us from the earliest days of science and medicine in Western civilization and from the Buddhist quest in the East. If that is a correct reading, it is perhaps the oldest iteration of the precautionary principle.

The third question, What will Nature help us do here? turns the one we most often ask on its head. The more familiar question, asked often in our land-grant universities and their laboratories, is, What can we do to help Nature?—"which has never yet needed our help," Jackson tells me, smiling, as if no one could doubt his premise.

Somehow, we have to get back to more natural systems. As we walk around the Land Institute, looking at the plants in his experimental plots, Jackson notes, "We are not asking how we can make things better, or imposing our notions on what might happen here. We are trying to learn from the place itself how to mimic the structure. . . . Maybe when we get that figured out, we will be granted the function. That is, we will be able to participate in the process in a way that it will function naturally —and yet revolutionize the agricultural system." Jackson clearly believes that is possible.

Possible, but not easy, and not soon. Part of Jackson's dilemma is that his project is long term (a real sign of hope in a culture that always prefers the short-term fix, the crash diet to the healthy diet), and what most financial backers want is a quick return on their investment. Our American inclination, often celebrated, is to want immediate results. Yet Jackson's long view is hopeful precisely because it envisions a long-term process; the long view is one healthy injection of an antidote we need for a cultural addiction to a quick—and often transitory or illusory—fix. "Even our sustainable agriculture practices are still too short term," in Jackson's view. "People want results, but they have wanted them too quickly, and to get quick results, they shortcut the process, so there is no return to a natural system." Instead of really changing the system, "we adjust it here and there, modify it a bit, but still take part in the larger farming and social systems." So, like Sisyphus, "We roll the rock up the hill, and it rolls back down on us."

In describing what he calls natural systems agriculture, Jackson distinguishes between the problems *in* agriculture and the problem *of* agriculture. In his critique of current systems, Jackson would likely include this book as part of the problem. That is, it deals with the problems *in* agriculture—genetically modified organisms, pesticides, marketing, trade, the loss of farms—rather than the more pressing and profound problem, the problem *of* agriculture. The Land Institute reports that it is the latter that it "has been addressing, almost since its beginning in 1976, mainly because we believe it may be our worst problem and paradoxically the one most easily solved. It is also the least recognized by the general public." How did this problem of agriculture come to be? The Land Institute offers a brief historical perspective.

> Before agriculture we were gatherers and hunters. We took but we did not plant. Once we started planting, soil disturbance was required *every year* because the plants that sustained this more settled way of human life were annuals, mostly from the grass family. Annual grasses

like wheat, rice, corn, rye, and barley account for 70 percent of all human calories. For convenience sake, we planted these annuals in monocultures. As we did, we created two edges on this agricultural sword: soil erosion as the consequence of the necessary *annual plowing*, and the pests that the *monoculture invites*. The first, over the long term, is more serious than the latter, but combined, they largely, but not exclusively define The Problem OF Agriculture.[6] [Land Institute's emphasis]

Aside from our utter dependence on oil to serve our industrial agricultural systems, there is the additional thrust of chemicals onto the scene and into our soils, groundwater, and air. "Farms have become among the most dangerous places to live," Jackson writes. "The epidemiological data mounts as we try to pinpoint the causes of many human diseases that exploded to commonplace during the chemical era." Jackson offers a local example of how pervasive the chemical problem has become: "The Kansas River provides water supplies to cities and communities, even though it is ranked as one of the top most polluted rivers in the nation. Since it flows through agricultural land, its pollution cause is mainly agri-chemicals rather than industry. . . . History does not yet know the full magnitude of the results of spreading chemicals on our planet's surface. We do know that many of our children in the agricultural Midwest cannot safely drink the water in their towns because of agri-chemicals." In this, Jackson buttresses the revelations made by Sandra Steingraber in her beautifully written landmark book, *Living Downstream: A Scientist's Personal Investigation of Cancer and the Environment*, and the further documentation offered by Theo Colborn, Dianne Dumanoski, and John Peterson Myers in *Our Stolen Future: Are We Threatening Our Fertility, Intelligence, and Survival? A Scientific Detective Story*. But Jackson is not content with looking at the history of chemical uses in agriculture. "Chemicals may harm or kill people, but chemicals eventually degrade." More important is soil—which Jackson calls "as much of a non-renewable resource as oil."[7] Of course, soil can regenerate itself, but when it is gone, it is gone for many generations before natural factors can replenish it as much as an inch.

Given these considerations, the Land Institute asks, "Can we mimic nature's solutions?" Its answer is yes. That confidence is based on the institute's observations of and research into the native ecosystems of the local tallgrass prairie and soils. The Land Institute has found that "the only communities that persist through evolutionary time are those that: (1) maintain or build their ecological capital, (2) fix and hold their nutrients, (3) are adapted to periodic stress, such as drought and fire, and (4) manage their weed, pest and pathogen populations." The Land Institute has also identified two "key elements of the structure of the tallgrass prairie" that are different from our current agricultural systems: "(1) the dominance of *herbaceous perennials* instead of annuals, and (2) the occurrence of these plants in *mixtures* instead of homogeneous monocultures" (Land Institute's emphasis). It is commonly known among botanists and farmers that "there are four major plant functional groups in the prairie: warm-season grasses, cool-season grasses, legumes and composites." So what the institute has adopted as its model for natural systems agriculture is "a 'perennial polyculture,' a mixture of perennial species from each of the four functional groups." And what the institute has been trying to find out is whether it can develop and test perennial plants that "can produce high seed yield without sacrificing vigor or perennialism, and determining whether perennial polycultures out-yield perennial monocultures in seed production."[8] If the answers to their questions are positive, that sounds to me like "Bingo!"

I can understand the Land Institute's confidence in and enthusiasm for its project, although, looking through the greenhouses and fields with their various test plots as Jackson explains what I am looking at, why it seems easy still escapes me. Nevertheless, the success of his work and the wonder of his vision can be noted in the institute's research and publications in refereed journals over a twenty-year period. Those many articles and papers and presentations at both scientific conferences and public meetings indicate that, as Jackson said in his Right Livelihood Award for 2000 acceptance speech, "Our hopeful message is now being more broadly sown. The message is that humanity fashion an agriculture as sustainable as the nature we have destroyed, an agriculture that rewards the farmer and the landscape more than their external supplier of inputs. An agriculture in which irreparable soil erosion ceases. An agriculture not dependent on fossil fuels or alien chemicals, an agriculture that honors the reality of the ecologic mosaic as it honors the cultural mosaic of men and women in local habitats."[9]

After I visited Wes Jackson and listened to him talk about his work,

I drove from Kansas on down into Oklahoma on my way to Texas, and so I had time to think about what he had said. One of the things that I most appreciated was a strain of humility unusual in a true believer who is totally committed to his work, as Jackson clearly is. It showed up in subtle ways, as in his comment "We have to mimic the structure to be granted the function." He does not say, "If we do this, *we can get* that," as if he were plundering a neighbor's larder. Rather, it's "We have to . . ." as in "We have to operate within the constraints of the natural structure." There is no bending or bullying nature to do our will here; we will do it nature's way. And if we do that, then we may "be granted the function." Jackson could almost be talking about grace rather than accomplishment. The benefits of the function are like a gift, a grace granted by nature to those who follow its course. That humility also shows in Jackson's insistence that, despite our science, we really don't know much. In public speeches, he urges his listeners to work from an "ignorance-based worldview" rather than one that is "knowledge-based." That would put our true position in its proper perspective and add a dollop of humility to a culture too proud of its knowledge and too lacking in wisdom. In presentations, Jackson occasionally quotes scripture, which contributes to his image as a prophet. A favorite passage has to do with the Genesis story of Eden. God put Adam down in a garden filled "with every tree necessary for food, and in the center was the tree of life and the tree of the knowledge of good and evil." In the last two hundred years, Jackson says, we have eaten too much of the tree of knowledge and too little of the tree of life. His lifelong project is to shift that relationship, not reducing knowledge but strengthening the tree of life, adopting it as our primary model. Then, perhaps, we may be granted the boon of a greater sustainable life.

I do not know whether Wes Jackson's vision will become reality or whether the promise inherent in mimicking natural structures will be kept. But I can see how it might be. And I can put a powerful lot of hope in that, in part because the present system is clearly destined to a short life. And I can put my hope in an agriculture that will be transformed for all our benefit if "the true grain comes ripe." William James declared that "the world can and has been changed by those for whom the ideal and the real are dynamically contiguous."[10] If anyone at the moment has the ability to see the ideal and the real as dynamically contiguous, it is the man who can be found standing in the middle of the Kansas prairie, surrounded by a perennial polyculture requiring only the great fundamentals of all life: soil, air, sunlight, and rain. If you should find him there, you can walk right up to him and say, "Mr. Jackson, I presume," just as if he were someone you had been looking for a long time.

Part VI

An Ecology of Hope

CHAPTER 15

Ours for a Short Time

PEGGY THOMAS

Driving along the shore of the Mississippi this morning, I move in and out of mist rising, a red-orange sunrise suffusing the sky wherever the drifting curtains of mist thin, part, and open to color. I am headed to see Peggy Thomas, who farms with Larry Gates just a mile and a half west of the great river below Kellogg, Minnesota. We begin where most farm and ranch visits begin—at the kitchen table over coffee and something baked and sweet.

Having a farm and making it work was not our original intent. We just wanted to live in the country, closer to a rural, more natural environment, and have a relatively small place. This one has 160 acres. We both had a natural resource background, so our concern was for the whole resource, not just farming or crops. We decided to see if maybe we could make some money, so I quit my job. That same summer we bought seven mixed-breed ewes to start. The ewes multiplied pretty fast, became forty lambs. . . . Then we added turkeys, and aimed to keep them pasture raised, staying away from medications, trying to respect each animal and bird's life, freedom.

As Peggy describes the farm operation, she reveals a purpose and a worldview that are different in quality and in kind from those of the larger culture. It is an old agrarianism made new by new knowledge and a deeper, more holistic understanding of nature and of farming's role in it and in the society.

Maximizing crop and animal production was never an aim. We always wanted to respect the animals, take care of the whole environment

—create a diverse, natural production. We're not simply trying to farm in the conventional way, but to manage the whole place toward biodiversity. Sustainability means that you can live on the land, live with the land, forever. Sometimes I do without "modern conveniences" so that we can demonstrate that such a system works. Maybe not many people are really willing to do that.

Even so, we never lost money. It's labor intensive and small profit because we have such small numbers. But I look at the value of my being home, of preparing our food, of having better, more flavorful, healthier food because I am fixing it or canning it myself.

Now we attend farmers' markets, direct-market all our farm produce and meats, and it's starting to be enough income. I enjoy bringing our produce and meats to the local farmers' markets, although I never pictured myself as a saleswoman. I don't mind selling my own good stuff, but if the tomatoes are blemished, uh-uh. . . . I can't sell them.

Peggy echoes other farmers in the region who are concerned about the broader farming enterprise regionally and nationally, not just about making a living on their own place.

We're really trying to show that this can work. We have about 160 acres, but only about 27 acres of that is cropland and pasture, so we are about up to capacity with animals. We want to be certified organic.

But Peggy holds that being certified is not as important as doing things right.

We raise everything that way so we might as well get certified. At the farmers' market in Wabasha last summer, not many people seemed to care if things are organic. The woman in the stall next to mine sold some of the same things I did, and we both sold out each day. Hers were not organic and I had a sign on mine to show they were, but not many people asked about it.

Like Mike Rupprecht, the Dietzes, and others I have talked to, Peggy thinks carefully about the future, about what needs to be done on the farm to benefit all of its elements and contribute to the larger culture.

We're part of the Hiawatha Sustainable Woods Co-op. Members have a stewardship plan to provide a variety of benefits.

Like others we have heard from, Peggy knows the priorities that will help her accomplish all the goals for the farm, and she and Larry understand the role of each element of the resources they have at hand.

First and foremost here, for us, is habitat protection and biodiversity. The cutting plan won't allow us to harvest any trees for the next five years. Larry loves the forest. It is a valuable resource too—helps provide for other components of the place, keeps water on the land, provides great habitat, and allows for a continuous wood supply. He likes to go up there in the forest and just hang out. He's trying to restore the oak savannah. That requires a lot of burning as well as clearing the place of that exotic European buckthorn.

We've been looking at how the land was before we came to this continent, and how it was done before, and we've concluded that most practice today is destructive: the quality of the food produced, animal welfare, . . . the environment, the lack of biodiversity . . . We've reduced the variety of fruits and vegetables down to about eight groups. People don't seem to know about the relationship between the health of the land [and] the health of people. Land health and food health are tied together, and it's the consumers' choice—and responsibility. If people watch TV (we don't have one), read the local paper (that never has stuff about food production in it), and go to the school sports events—then they have no idea about what's really going on. We've got a whole society that doesn't know, maybe doesn't want to know. If they don't know the real issues, then the land is going to get abused. I feel good about how Larry and I feel about the land, and doing good by the land. It's only ours for a short time. I don't know how many farmers feel that way, but it's not only the farmers' responsibility but everybody's to take care of the land. If they don't, then I just don't know what to do next.

One drawback to raising animals is that you can't get away. I feel such an incredible bond with the animals. I hesitate to leave them with anyone—I let Larry take care of them. I trust him. We try to arrange for the most humane slaughter of our animals. Our lambs range between 80 and 110 pounds, and you get a hanging carcass about half that. The turkeys dress out between 15 and 40 pounds.

I'm encouraged by the fact that I have friends who make a sustainable system work. Mardo and Jack have four hundred acres, raise beef cows, and have an organically certified CSA [community-supported agriculture]. We're looking at quality of life issues, and we're looking at buying the neighbor's place. It has a hundred more acres of cropland. We could add some beef cows to rotate with the sheep. We could increase revenue, but not at the expense of keeping water on the land and keeping the biodiversity. For example, we would want to make sure that we don't have erosion problems that damage the creek.

There's got to be a profit in it in some configuration of activities, balancing this and that, and we just have to find what it is.

Well, let's go take a little walk, look at the sheep.

We head up to the barn. The limestone foundation has been stuccoed and the rafters have been rebuilt to make use of the old barn. The sheep are all inside, where Peggy feeds them. She has names for all of them. The sheep move almost noiselessly; a quiet rustle of hay is the only sound, the air inside warm compared to the zero temperature outside. Peggy watches the sheep as they feed, checking for anything that might bother one. Finding them all fit, she comes back and leans against the bottom door. I know her smile is not only for me, but for (and from deep within) herself as well, her own sheer pleasure in her animals, too radiant to contain.

CHAPTER 16

An Ecology of Hope

Where lies hope? The phrase "ecology of hope," which serves as the title of the final part of this book, comes from a fine book of that name by Ted Bernard and Jora Young. I think there is an ecology of hope just as there is an ecosystem for marmots, mussels, or mallards. That ecology is formed from the constituent parts of a sustainable culture, rather than a sustainable agriculture or sustainable communities. The list of dangers that threaten a sustainable culture is impossibly long, but within each threat, a hopeful aspect grows. For me, the ecology of hope is composed of six notions that revolve around a central value that the farmers I have talked to share: diversity. Farming is one ecosystem in which these six are active elements in our hope for a sustainable future. All six are interrelated and mutually dependent; all six are available to every one of us who live in this region and elsewhere; all six express themselves in our agriculture; and all six, working together, offer us a hope far greater than the sum of their parts.

Our first hope lies in the diversity of our farming practices. The range of practices on the variety of farms in the region provides an expansive laboratory in which various practices can be observed, their strengths and weaknesses noted. That diversity of practice makes available a large pool of choices about how we farm, no matter where we live.

The second hope lies in a sustainable environment. Biodiversity is one of the keys to that healthy ecosystem; indeed, it is essential to the survival of any culture at all. Following the best practices among those we know are practiced in the area, it is within our reach. That is not to say it will be easy, but it is still possible, and with folks like these alongside, we won't have to work alone.

A third hope lies in the land itself, in our southeastern Minnesota karst topography, which inhibits some of the more destructive aspects of current agriculture. Yes, it encourages others. As we have seen, erosion comes easily to the steep hills and thin soil, but grass and pasture and hay can limit erosion, and there are examples for us to see. That limestone base that creates our steep hills also prohibits three- or four- or five-thousand-acre fields of corn and beans, a practice that without animal husbandry and its attendant pastures and hay would lead to massive erosion. Even with smaller fields of corn and beans, even on the slightly rolling prairies to the west of us, we see that erosion.

Our fourth hope lies in our capacity to create social justice. Wherever injustice exists in a society—poverty, racism, too great a gap between rich and poor, the arrogance of power, contempt for the religion of others—there are the seeds of that society's destruction. We have the capacity to diminish the inequities that exist in our society. Acting to achieve that justice is another available choice, one that many have already made and are working on.

A fifth hope stems from a kind of diversity that I've never heard mentioned in discussions on sustainability: idea diversity. Thinking our way out of our present dilemma will require a large resource pool of good ideas. The conversations reported here reveal that farmers, university faculty members, students, and other citizens have myriad good ideas, a notion we will follow up in the next section.

Our sixth hope lies in a right spirit. That is not the same as the "right stuff," though it will take some of that kind of courage to work our way into a more secure future. I have come to believe that right spirit is already innate in many of our farmers and other citizens. That is, it lies in their good stewardship, idealism, frugality, flexibility, altruism, and apparently infinite inventiveness.

I have found all six of these notions alive and well among the farm families I've visited. I know that in some citizens they lie dormant, and I suppose that, given the nature of the world and our own human cussedness, in some folks they are dead as a hammer.

Most of these seeds of hope seem obvious and need no further explication from me, but I want to address two of these signs of hope at greater length. The first is our diversity of farming practices—which is a reflection of a great diversity of ideas held by farmers who are "farming for the farm," as Elton Redalen, whose story serves as an epigraph for this book, says. The second is revealed in the innate character of our farmers. Peggy Thomas also articulately expressed both of these in our interview in chapter 15.

A DIVERSITY STORY

At the beginning of this project, I heard much about "conventional farmers" and, sometimes, "traditional farmers." Now I'm not sure there is such a creature. Sure, there are lots of folks doing large-scale farming, raising corn and beans, raising hogs, or dairying, and from a distance, their practice may seem traditional or conventional. There is nothing new about raising corn or beans, but our own regional history indicates that raising them extensively, on a large scale, is very recent. There is nothing traditional about it. It is conventional only in the sense that it seems to be what many folks, on the advice of experts, have done recently. There are ways to grow larger fields of corn and beans that simply exploit resources and ruin the soil, and there are other ways to use more sustainable practices. A wide variety of both approaches are in use throughout the Midwest. The same could be said about the way we do large-scale poultry, hogs, or cattle. There is nothing traditional or conventional about large hog or beef confinement operations that require antibiotics, result in easily broken legs, and turn our air into methane. Those efforts are all very new on the scene. Some think they are the wave of the future, but as oil diminishes, they are among the least sustainable practices.

Further, I have yet to talk with any farmer who does not put his or her own personal twist on the farm's operation, its way of coping with the current market, and its strategies for dealing with the land, livestock, or crops that it has available for use. If everyone does it differently, what is conventional? When I've asked farmers about that, suggesting that I have yet to meet a conventional farmer, the response has generally been recognition—and often laughter, especially among farmers others might see as conventional. No one has suggested that my assumption is wrong.

Why is that lack of convention a hope for us? When I talk with Dean Harrington, president of First National Bank in Plainview, Minnesota, he describes several obstacles to positive change, and also the signs of hope he sees that may finally help us develop a viable future. As we follow his thinking, the hope that resides in our wide range of farming practices comes clear.

There's a kind of fatalism that is awfully hard to overcome. We're losing dairy farms. . . . In the lending side of it, it's really important to look at that future, and we're doing it now, and we're doing it specifically with one enterprise and that's the dairy business. We can see that we've had a definite decline and we can measure the number of dairy customers that we have. That's an easy thing to get, and we can see over the years that's going down. . . . The response is pretty much that it's inevitable, the attitude that, "Well, you know, that's the way things go, and if you aren't efficient and growing, then you'd better get out and make room for somebody who can do a better job." That's kind of the whole attitude. If we were talking about this at a board meeting, we'd all give a collective sigh and say, "Yeah, well, but . . . that's the world, you know." And then it would be back to business.

If we wish to find ways to act that will go against trends, we need to find the courage and conviction that will lead us to say, "We're going to take a different route and see if we can change that future." Dean believes he is now beginning to see a movement in that direction among his bank's board of directors—a sign of hope.

Now I think there's enough erosion that those of us who have been pretty complacent about it are starting to say, "Well, wait a minute, why does anybody need us in the dairy industry if it's going in a direction that needs fewer farmers, or goes to larger ones who aren't dependent on the local suppliers like the bank, and what can we do about it?" So I think there's a little opening now for that kind of discussion. . . .

I was trying to think, well, what's changed in the years since I've been in the banking business and in the community? And I wrote down without even thinking of it, "grandiosity" and "a kind of ambition." It seems that we want to reach for more than we can grasp, not just in the rural areas, but it's probably generally true.

Paul Rogat Loeb, writing in Soul of a Citizen: Living with Conviction in a Cynical Time, *would agree with Dean's assessment. He notes that we spend more and more time at work, in part because so many families need two incomes to get by. "There are various reasons, of course," he says, "but none more relevant than the fact that average wages buy less than they did in the early 1970's." But then Loeb wonders, "Would such pressures be so acute if we hadn't enshrined 'Greed is good' as a core cultural ethic, justifying all manner of destructive economic and political decisions?"[1] Dean's assessment is a bit broader than Loeb's, and it has to do not only with grandiosity and ambition, but also with the scale of things.*

We've automatically focused on the larger scale and that it just takes more to satisfy whatever our goals are. We see it in the finance world, in the venture capital business. We want to average 30 or 40 percent return, and who can do that? That's ridiculous to me. But how to find something going forward that will put it in a different context is what I'm kind of struggling with in the bank. And we talk about it in the theater, at our local rep theater, the Jon Hassler. How do we make sure that we don't overreach what we can really accomplish? That is one key area.

The dynamic at work in this assessment of our cultural stance toward change is that we have become accustomed to the scale of the larger social and political scene and to its inevitability.

The whole lack of local confidence is, I think, another general area. I just see that we concede so many things to the broader scale. We concede local decisions to regionalism, and regional decisions to national organizations, and national to global, and it just seems like we've kind of given up locally—or that we can't do things that we used to just assume we could do. Part of it is, I think, that the decision makers are less likely to live in the community their decisions affect. We've seen that in business and in education, and I think that takes a toll on local confidence. . . .

If we could increase that confidence and increase some skepticism toward the way things are going now, the current trends . . . I think of the people who look at the thundering herd going one direction and tend to be skeptical of that rather than joining the herd immediately —and sometimes you have to join the herd to keep from getting trampled—but if you ask a few more questions and look for a few more solutions, that can make a big difference in the local and regional scene. I think you've seen that. Apparently there are a lot of seeds out there that have been planted, and some of them are beginning to sprout.

This is where Dean sees a glimmer of hope in the diversity of unconventional approaches to farming that are now taking root in our region and elsewhere.

You commented on how hard it is to describe a conventional farm. If there is more individualism built into those operations, to get those enterprises, those practices, brought into a more prominent light, say there's a big corn and bean operation and there's also a guy selling birdseed, . . . the bank would tend to look at the corn and beans operation and just skip over the birdseed. Yet that might be where the mistakes are, that we don't look at those viable parts of that business and try and help those sustain themselves. But having the confidence to do that is hard. The more examples that you pick up as you work into that, of people who are not the radical fringe but the people—like Bill McMillin, who is a very highly regarded farm operator in the county, and people tend to learn from Bill, and see that he has taken himself out of the mainstream, yet he's part of the community—those are the kind of examples I find useful.

I like that notion that maybe we need to ask more questions than we have thought to. Maybe what we're looking for in a lot of areas of our civic and agricultural life is not policy but the best questions that we might ask about the directions in which the culture is being manipulated, often by forces hidden from view. I tell Dean about a lecture David W. Orr, a professor at Oberlin College and director of its Center for Sustainability, gave at the Land Institute in 1999. Orr noted, "It is time for us, plainly, to be asking larger questions, developing bolder visions, and thinking in terms of centuries, not years."

Right. And how to get a longer-term view worked into our plans. We're going to be all right till the end of this year (*Dean laughs*) . . . and next year probably . . . but ten or twenty years from now, I wouldn't be so confident if the present trends continue.

That comment surprises me. I reply, "Well, you've got a long history here in this bank, one hundred years, and I assume everyone is going to want to continue."

Well, I think so, unless there's a fatalism takes over: "Who are we to try and stay independent when the whole world isn't?"

I tell Dean that one of the agrarian virtues crucial for farming and for the future of our culture is represented, for me, by Phil Abrahamson, a Fillmore County livestock grower. Phil exemplifies the independence innate in many farm folks. Once, he told me that he had attended a meeting years ago where powerful corporate, academic, and government representatives made clear that farming's future was to get big or get out. "There's no room for the small guy." Cheap food was the goal, commodities were the means to achieve it, and they could be raised more produc-

tively and efficiently on a large scale. When Phil got home and got out of the pickup, he told me, his thought was, "No, I'm not going to do that." To take that stance, in the face of that entrenched power, is not merely independent but courageous. I believe that willingness to stand up in front of what appear to be overwhelming forces shows up frequently in southeastern Minnesota.

That's the kind of skepticism I'd like to get at.

It's the healthy skepticism about an inevitable future that leads to a diversity of ideas, to creative thinking about "what might work on this place," and to that willingness to try something other than what everyone else is doing. "The limiting factor, I believe," writes Orr, "is not so much resources or money as imagination." No wonder that diversity of ideas and practice gives Dean hope. Following the conventional is what got us into the situation we now face. The more diverse the unconventional ideas and unconventional practices are that work, the better pool of resources we have to draw from. That better pool can help us create a future that unfolds more by design and less from so-called inevitability. Orr says of his students in a course on sustainable agriculture, "They've learned to be unskeptical, which is to say unscientific, about the craft of science itself. This is the worst kind of naïve positivism, and it places a great deal in jeopardy. We need a science of a more integrative type, one that connects knowledge from different fields and perspectives." Orr outlines the diversity of ideas that will change the course of the future for us in his description of the old agrarianism—all we need do is put his verbs into the present tense. The agrarian farmer can't till the land by drawing on a single scientific discipline or by using only verified knowledge. Intellectual breadth is a matter of necessity. Intuition and tradition must be instinctively mixed with empirical knowledge. Responsibility is undivided.[2]

SINGLE IDEAS AND A NARROWED FUTURE, DIVERSE IDEAS AND A BROADER FUTURE

There is a continuum between good farming practice, which fosters diversity, and poor farming practice, which reduces diversity. What interests me is that diversity in good practice directly correlates with diversity in the ideas and the language we use in our conversations about agriculture. Those diverse ideas and rich language are hallmarks of the imagination that Orr calls for. The lack of diversity in poor practice directly correlates with a lack of diversity in ideas and the lack of shoptalk in those farmers' representations of what they do.

An outline of the simplest, briefest, least controversial expressions of good practice might look like this: contour strips; reduced or no pesticides, herbicides, or chemical fertilizers; grass waterways; some grazing animals and pasture; long rotations, careful management, and detailed observation. The operation is multipurpose, multi-idea, and multifunctional. The family is resident on the farm, "farming for the farm." The primary question that family asks is, How do we make this place work? The question itself implies an innate diversity of ideas and of farm functions: it has to work for the soil, for the animals—both domestic and wild—for the crops, and for the people on the farm. It also has to be economically profitable. There is a holistic view here, and it requires something akin to the intelligence of Einstein and the balance of the Flying Wallendas to maintain the integrity of the soil, the health of plants, animals, and humans, and the concern for songbirds and other wildlife, and to balance the very complex mix of diversity and productivity on the farm so that it is environmentally sound and economically viable.

Kentucky author, philosopher, and farmer Wendell Berry holds that a very similar primary question is "the discipline of all agrarian practice." He asks, "What is the best way to use land? Agrarians know that this question necessarily has many answers, not just one. We are not asking what is the best way to farm anywhere in the world, or everywhere in the United States, or everywhere in Kentucky or Iowa." Reading those words of Berry's, I hear echoes of Donna, Peggy, Larry, Lonny and Sandy, Ed, Art, Ralph, Ron and Ron, Dave, Dennis, Gene, Mary, and others you have met in these pages, each saying it one way or another like a mantra: "I'm not saying everyone should do it this way. We just think this is the best way for us on this place." Berry concludes, "The standard cannot be determined only by market demand or productivity or profitability or technological capability, or by any other single measure, however important it may be. The agrarian standard is, inescapably, local adaptation, which requires bringing local nature, local people, local economy, and local culture into a practical and enduring harmony."[3]

One problem with federal farm bills and world trade agreements is the false assumption that there is a single farm system that ought to work for the whole country, and that America's farm system is an appropriate model for the whole world. There is also a mistaken assumption that agriculture is an industry like the manufacturing and technology industries. There is a monoidea here, a failure of imagination, a loss of idea diversity that wreaks havoc on our agrarian enterprise. I can hear Mike Schuth muttering, "Sh—, it isn't even a good system for the whole of Wabasha County, let alone the whole country." Let alone the whole world!

If, on the other hand, one were to outline the simplest, least controversial expressions of poor practice, they might look like this: fence-to-fence cultivation, large-scale monocropping of commodities, large-scale confinement feedlots, lack of biological diversity, absentee landlords or absentee renters, and heavy dependence on chemicals and federal farm supports, resulting in "farming for the farm bill instead of the farm." The primary question that distant landlord or his or her farmer renter asks is, How do I make this place pay? Both the question and the answer imply a single idea and a single function: maximum income from maximum commodity production.

I suspect there are a few folks in every region whose operations are at the farthest edges of this continuum. Most farmers I've met in southeast Minnesota, fortunately for all of us, fall somewhere in between, using a mix of practices that come from personal observation, fertile imaginations, and diverse ideas about what will work. Fewer ideas and less independent thinking are reflected in poor practice. Gyles Randall of the University of Minnesota Southern Research and Outreach Center in Waseca told me, "The commodity growers' choices are pretty easy. They pick one of about three different kinds of fertilizers, often chosen for them at the suggestion of a salesman or somebody else. It takes three weeks to plant and three weeks to harvest, and all they have to do in the meantime is wait."

Such farms tend to become single-idea places: getting the largest number of acres of the highest-paying commodities possible into the base acreage in order to reap the highest income from the farm bill. Miguel Holle, director of a Peruvian agricultural nongovernmental organization, said during the agriculture e-conference described in chapter 11, "The richer North is lagging behind poorer peasant farmers" in

understanding these issues. "Our most common thinking is monofunc-
tional, and the reality of small or poor or traditional or marginal farmers
is multifunctional with quite a complexity."

LANGUAGE AND STORIES

Larry Gates, the farmer and hydrologist, talks about "being able to do
something in a positive fashion," but then he adds a sentence that to me
is striking: "And I have to think about how to represent what I do, and
how to talk about it." I think that statement is a crucial one, and one that
not many folks think about—it is important to find a way to articulate
things. "There are some sustainable farmers that have a language that's
rich and a way of talking about what they are doing that is rich, and part
of that richness is in terms of language," says Larry.

We present ourselves in our actions; we re-present ourselves in our
words, our stories. When we are scrupulous and direct, that re-presentation
may make our lives clearer to us at the same time we reveal ourselves to
others. To present and re-present are our best possible ways to grow in
understanding, to learn from the past, and to find our way into a new and
possibly better, even more rewarding, future. If our language and our stories are as important as I think they are, then perhaps the worst pollution
of the last fifty years has been language pollution, the use of political,
media, corporate, and especially public relations language to convince
rather than reveal, to sell a position, an idea, or a thing rather than to illu-
minate reality. So thinking about how we re-present ourselves is crucial—
both the story we tell ourselves and the story we tell others. If the story I
tell myself is that I am the biggest, meanest guy in the county and I'll get
mine, by god, or know the reason why—guess what? That's what I become. If I tell myself a more positive story about who I am, I have an opportunity that the first story short-circuits: the possibility of becoming
more useful to the community, of leaving the place at least as good as, if
not better than, it was when I arrived, despite the messes, pain, and grief
I inevitably create while here. So Larry is right, and he is thinking about
the right stuff when he worries about how to re-present himself.

I have heard this, too, in the language of the thoughtful farmers I've
talked with. There is a kind of linguistic diversity that is parallel to the
diversity we want to encourage in the biological world. Perhaps there is

a direct tie between diversity in our language and the diversity of micro-organisms in our soil, or the diversity of the crops or livestock we raise. Whether it is the soil and our practices that diversify our language or our language that leads us to diversify our crops is immaterial. The interplay is a loop or, perhaps, a Möbius strip, taking its own particular twist, all of it running into itself, influencing everything along the way. So when the way we talk, and the language we use, stops being rich and diverse, everything suffers, including the biological world. Alas, the most pervasive language of our time is not English or French or Chinese. It is the public relations language of sales, which has but one purpose. Some still argue in order to arrive at the truth or an understanding; too many of us now use our public discourse only to sell, or to persuade to our side, to "get a win." Such discourse is the ultimate sophistry, where to win means we lose track of reality.

If we do not stand against the uniformity of culture, against the monomind that the culture wants to inculcate, then the language we use, the arts we create or perform, and the literature we write become more and more like monocrop commodity agriculture. If we do not stand for diversity in the culture, and acceptance of that, then our creativity in every area dwindles to become more and more the same, duller and duller, until our arts and crafts, including the art and craft of agriculture, are hardly worth thinking about. As Spandrell says in Aldous Huxley's *Point Counter Point*, "The wages of sin is boredom."

It makes me wonder if the way we farm doesn't either set free or impose restraints on everything else in the culture. The rhetoric of the farm legislation is always the same, always "Here is the way to protect the family farm." And the farm bills themselves pass by in their monotony, supporting agribusiness, transnational corporations, and commodity producers while driving small farmers out of business and depleting small communities, social capital, and biodiversity. Maybe everything in the culture becomes monogray when agriculture becomes monocrop. The borders all blur in the pale shadow of blandness. We become a culture without fencerows, without habitat for new or exciting ideas. Even the distinctions between morality and immorality, compassion and exploitation become a blur, for the rhetoric some of our legislators present, or re-present, about our farm bills is surely deceitful. Perhaps one of the reasons we get monopolitics instead of a genuine two-party system and loyal opposition is that we suffer from a "mononucleosis" of all our creative juices that stems from a monotonously federalized agriculture. But "men are like plants," J. Hector St. John de Crèvecoeur wrote in 1782; "the goodness and flavour of the fruit proceeds from the peculiar soil and exposition in which they grow."[4] The differences between us are as fruitful to the whole as the commonalities. If that is true, though the farm bill certainly pushes us toward monoculture, then agriculture has a function in the culture that is way out of proportion to the number of people doing it, or its economics, or farmers' political power, or anything else. In that case, how we re-present agriculture and our practice of it, how we talk about it, as Larry says, is of paramount importance. And perhaps, then, the more diverse and colorful our agriculture, the more diverse and colorful our larger culture.

Further, that rich and varied language is revealed in the richness and diversity of the stories we tell when farm folks get together. One of the great privileges of working on this book has been listening to the stories many of our farmers tell, not just when they are talking to me but also, especially, when they talk to each other. They are full of stories, some funny, some poignant, some teasing, some outrageous. Nearly all their talk is farm stories or stories that represent some critique of the larger culture based on the way it affects farming. It seems there is nothing they'd rather talk than shoptalk. In a long evening of conversation at Ralph Lentz's farm, with the Rabes, the Minars, the Thickes, Larry Gates, and Peggy Thomas all talking after a bountiful meal grown entirely by themselves, perhaps with some fish caught in Ralph's pond, I have never heard mention of the Vikings or the Timberwolves, Yankees or Twins, or the doings of the latest pop music or Hollywood star. It's not that they are unaware, for they do talk about critical issues in the society, including war, the economy, and education. But their own work, their own soil, their experiments and observations, their own lives are more interesting to share than the fantasy worlds of Hollywood and sports. "Get a life," people sometimes say, but these folks don't need to. They have carefully and deliberately created real lives, filled with vital interests and puzzles enough to think about and brood over for a lifetime. They have no need for fantasy distractions to fill a void in their interest in the world right around them. I don't know whether or not farmers outside the region, who are really into giant fields of corn or beans and have no plans to do anything else, talk that way, but if Gyles Randall's earlier description is right, I suspect there are more stories about golf and the NFL among industrial commodity producers, less shoptalk and fewer stories about farming itself.

It's hard to overestimate the power inherent in those endless farm stories one hears where the practices and the stories are many. "Statistics can give us a sense of the magnitude of our problems, and help us develop appropriate responses," writes Loeb in *Soul of a Citizen*. But they "can't provide the organic connection that binds one person to another," which is the catalyst that ultimately leads us to action. That catalyst is most often a story. Loeb quotes Scott Russell Sanders writing about the power of stories in the *Utne Reader*: "They link teller to listeners, and listeners to one another," and they "give us images for what is truly worth seeking, worth having, worth doing." If the farm stories we are telling are healthy stories—healthy for the soil, for the animals, for our families and our communities—then we can take great hope in their power. Loeb continues, "In a time when we're taught that our actions don't matter, stories carry greater weight than ever. They teach us, Sanders suggests, 'how every gesture, every act, every choice we make sends ripples of influence into the future.' Indeed, it's no exaggeration to say that the stories that gain prominence in the public dialogue will significantly shape public policy."[5]

Fortunately, many of our farmers are independent thinkers who use common sense and trust their own observations and their own thinking. Dairyman Gene Speltz, thinking about farming, muses, "Common sense has probably a major role in farming and all of agriculture. Common sense is the big thing." Such farmers refuse to accept narrowed options. They also avoid aligning too closely with the radical fringe of anything. They represent the real hope for the creation of a sustainable culture and the survival of small- and medium-size farms and rural communities, for our greatest hope lies not in any legislation but in the character and qualities inherent in farmers themselves.

THE FARMERS' VALUES STORY

The farm folks I've talked to have left me with powerful impressions. Others, listening to the same comments, might come up with a different list of the characteristics of farmers in this region, or a greatly expanded one, but this is how I heard it. This is not what I intended to write when the project began; it came into focus without any bidding on my part. What interests me about the folks I interviewed surprised me. I started out looking for commonalities in good practice; what I found was a commonality in values. I do not mean to imply that other folks do not hold these values. But these values imposed themselves on me as I talked with farmers. Perhaps if I were working with ironworkers, something similar might have emerged. I've used "they" as the primary pronoun in what follows; there is of course a wide variation from one farmer to another, but they all seem to have these values to one degree or another. These are the characteristics that struck me, and they all offer great hope.

Their Intentionality

Many have picked their own lives up by the scruff of the neck, turned them 30, 60, 180 degrees, and set them down to face in a new direction. They are living with purpose. None of them are drifting through life with values congruent with those of the general culture. One, after thirty years as an auto technician, became a farmer for specific reasons that are very clear to him. Another, after a successful life as a consultant, is raising flowers. Yet another, after thirty years of teaching farming according to the best science and the methods promulgated by the university (except getting bigger and bigger), took a fresh look at that knowledge and started farming according to what works on, and for, his place, his land, his animals. Some chose to stay on the home place and make it work although they could have made more money elsewhere. Others left the home place to make better money—and then learned that their work was not as satisfying as farming and chose to return home. All of these farmers' goal appears to be a meaningful life even if it may be less profitable. All of them have decided to try "unconventional" strategies that are appropriate to their particular farms.

Their Deliberateness

Deliberateness is intentionality with an added dimension. Here it has two senses: They are being very deliberate about their lives, and they deliberate about things—not only about what they are going to do and how they will do it, but why they are doing it. They don't tackle projects without thinking them through, and they don't adopt them till they have experimented with them. One man, farming a place new to his family, said, "We've got the wildlife area now, some row crops; we've done tree planting and prairie restoration, and we've set up a whole farm ecosystem. It's all necessary. . . . Now we are creating 'live food' full of enzymes and are building 'live soil' because you get a healthier crop from healthier soil. You get more off less land, too, so there is room for wildlife."

Elton Redalen, a lifelong dairy farmer, is deliberate in his farming. He talked about how his father came to be intentional and deliberate in his conservation practices, and how his father's influence has kept him conservation minded. "My dad, way back in the thirties—after a hard rain, terrible washing—Pa came home and said, 'We have to do something. We're going to Triple A.' And through '37, '38, '39, we ploughed around the hill, began rotation, not just of corn and beans, but corn, oats, hay—three or four years of hay. We're not going to raise beans because it loosens the soil." Later, driving around his farm in his pickup, he pointed out the way the contours lay, the strips to hold moisture and prevent runoff, the holding pond that helped stop the erosion on his steep hillsides and provided water for his cattle. "See how that's helped heal the land?" he asked.

Their Tenacity

Those farmers who have hung on to their places for three, four, or even five generations are clearly tenacious. They, and their forebears, represent what a number of older farmers I've talked with call rural values. Some would say they are the values of an important new agrarianism. "They [those old-timers] were frugal, and persistent," said one farmer in a public meeting in Lanesboro, and the others nodded their heads in agreement. "They knew how to hang on when things got really bad."

Frances Merwood, a dairyman near Lanesboro, said, "I always made my most progress when times got tough. You have to be in it for the long haul. There will be ups and downs. The weather . . . We worked awfully hard. It was a real struggle. The eighties were awful, but I made the payments. When you get your back against the wall and a lifetime of work at risk, you become a better businessman. Timing is part of it, luck is part of it. Nineteen eighty-two was a cold, wet summer, poorest crops I ever saw. Snow knocked the corn down. Clouds every day over this corner of the state. Perseverance, work hard . . . got to manage good. It's a family thing."

Migrant workers reflect the same tenacity. When asked how she saw the future, Consuelo Reyes of Centro Campesino in Owatonna replied, "I just see a lot of struggle." Victor Contreras, codirector of Centro Campesino, echoed Consuelo's view. But both keep up hope. Consuelo explained, "My hope is here, in the organization. The support we get from one another, the help we give each other, helps. People really work to help each other." Victor added, "Talking to people, seeing people react, being able to support people in their rights. I get the energy from my heart." It is their hope that strengthens their tenacity and allows them to persist in the face of struggles that must be won over and over. Consuelo summed it up: "People ask us why we stay. But our workers come from the poorest parts of Texas and Mexico. They come hoping to make a little money so they can send it home to their families, and because they want their children to have a better life. They think they will find it here, and so, in spite of the difficulties, we stay. We have to sacrifice something for the sake of our children. . . . We have to keep working."

Dean Harrington says the strength of these farmers and farmworkers is in "coping. Quitting is not an option, and never occurs to them. They have no exit strategy. . . . The ethic that is there is very powerful."

Their Daring

They have made huge changes in their lives, often giving up successful careers, risking everything to find a more satisfying life for themselves and their families, but also to improve the life of the land and the life of the society. Whether they are new to farming or old-timers, they are making changes that involve learning to find their way in uncharted terrain. They do not follow any conventional wisdom. They are entering farming—or persisting for another generation—at a time when many others are getting out because there is no future in it.

Like Huck Finn, they have "lit out for the territory," not knowing for sure what they will find, knowing there are no guarantees, but learning to trust their own instincts and believe their own observations. "You have to be careful about what you hear," said one, talking about the advice of experts, "because what you hear doesn't always work." Another said, "Last year I quit my job, so now we have to learn fast!" His principal teachers are other farmers pursuing similar goals and holding similar values. He also learns from his own experiments with the land.

The stakes are clearly high. These folks are out there, on the edge of culture, both learning and demonstrating what the conventional culture does not believe and does not want to believe, so they are constantly facing resistance, trying to overcome obstacles the culture puts in their path. "I want to prove this can be done," several said. "Not just for myself, or that I can do it, but also so others can see that it can be done, and take encouragement and stay with it." Among the obstacles, as they see them, are the federal government, the major land-grant universities, the state government, the world's largest and wealthiest transnational corporations —and sometimes their more conventional neighbors and the folks in town who have no idea what their lives are like, no conception of the issues they wrestle with every day.

Their Attentiveness

They are always examining things, paying attention all the time. They do not believe what the experts tell them without testing to see whether it is true. "I spend a lot of time down on my knees looking at grass," a Colorado rancher and rotational grazer told me. "You have to pay attention all the time." They try things and see. Ralph Lentz looked at his place and set up an experiment. The part that worked flew directly in the face of conventional wisdom and expert opinion. Experts said it couldn't work. Ralph knew it was working because he was paying attention and could see it. "Come look at it," Ralph always replied to skeptics. Finally one did, and he told his colleagues what he had observed, and now they bring other farmers, agency personnel, and faculty members and ag students from the university to see. That "seeing" led to a permanent bond between farmer and bureaucrat, though argument is still one means they use to figure out what is really going on.

The farmers I've talked to study and learn and test some more. If you ask them how they learned to farm, they will say, "Haven't yet," cheerfully —even though the land has been in their family for five generations. They know that they are not experts but believe that, if things are going to change in this drifting mass culture that is becoming ever more homogenized, ever more docile, ever more under the control of corporations and governments indifferent to them, they are going to have to change themselves. Indeed, each of us has to do the changing the culture needs if any of us are to have a future with some measure of freedom, dignity, independence, and hope.

Their Intelligence

I admire the brains they marshal to think about things—the intelligence with which they analyze their own lives, the society in which they live, and the land they occupy (and their neighbors' as well!). And I admire the understanding they have of the direction of the larger world culture that looms over us all so threateningly. They don't always agree with one another, and I don't always agree with them, but they sure are thinking hard.

I am not alone in seeing this. Deborah Allan, one of the soils scientists at the University of Minnesota who works with farmers in southeast Minnesota, has come to respect farmers' intelligence as I do. "They're incredibly intelligent and perceptive. I think that what they are is, they're innovators." Deborah believes that the best research ideas come not from university faculty members but from farmers. "I think there are more researchers that would be very interested in doing research on alternative systems and methods. But we're not the ones in a position to innovate them. It's not something you can look at academically and say, 'Well, I'll bet if you tried this . . .' It's the guys out there trying stuff, who are the innovators, that have to feed us the ideas that we can check: is this really doing what they think it's doing, or not? They may not agree with what we find out, but they are the ones that have to give us the information to research. They're the idea people."

Their Flexibility

The tentative nature of their efforts and the modesty of their claims to success show through their conversation. Lonny Dietz, the organic farmer who raises vegetables on the high ridges above the Whitewater River, said, "Next step is to get the vegetable production down, maybe expand into bigger fields—maybe ten acres of carrots because there's a market out there, but we've got to get the small system down first." He continued, "I was lucky more than smart. I didn't have a clue at first." But he tried this, tried that, talked to folks, observed the good practice of others, and finally was developing a system that seemed to work. "Just started doing greenhouse," Lonny said. "Tried strawberries the first year, but they didn't work so well the second year. Now we do early crops. We have six acres of vegetables outside." That idea—that he and his wife, Sandy, are developing a flexible system, learning as they go, experimenting, observing closely, trying again, creating a system that works for them and their land—permeates every visit with these farmers.

Others have been at it for a couple of generations or more. They have a system, but there is always room to fine-tune, to try something new—something that seemed to work for a neighbor or a workshop leader at a regional meeting, or something they saw during a field day on a distant farm. "We had forty goats," said Lonny. "My wife wanted to do some weaving." Sandy added, "But the market for mohair just went"—she turned her thumb down and continued—"so now we have just two, a couple of geriatric cases." She did not seem to be put out about the change, was apparently willing to roll with the punches, and it was clear she would soon find another outlet for her creativity.

Their Altruism

One striking commonality is the sense that these folks live not only purposefully but for a purpose or purposes beyond themselves. They want to make a living, yes, absolutely, but they also want to make a contribution to the land or the larger society somehow, too. Larry Johnson, who raises organic flowers on the edge of Winona, sells them at local markets. He believes that "this small-scale farming is a way to mix working with people and the conservation issues I'm concerned about, and yet be part of the whole process at the same time. . . . There is the beauty of plants and flowers; they make people smile, they soften the mood." Later he added, "I spent a lot of time for a while cleaning up groundwater. I thought, somehow we've got to get ahead of the curve on this. And seeing a lot of sad, angry, upset people got to me after a while, so I just

thought, I have to do something about this." He studied holistic management with Allan Savory, became certified, and now helps farm families
plan their way into more workable systems.

Thinking about the environment is a constant for most of the folks
I've talked with. Vic Ormsby, a vegetable farmer near Winona, said, "My
question is, How do we maintain stewardship over the long term? We
have this private ownership paradigm in our western European tradition, and it works against stewardship. We know state ownership doesn't
work either. Community ownership is a better model, but there isn't
much experience with it."

Bonnie Haugen, the rotational grazer near Mabel, pointed out some
ways in which local farmers, food, and the world are related.

> Years and years ago, someone was reminding people that if you give
> people—he was speaking at one of the Bread for the World luncheons
> —if you give a person a gun, you might have a friend for a day, and if
> you teach them how to grow their bread or the grain, you're likely to
> have a friend a lot longer. I think, socially, that's a lot of what our
> whole nation's problem is—you have to balance your security against
> your friends', but too often it's too fast and too easy to lean towards
> just the security. We might be in different countries, but we are still in
> the same world. We are still sharing the whole air and water and all of
> that. It's a difficult thing.

Altruism extends especially to other farmers, both current and aspir-
ing. One organic farmer said, "I look at my work as generational. . . . I like
working with young farmers." Ron Pagel, a dairy farmer, spent a full day
last fall talking with students at the middle school's farm day. He seemed
to know just how to get them interested. The reward he gave for attention
and right answers to his questions was a plastic Holstein cow, casually
pitched across the room to successful students, who caught them, laughing. Soon everyone wanted one and the discussion became lively.

Pam Benike, a Dover cheese maker and one of the founders of the
Southeast Minnesota Food Network, told me while she made cheddar,

> The market for our cheese, just from word of mouth, being at farmers'
> markets, selling to restaurants, is growing very quickly. People like our
> product. I figure we could add another container to pasteurize the
> milk and make the cheese without adding anything to this room, and

double our production in the next two years. Yes, that would double our income. But I've been involved in the food producers network, and I also get two or three calls a week from farmers who would like to change to a more sustainable system, but they don't know how, or they are afraid of a loss of income. I'm trying to get a mentorship program going for them. All that takes a lot of time. I decided that those activities are really important for a lot of folks, so if we have to put off the new vat and doubling production for three or five years, that's what I'd better do.

Ron Scherbring, the Rollingstone farmer, has an idea where these farmers' altruism comes from.

We're probably the last group of people in America that is so ingrained with the religious background. You see a lot of these people go to their churches—whichever denomination, it's irrelevant. They will go to their churches and be a part of their schools.

Religion teaches you to be humble and respectful. So you don't go out there and crawl all over somebody else's failures. You like to see everybody come along, and that's something that needs to be reflected in your farm story. . . . You never bragged, you never flaunted money around—humility was a virtue and even though you thought you were doing good, you thanked the Lord and prayed and you moved in those manners. And, yes, that's done in other businesses, but I'm just saying it's important in agriculture.

Scherbring may be right. I've been impressed at how frequently religion comes into the conversation with no prompting on my part.

Their Carefulness

They are not careful in the sense of cautious, for what they are doing is daring, but in the sense of being filled with care for something beyond themselves. In the case of these folks, it is care for the health of their children, the health of the land, the health of their neighbors and the general society, and the health of the larger culture. "I just hate to see farmers going under," said Larry Johnson. "If I can make this work, maybe others will see it and hang in there, and have a better life too." Art Thicke, showing other farmers his winter grazing system, said, "It's easy to be good for the environment. We're having a good life here, improving the soil and the grass, and our cows are healthy. The songbirds are back." Art talks so much about the pleasure of walking with his wife on summer evenings, listening to the songbirds, that sometimes it is hard to get him to talk about farm issues. Lonny Dietz says happily, "We've got a couple of eagles, red tail hawks, owls, songbirds. We've restored four acres of native grasses. Now we want to get some forbs in." Mike Rupprecht points out that "the land approves" of his efforts; the indicator for him, too, is the return of the songbirds. These farm families rejoice in songbirds because songbirds are pleasing in themselves, and because their presence tells farmers something hopeful about the health of their land.

Their Tendency toward Exuberance, Joy, Pleasure, Hope

"These farmers do not gripe a lot or sit around bitching and moaning and blaming others for their plight," said one fellow. I have found they smile a lot; they laugh; they josh and tease. Instead of taking vacations (as Sandy Dietz and Peggy Thomas pointed out, it's hard to get away from vegetables and animals), they get together for supper, and it becomes a party. They marvel, awestruck, then shake their heads and laugh at the foolishness and ignorance of the institutions—like their universities and their governments—that they should be able to depend on but that fail them frequently, and that too often condescend to them. "This isn't top-down work," said one farmer. "The feds, the university, the state, they just don't get it."

These farmers look forward to the future despite the obstacles to their success—and maybe because of them. "I'm going to the organic conference this weekend," said one farmwife; "that's a given. You can feel the energy and the hope, the optimism. You have to have that. It nourishes you." They are definitely not Pollyannas about their lives and the high stakes that are on their tables, and sometimes the strain shows. Nevertheless, they have confidence. "Don't tell me I can't do this," said one woman with a bright laugh.

This optimism is characteristic not only of the independent farmers I've talked with but also of farmworkers, such as the members of Centro Campesino in Owatonna, who work so hard for the simplest elements of social justice, who daily risk injury in our processing plants, and have no job security whatsoever. Yet, talking with this gringo, Consuelo and her colleagues smiled frequently, and their laughter transformed the room.

I am not trying to make these folks out to be saints; they would blanch at the idea. There is no reason that you can't be altruistic and attentive and still be a crank or a curmudgeon at the same time. We ponder why it is that the innocent suffer and why there is evil in the world, but it may reward us even more to ponder the good. Notice that there is not a single "Thou shalt not" among these values. As one farmer (who may or may not be a curmudgeon) said, "Too many people want to talk about what to avoid instead of what we want to accomplish." Fortunately, we can talk about what we want to accomplish here. We have a rich pool of good ideas and good practices to draw on.

THE STORY OF HOPE

What do these characteristics and values indicate beyond certain attitudes and personal character? For me, they confirm the truth of what Eric T. Freyfogle writes in *The New Agrarianism: Land, Culture and the Community of Life*:

> With no fanfare, and indeed with hardly much public notice, agrarianism is again on the rise. In small corners and pockets, in ways for the most part unobtrusive, people are reinvigorating their ties to the land, both in their practical modes of living and in the ways they think about themselves, their communities, and the good life. Agrarianism, broadly conceived, reaches beyond food production and rural living to include a wide constellation of ideas, loyalties, sentiments, and hopes. It is a temperament and a moral orientation as well as a suite of economic practices, all arising out of the insistent truth that people everywhere are part of the land community, just as dependent as other life on the land's fertility and just as shaped by its mysteries and possibilities.

Just as important as the methods and economics of farming from ancient times till now are "the ways that farm life has figured in people's social and moral imagination," writes Freyfogle. As evidence for the emergence of the new agrarianism, he cites current features of America's farm story that have been reported here: a rise in the interest of urban people in the way their food is raised, as expressed in the success of farmers' markets and community-supported agriculture; "in the woodlot owner who develops a sustainable harvesting plan for his timber, aiding the local economy while maintaining a biologically diverse forest"; in the many efforts to restore watersheds in such a way that the river's health is the primary goal; in the farmer who radically reduces "chemical use, cuts back subsurface drainage, diversifies crops and rotations, and carefully tailors farm practices to suit the land"; "in the faith-driven religious group that takes seriously, in practical ways, its duty to nourish and care for its natural inheritance"; and in urban citizens who work to transform empty lots into gardens, lobby their government to provide green space, conserve land, and work in other ways "to translate agrarian values into daily life." Freyfogle mentions other signs, but the picture is clear. "Agrarianism is very much alive and flourishing in America today, in ways both new and old and in diverse vocations and avocations. . . . One could not call it a major element of contemporary culture, yet once aware of agrarianism, one stumbles on its outcropping at many a turn."[6] That has certainly been my experience in this region.

Working against that resurgence of values are some of our culture's most enduring romances: our romance with speed, with efficiency, with production, with income, and with the accumulation of "toys" to prove our worth. Our hope has always lain in the longheadedness that Steven Stoll referred to, the longheadedness it takes to be a really good farmer, good husbandman, good citizen of a democracy. "Good things often take time," writes David Orr. "There is no fast way to develop good character, to sink roots, or to acquire wisdom. There is no fast way to build decent communities or to repair an eroded field."[7] If Peter Rosset of Food First is right, efficiency as we currently think of it in agriculture may be mostly myth, and a more precise view of it would help us get down the road toward a more sustainable agriculture. The religion promulgated in those churches that Ron Scherbring insists are still influential in our region holds that we should not lay up for ourselves "treasures on earth," that we need to "consider the lilies of the field" and pay attention to our stewardship of the land.

Orr's students at Oberlin College worked to convert a school parking lot into a garden that could produce food for the school cafeteria. In a report to the board of the foundation that provided funds for the project, one student said that what he learned was that "asphalt isn't forever." That, it seems to me, is the insight that Dean Harrington, the farmers I've visited with, and many other citizens as well are striving toward, and that many have come to live by. Orr sums up the possibilities in such a notion with customary clarity: "Now, if a young student can grasp this truth, perhaps others of us can see that decaying, sprawling, crime-ridden cities are not forever, any more than degraded rivers, eroded fields, abandoned farms, and dying small towns are forever. Better alternatives are available. It is up to us to bring them into being."[8] For those of us who live in the Midwest—and for farmers wherever they are, working at whatever scale—it is a good reminder. The expansion of soybean acreage and the decline in the number of small farms is not forever. Plowing straight up the hill to plant our corn is not forever. Erosion is not forever. Farm bills and trade agreements that steer us in the wrong direction are not forever. The dominance of agriculture by three or four humongous international corporations without loyalty to any community or any nation is not forever.

Perhaps one way to bring about these better alternatives sooner is to turn the characteristics of farmers in southeastern Minnesota and throughout farm country into a prescription for a healthy, sustainable life. We are not called to obey these values as if they were commandments, but to live toward them. If we do, health and sustenance will be restored for the soil, plants, animals, and humans. "In the course of our practice," poet Gary Snyder says, writing about the precepts of Buddhism, "we will not transform Reality, but we may transform ourselves. Guilt and self-blame are not the fruit of our practice, but we may hope that a *larger view* is" (Snyder's emphasis).[9] The match here is not perfect, for Snyder's Buddhist precepts are not directly parallel to these agrarian values. Still, we know we may change ourselves and our practice, whether it be agricultural practice, personal growth, or achieving a broader vision of our social concerns.

The origin of the values I have heard from the mouths of our farmers has become a kind of mystery for me to ponder. What gives rise to them? The church that so many mention? Sometimes, perhaps. Fathers and mothers, grandfathers and grandmothers? Yes, often. A neighbor's good practice? In some cases. More broadly, are these values simply embedded in the local culture of the region, and by osmosis leak into the minds and hearts of those who till the soil or graze it? I think that cannot be the case, because these values are held far beyond the region and beyond the culture. As Philip Snyder, the farmer who loves potatoes, noted, "The Hmong are good farmers too." And it must be something even greater than our churches, for as Mike Rupprecht pointed out, churches rarely mention the importance of stewardship.

I choose to think that it is something in the soil, the plants, the animals and birds both domestic and wild, that we live with. They call us to respond. Those who hold these values do so in response to the land they work, and they work it in the way they do because they watch it with such attention. As they fit their practice to the land they live on, they can see it change and improve. Such attention is an investment; such an investment leads to good care; and such care leads to good practice. In their good practice, they discover what seems like "the right thing to do," as Dennis Rabe said. Living sustainably is less a burden we must bear than an ethical, spiritual, and intellectual response to what we see around us. Sustainability, then, may have more to do with our capacity to respond fully than with our acceptance of responsibility.

There are a variety of healthy futures available to us. "It is," as David Orr says, "up to us to bring them into being." But where does the vision for a sustainable culture reside? It is not among the visions shown us on TV of how the culture ought to look: sleek, rich, ostentatious, and utterly self-conscious in that "It's all about me" way that the culture approves now. No, I am not asking for extreme asceticism or self-denial, but a recognition that a ten-thousand-square-foot house is not a dream realized but a failure of imagination. It shows us not how to change the culture, or how we are superior to it, but how we submit to it. It is not a vision of how the future might unfold but a model of how it has, how our history has overtaken us, and how a sustainable future has been put off.

So who is showing us that the culture can be changed? Art and Jean, Mike and Jennifer, Sue and Dennis, Peggy, Larry, Ralph, Dave and Dan, and all those others whose names have come up in this narrative. These are not people you know or have ever heard of, unless you live in the region. I could go on naming names: while we had him, there was that entrepreneur of ideas and realizer of dreams, Dick Broeker; we're lucky to still have our community-minded banker, Dean Harrington . . . But what I like about my short list is that you can add the names of those you know, and others can name the folks they know, and soon we will have a whole army of names that can move our future from ever-increasing oppression and poverty toward sustainability—which is the home of the only enduring freedom.

What do these folks show us? That the future does not depend on chemicals. That Monsanto and Archer Daniels Midland are not as pervasive and powerful as they seem; that they can be gotten around. That the good life is not one that consumes but one that creates, one in which the imagination can replenish itself and refresh, renew, restore what is essential: our soil, our streams and grasses, an all-pervasive justice, freedom . . . and good food and healthy stories we can grow into.

These folks are as creative and artful as anything that hangs in a New York gallery. If, as Curtis White claims in *The Middle Mind: Why Americans Don't Think for Themselves*, it is characteristic of both art and imagination to always insist that alternative worlds and alternative futures are possible, then these farmers and their allies in the small towns around them represent some of the most creative imaginations in the country.[10] They are not artisans but artists. What they create is not "product," as our culture's lingo has it these days, but produce. It is healthy, colorful, beautifully sculpted, and richly diverse, each squash or cucumber or head of lettuce, cow, sheep, hog, or chicken unique. As such, it stands as an inspiring alternative to the products of our international agribusiness giants, whose every chicken's breast weighs exactly one-quarter pound and fits a fast-food bun perfectly —and tastes just as much like cardboard as every other fast product of a culture growing ever more gray and tasteless in its business, its politics, and its oppression.

The new agrarian values held by the farmers you've met in this book, and by those who work with them, are smooth stones for our slingshots against the Goliaths that are now stomping around our world, more vulnerable than they appear. Those farmers and their values represent an ecology of hope that can help us create a sustainable culture and feed us mighty well as we work away at that project. The very concept of sustainability that we currently hold began with them, not agribusiness, government, or academe. Those institutional imaginations are often severely limited by the necessity for profit and power. Without these farmers and their friends in our small communities and great cities, without agriculture, we have no culture at all, no matter what New York,

Los Angeles, Paris, or London might do. With them, we may yet create a sustainable culture that can save us all.

Now that's worth a celebration! Come join us—say, for supper at Ralph's place? Four or five couples will be happy to meet you, sitting outside on the back porch. They'll all be glad you joined us. We'll have a few fresh fish from the pond, just-picked vegetables from the garden, salad foraged from the pasture or the fencerow, grass-fed beef on the grill, cold regional beer from the cooler between us, and great stories around a table where the arguments are loud, the observations astute, the teasing and laughter frequent and ringing, and everyone gets a hug before heading home.

Notes

CHAPTER 1. FUNDAMENTALS

1. Lao Tzu, *Tao Te Ching: The Definitive Edition*, trans. Jonathon Star (New York: Tarcher/Putnam, 2001), 21; Lao Tsu, *Tao Te Ching*, trans. Gia-fu Feng and Jane English (New York: Vintage, 1987), 10.

2. J. R. R. Tolkien, *The Fellowship of the Ring* (New York: Houghton Mifflin, 1994), 241.

3. Richard Fortey, *Life: A Natural History of the First Four Billion Years of Life on Earth* (New York: Knopf, 1998), 16.

4. Steven Stoll, *Larding the Lean Earth: Soil and Society in Nineteenth-Century America* (New York: Hill and Wang, 2002), 37.

CHAPTER 2. HISTORIES

1. Recounted in James Banks, as told to Irving Wallace, *Wing of Scarlet* (Washington, D.C.: Pioneer, 1946), 31–32.

2. Ibid., 36, 40.

3. For a summary of American Indian cultural development in the region, see Frederick L. Johnson, *Goodhue County, Minnesota: A Narrative History* (Red Wing, Minn.: Goodhue County Historical Society Press, 2000), 3–19.

4. W. H. Mitchell, *Geographical and Statistical Sketch of the Past and Present of Goodhue County, Together with a General View of the State of Minnesota* (Minneapolis: King's Book and Printing House, 1869), 13–16.

5. Reverend J. W. Hancock, diary for 1869, Goodhue County Historical Society Museum, Red Wing, Minn.

6. Steven R. Hoffbeck, *The Haymakers: A Chronicle of Five Farm Families* (St. Paul: Minnesota Historical Society, 2000), 19–45.

7. Ibid., 49–73.

8. Ibid., 77–105.

9. Ibid., 109–35.

10. Ibid., 139–72.

11. H. C. Hinrichs, "As I Remember: A Treatise on Early Rural Life in Goodhue County, Minnesota" (unpublished manuscript, Red Wing Public Library, Red Wing, Minn., 1982).

12. U.S. Department of Agriculture, *Soils and Men: Yearbook of Agriculture, 1938* (Washington, D.C.: U.S. Government Printing Office, 1938), 271–72, 275, 276.

13. Ibid., 272, 274, 275.

14. Eric T. Freyfogle, ed., *The New Agrarianism: Land, Culture, and the Community of Life* (Washington, D.C.: Shearwater Books, 2001), xxiii.

15. Don Worster, "The Wealth of Nature," in Freyfogle, *New Agrarianism*, 167–68.

16. Freyfogle, *New Agrarianism*, xxxii.

17. Willard W. Cochrane, "A Food and Agricultural Policy for the 21st Century," November 16, 1999, http://www.tradeobservatory.org/showFile. php?RefID=29697.

CHAPTER 3. TWO VIEWS, ONE FARM: VANCE AND BONNIE HAUGEN

1. GM-Free Ireland Network, "Illegal GM Maize Should Be Returned to USA," news release, May 28, 2005; Friends of the Earth, "Illegal Maize Found in Japanese Imports," news release, May 30, 2005; Friends of the Earth International, "World Food Programme and United States Accused of Ignoring Concerns of Central American Society," news release, February 18, 2005. For more on StarLink and Mexican corn, see Christy Harrison, "Genetically Modified Foods," March 3, 2004, http://www.askquestions.org/details. php?id=30. For a counter point of view, see International Foundation for the Conservation of Natural Resources, "Greenpeace and Biotech: Truth or a Deliberate Scare," Biotech Web site, June 2, 2002, http://biotech.ifcnr.com/ article.cfm?NewsID=279.

CHAPTER 9. FARMING CONNECTS US ALL

1. Lucretius, *On the Nature of the Universe*, trans. R. E. Latham, rev. John Godwin (New York: Penguin, 1994), 65–66.

2. U.S. Department of Agriculture, *Agricultural Statistics* (Washington, D.C.: National Agricultural Statistics Service, USDA, 1997–2000).

3. Peggy J. Cook, "Rural Areas Show Signs of Revitalization," *Rural Conditions and Trends* 7, no. 3 (February 1997): 3, 7; *Center for Rural Affairs Newsletter*, "New Data on Rural Poverty," September 2002, 1; U.S. Department of Agriculture, *2002 Census of Agriculture* (Washington, D.C.: National Agricultural Statistics Service, USDA), tables 1, 4, 5, 43, 44.

4. *Agri News*, November 14, 2002; U.S. Department of Commerce, Bureau of Economic Analysis, "Personal Income and Outlays," news release, November 1, 2002.

5. Truman quoted in Peter Forbes, *The Great Remembering: Further Thoughts on Land, Soul, and Society* (San Francisco: Trust for Public Land, 2001), 26, 28.

6. Land Stewardship Project, *Farming in Goodhue County (Wells Creek): Row Crops Increase and Farms Consolidate* (White Bear Lake, Minn.: Land Stewardship Project, August 2002), 1; 1950 figures provided by Douglas A. Hartwig, director of Minnesota Field Office, National Agricultural Statistics Service, USDA, July 20, 2006; 1987 figures from U.S. Department of Agriculture, *1992 Census of Agriculture* (Washington, D.C.: National Agricultural Statistics Service, USDA), vol. 1, pt. 23, chap. 2, table 6.

7. *New York Times*, February 13, 2003; *Rochester (Minn.) Post-Bulletin*, February 13, 2003.

8. See Peter M. Rosset, *The Multiple Functions and Benefits of Small Farm Agriculture in the Context of Global Trade Negotiations*, Food First Policy Brief no. 4 (prepared for the FAO/Netherlands Conference on the Multifunctional Character of Agriculture and Land, Maastricht, the Netherlands, September 12–17, 1999).

9. Ibid., 6–7.

10. Ibid., 10; Dean C. Ludwig and Robert J. Anderson, "A Model of Indigenous Revival for U.S. Agriculture," *Journal of International Food and Agribusiness Marketing* 4, no. 2 (1992): 35, quoted in Rosset, *Multiple Functions*, 11; Darryl E. Ray, "Export Values Up, Net Income Down: Which Is More Important?" *MidAmerica Farmer Grower*, September 13, 2002; National Farmers Union, "The Effects of Export-Oriented Agriculture on Canadian Farm Families, Canadian Consumers, and Farmers around the World" (submission to the Canadian House of Commons Standing Committee on Foreign Affairs and International Trade, April 26, 1999); *Western Producer*, "Farm Exports, Farm Income—the Gap Widens," September 14, 2000.

11. Ludwig and Anderson, "Model," 35, quoted in Rosset, *Multiple Functions*, 11.

12. Lincoln quoted in Corporate Agribusiness Research Project, "Quotable Quotes," http://www.electricarrow.com/CARP/furrows/quotable.html.

13. Rosset, *Multiple Functions*, 9–10; Willis L. Peterson, "Are Large Farms More Efficient?" Staff Paper P97-2 (Department of Applied Economics, University of Minnesota, St. Paul, 1997), cited in Rosset, *Multiple Functions*, 10.

14. Richard A. Levins, "An Essay on Farm Income," Staff Paper P01-1 (Department of Applied Economics, University of Minnesota, St. Paul, April 2001), 18, 19.

15. U.S. Department of Agriculture, *Soils and Men*, foreword.

16. U.S. Department of Agriculture, *Quick Stats: U.S. and All States Data—Crops* (Washington, D.C.: National Agricultural Statistics Service, USDA, 2005); Dale Lattz, "Costs to Produce Corn and Soybeans in Illinois —2005," *Farm Economics Facts and Opinions*, June 1, 2006.

17. Tad W. Patzek, "Thermodynamics of the Corn-Ethanol Biofuel Cycle," *Critical Reviews in Plant Sciences* 23 (2004): 519–67; American Coalition for Ethanol, news release, July 19, 2005; David Pimentel, "Ethanol Fuels: Energy Balance, Economics, and Environmental Impacts Are Negative," *Natural Resources Research* 12, no. 2 (2003): 127–34.

18. R. F. Rockwell, K. F. Abraham, and R. L. Jefferies, "Tundra under Siege," *Natural History*, November 1996, 20–21.

19. Ibid.

20. Ibid.

21. Kala Mehta, Susan M. Gabbard, Vanessa Barrat, Melissa Lewis, Daniel Carroll, and Richard Mines, *Findings from the National Agricultural Workers Survey (NAWS) 1997–1998: A Demographic and Employment Profile of United States Farmworkers* (Washington, D.C.: Employment and Training Administration, U.S. Department of Labor, March 2000): vii–viii; Victor Contreras, Jaime Duran, and Kathryn Gilje, "Migrant Farmworkers in South-Central Minnesota: Farmworker-Led Research and Action for Change," *CURA Reporter*, February 2001. Also see U.S. Department of Labor, *A Profile of U.S. Farmworkers: Demographics, Household Composition, Income and Use of Services*, Research Report no. 6 (Washington, D.C.: Office of the Assistant Secretary for Policy, Office of Program Economics, U.S. Department of Labor, 1997), and U.S. Department of Agriculture, *Farm Labor* (Washington, D.C.: National Agricultural Statistics Service, USDA, 1995–2006).

22. Kate Bronfenbrenner, *Uneasy Terrain: The Impact of Capital Mobility on Workers, Wages, and Union Organizing* (Ithaca, N.Y.: Bronfenbrenner, September 2000); Velasquez quoted in *Multinational Monitor*, "The Power

of Organizing: Securing Farmworkers' Rights," May 1993; Contreras, Duran, and Gilje, "Migrant Farmworkers."

23. National Coalition for Agricultural Safety and Health, *Agricultural Occupational and Environmental Health: Policy Strategies for the Future*, 1998, http://www.public-health.uiowa.edu/AgAtRisk/; U.S. Department of Labor, *Agricultural Operations* (Washington, D.C.: Occupational Safety and Health Administration, U.S. Department of Labor), cited in Eric Schlosser, *Fast Food Nation: The Dark Side of the All-American Meal* (New York: Perennial, 2002), 328 n. 173.

24. Alice Larson, *Migrant Health Issues: Environmental/Occupational Safety and Health*, Monograph no. 2 (Buda, Tex.: National Center for Farmworker Health, October 2001); Rupali Das, Andrea Steege, Sherry Baron, John Beckman, Ximena Vergara, Patrice Sutton, and Robert Harrison, *Pesticide Illness among Farmworkers in the United States and California* (Oakland: California Department of Health Services, July 2002); Dina M. Schreinemachers, "Birth Malformations and Other Adverse Perinatal Outcomes in Four U.S. Wheat-Producing States," *Environmental Health Perspectives* 111 (2003): 1259. Also see Fawn Pattison, *Examining the Evidence on Pesticide Exposure and Birth Defects in Farmworkers: An Annotated Bibliography, with Resources for Lay Readers* (Raleigh, N.C.: Agricultural Resources Center and Pesticide Education Project, May 2006).

25. Charles M. Benbrook, "Genetically Engineered Crops and Pesticide Use in the United States: The First Nine Years," AgBioTech InfoNet Technical Paper no. 7 (October 2004), http://www.biotech-info.net/technicalpaper7.html; U.S. House Committee on Governmental Reform, *Human Pesticides Experiments*, 109th Cong., 1st sess., June 16, 2005; "Facts on Farm Workers in New York State," Coming Up on the Season, College of Human Ecology, Cornell University, October 2001, http://www.farmworkers.cornell.edu/pdf/facts_on_farmworkers.pdf; League of Women Voters of Oregon Education Fund, "Farmworkers in Oregon," 2000, http://www.open.org/~lwvor/Farmworkers2.htm. Also see Charles M. Benbrook, "Troubled Times amid Commercial Success for Roundup Ready Soybeans: Glyphosate Efficacy Is Slipping and Unstable Transgene Expression Erodes Plant Defenses and Yields," AgBioTech InfoNet Technical Paper no. 4 (May 2001), http://www.biotech-info.net/troubledtimes.html.

26. Ann Ziebarth and Jaehyun Byun, *Migrant Worker Housing: Survey Results from South Central Minnesota* (Centro Campesino, University of Minnesota Center for Urban and Regional Affairs, and Hispanic Advocacy and Community Empowerment through Research, 2000), 64.

27. James J. Kielkopf, *Estimating the Economic Impact of the Latino Workforce in South Central Minnesota* (Mankato: Center for Rural Policy and Development, Minnesota State University, September 2000), 3.

28. Ibid., 13.

29. Neil E. Harl, "The Structural Transformation of the Agricultural Sector" (paper presented at the Master Farmer Awards Ceremony, West Des Moines, Iowa, March 20, 2003); Darren Hudson, "Contracting in Agriculture: A Primer for Farm Leaders," Research Report 2000-007 (Department of Agricultural Economics, Mississippi State University, Starkville, August 2000), 2–3, 17; Nigel Key and James MacDonald, "Agricultural Contracting: Trading Autonomy for Risk Reduction," *Amber Waves*, February 2006; Pat Stith, Joby Warrick, and Melanie Sill, "Boss Hog: North Carolina's Pork Revolution," *Raleigh (N.C.) News and Observer*, February 19, 21, 22, 24, 26, 1995; Brian DeVore, "Selling the Farm Down Contract Creek," *In Motion Magazine*, July 27, 1999; David Moeller, *Livestock Production Contracts: Risks for Family Farmers* (St. Paul, Minn.: Farmers' Legal Action Group, March 2003). For another view of contracts, see Allen Harper, "Contract Hog Production as an Alternative Farm Enterprise," *Livestock Update*, April 2005.

30. Joby Warrick and Pat Stith, "Boss Hog 2: Corporate Takeovers," *Raleigh (N.C.) News and Observer*, February 21, 1995.

31. Northern Plains Sustainable Agriculture Society, "Agriculture at the Crossroads" (position paper), http://www.npsas.org/Crossroads.html; *Agribusiness Examiner*, November 11, 2003.

32. Charles M. Benbrook, *When Does It Pay to Plant Bt Corn? Farm-Level Economic Impacts of Bt Corn, 1996–2001* (special report for the Institute for Agriculture and Trade Policy, Minneapolis, December 2001), 2; William P. Cunningham and Mary Ann Cunningham, "Food and Agriculture: Additional Case Studies," *Principles of Environmental Science*, 2nd ed., Online Learning Center, McGraw-Hill Higher Education, http://highered.mcgrawhill.com/sites/0072919833/student_view0/chapter7/additional_case_studies.html; rBST Internal Review Team, *rBST (Nutrilac) "Gaps Analysis" Report* (Ottawa, Ont.: Health Protection Branch, Health Canada, April 1998).

33. Mark Shapiro, "Sowing Disaster: GMO's and Mexican Corn," *Nation*, October 28, 2002; Zac Goldsmith, "This Should Be the End for GM," *Observer*, October 19, 2003; Simon McRae, "No GMO Liability? No GMO Releases: Why New Liability Laws Are Needed for GMO Food and Crops" (Friends of the Earth submission to the European Commission's "White Paper on Environmental Liability," June 2000), http://www.foeeurope.org/press/liabilitybriefing.PDF; Greenpeace, *Grains of Truth: False Promises of Genetic Engineering* (Amsterdam: Greenpeace International, September

2001); Myrto Pispini, "Compensation Claimed: Greek Farmers Sue the Seed Companies; Pioneer and Syngenta Are Responsible for Genetic Contamination," *Transregionale,* http://www.gmo-free-regions.org/fileadmin/files/ Transregionale-site5.pdf; GM-Free Ireland Network, "Illegal GM Maize"; Friends of the Earth, "Illegal Maize Found"; Friends of the Earth International, "United States Accused of Ignoring Concerns"; Jeffrey M. Smith, *Seeds of Deception: Exposing Industry and Government Lies about the Safety of the Genetically Engineered Foods You're Eating* (Fairfield, Iowa: Yes Books, 2003), 84–85, 101–2.

34. William Walker, "Monsanto's PCB Scandal: Dirt-Poor Residents Seek Compensation in Alabama Town That Was Secretly Poisoned for Decades," *Toronto Star,* February 17, 2002; BBC News, "The Legacy of Agent Orange," April 29, 2005; Principi quoted in *Agent Orange Review,* "IOM Identifies Herbicide Link with CLL, Principi Extends Benefits," July 2003, 1; Committee to Review the Health Effects in Vietnam Veterans of Exposure to Herbicides, *Veterans and Agent Orange: Update 2000* (Washington, D.C.: National Academy Press, 2001).

35. For a small sample from the British press, see *Guardian,* "Bush's Evangelizing about Food Chills European Hearts," June 2, 2003; *Independent,* "Think of the Benefits, Mr. Meacher: An Apple, Say, That Helps People Stay Slim," June 29, 2003; *Independent,* "Not a Modified Bean in the House: The Villagers Who Drove GM out of Town," June 29, 2003; *Independent,* "Science? The Public Is Right to Smell a Rat," June 29, 2003; *Independent,* "Economic Benefits of GM Crops Small, Study Concludes," July 12, 2003; *Independent,* "No Support from the Public, No Evidence, No Case for GM," October 19, 2003; *Independent,* "Science Backs Consumers' Rejection of GM Food—Are You Listening Tony?" October 19, 2003; *Guardian,* "Bad for the Poor and Bad for Science," February 20, 2004. It was not unusual for either paper to run multiple stories about GMOs in the same issue, as on June 29 and October 19, 2003. Articles were sometimes favorable, sometimes not, but they always raised questions for public consideration.

36. Smith, *Seeds of Deception,* 39–44; rBST Internal Review Team, "*Gaps Analysis*" *Report*; Smith, *Seeds of Deception,* 5.

37. Smith, *Seeds of Deception,* 5–44.

38. Ibid., 77–180; Mae-Wan Ho and Lim Li Ching, *The Case for a GM-Free Sustainable World* (London: Independent Science Panel, June 2003); "Open Letter from World Scientists to All Governments Concerning Genetically Modified Organisms (GMOs)," 1999–2000, http://www.i-sis.org. uk/list.php. Also see Mae-Wan Ho, witness statement (Chardon LL Hearing, Novotel, London, October 26, 2000), http://www.i-sis.org.uk/chardonLL-

transcript.php. Pusztai also testified at this hearing on a GM fodder maize product for cattle.

39. Neil King Jr., "EU Ban on Biotech Products to Face Challenge from U.S.," *Wall Street Journal*, May 9, 2003; Reuters, "EU's Nielson Blasts U.S. 'Lies' in GM Food Row," January 20, 2003; Michael McCarthy, "U.S. Firms 'Tried To Lie' over GM Crops, Says EU," *Independent*, October 14, 2003.

40. Mark Muller and Richard Levins, *Feeding the World? The Upper Mississippi River Navigation Project* (Minneapolis: Institute for Agriculture and Trade Policy, December 1999); Frederick Kirschenmann, "Feeding the Village First" (position paper, Northern Plains Sustainable Agriculture Society), http://www.npsas.org/Feeding.html.

41. Richard Strohman, "Crisis Position," in *Safe Food News 2000* (Fairfield, Iowa: Mothers for Natural Law, 2000).

42. Joe Cummins, "The Fate of Food Genes and the DNA CpG Motif and Its Impact" (paper presented at the Toxicology Symposium, University of Guelph, Guelph, Ont., March 3, 2001).

43. Richard Lacey, witness testimony, *Alliance for Bio-Integrity v. Donna Shalala*, civil action n. 98-1300 (U.S. District Court for the District of Columbia). A transcript of the testimony is available at http://www.saynotogmos.org/scientists_speak.htm.

44. rBST Internal Review Team, *"Gaps Analysis" Report*; Royal Society of Canada, *Elements of Precaution: Recommendations for the Regulation of Food Biotechnology in Canada* (Ottawa, Ont.: Royal Society of Canada, January 2001), viii–ix; Smith, *Seeds of Deception*, 77–105; Midwest Sustainable Agriculture Working Group, "Position Paper on Genetic Engineering" (Center for Rural Affairs, February 2000), http://www.cfra.org/resources/msawg_ge.htm. For the complete text of Monsanto's Nutrilac label, see Gabriel Hegyes, "BST Warning Label," e-mail to Sustainable Agriculture Network Discussion Group, February 5, 1994, http://www.sare.org/sanet-mg/archives/html-home/3-html/0312.html.

45. Charles M. Benbrook, *The Bt Premium Price: What Does It Buy?* (special report for the Institute for Agriculture and Trade Policy, Minneapolis, February 2002), 7, 4. Also see Benbrook, *When Does It Pay*.

46. *Agri News*, "Biotech Crops Becoming More Common," January 23, 2003; *Agri News*, February 20, 2003.

47. Mark Ritchie, *The Loss of Our Family Farms: Inevitable Results or Conscious Policy?* (Minneapolis: League of Rural Voters, 1979), 5–9.

48. Ibid., 12.

49. Ibid., 18.

50. *Mencius*, trans. D. C. Lau (New York: Penguin, 1970), bk. 2, pt. A, 75.

51. Ibid., bk. 1, pt. A, 51–52.

CHAPTER 10. AGRICULTURE AND COMMUNITY CULTURE

1. Mary Evelyn Tucker and John Berthrong, "Setting the Context," introduction to *Confucianism and Ecology: The Interrelation of Heaven, Earth, and Humans*, ed. Mary Evelyn Tucker and John Berthrong (Cambridge, Mass.: Center for the Study of World Religions, Harvard University, 1998), xxxvi–xxxviii; William Irwin Thompson, *Coming into Being: Artifacts and Texts in the Evolution of Consciousness* (New York: Palgrave Macmillan, 1998), 23–30; *Parmenides and Empedocles: The Fragments in Verse Translation*, trans. Stanley Lombardo (San Francisco: Grey Fox Press, 1982), 38, 100, 102.

2. Martin Rees, *Just Six Numbers: The Deep Forces That Shape the Universe* (New York: Basic Books, 2000), 7. For a discussion of *tian* and *tianli*, see Roger T. Ames and Henry Rosemont Jr., *The Analects of Confucius: A Philosophical Translation* (New York: Ballantine, 1998), 46–48, and David L. Hall and Roger T. Ames, *Thinking Through Confucius* (Albany: State University of New York Press, 1987), 201–16.

3. These figures are based on my experience serving on the board of directors of the Child Abuse Board and the Center for Children and Families in Anchorage, Alaska, in the 1970s.

4. Roy Beck, *The Case Against Immigration: The Moral, Economic, Social, and Environmental Reasons for Reducing U.S. Immigration Back to Traditional Levels* (New York: Norton, 1996): 118, 105, 109–10; Mark A. Grey, "Meatpacking and the Migration of Refugee and Immigrant Labor to Storm Lake, Iowa," Changing Face, Migration Dialogue, University of California, Davis, http://migration.ucdavis.edu/cf/more.php?id=154_0_2_0.

5. Beck, *Case Against Immigration*, 109–10; Poverty & Race Research Action Council, "Shattered Promises: Immigrants and Refugees in the Meatpacking Industry," *Poverty and Race*, November 1992, http://www.prrac.org/full_text.php?text_id=644&item_id=6514&newsletter_id=5&header=Search%20Results.

6. Schlosser, *Fast Food Nation*, 170–90; Poverty & Race Research Action Council, "Shattered Promises"; Rochelle L. Dalla, Sheran Cramer, and Kaye Stanek, "Economic Strain and Community Concerns in Three Meatpacking Communities," *Rural America*, Spring 2002.

7. Poverty & Race Research Action Council, "Shattered Promises"; Dalla, Cramer, and Stanek, "Economic Strain."

8. Beck, *Case Against Immigration*, 105; Schlosser, *Fast Food Nation*, 160–64.

9. Poverty & Race Research Action Council, "Shattered Promises"; Schlosser, *Fast Food Nation*, 162–63.

10. *Dallas Morning News*, January 15, 2000.

11. *Mitchell (S.D.) Daily Republic*, August 25, 1999.

12. Ibid.

13. Ibid.

14. Associated Press, "Equipment Sales Slow across U.S.," *Agri News*, March 27, 2003.

15. *St. Paul (Minn.) Pioneer Press*, "Harvest of Risk," June 27, 1999.

16. Stoll, *Larding the Lean Earth*, 17–20.

17. Ibid., 34, 37, 30.

18. Ibid., 26.

19. Ibid., 48.

20. U.S. Government Accounting Office, *Food Stamp Program: Various Factors Have Led to Declining Participation*, GAO/RCED-99-185 (Washington, D.C.: GAO, July 1999), 12; Food Research and Action Center, *Community Childhood Hunger Identification Project: A Survey of Childhood Hunger in the United States* (Washington, D.C.: FRAC, 2001); Food Research and Action Center, "Food Stamp Participation Jumps in July 2003 to More Than 22 Million Persons; Is 5.15 Million Persons Higher Than in July 2000," Current News and Analyses, October 9, 2003, http://www.frac.org/html/news/fsp/03july.html; Food Research and Action Center, "Food Stamp Participation Increases in September 2004 to Nearly 25 Million Persons," Current News and Analyses, December 8, 2004, http://www.frac.org/html/news/fsp/09.04_FSP.html.

21. U.S. Department of Agriculture, *Household Food Security in the United States, 1995–1998 (Advance Report)* (Washington, D.C.: Food and Nutrition Service, USDA, July 1999); Mark Nord, Margaret Andrews, and Steven Carlson, *Household Food Security in the United States, 2002*, Food Assistance and Nutrition Research Report no. FANRR35 (Washington, D.C.: Economic Research Service, USDA, October 2003); Barbara Cohen, James Parry, and Kenneth Yang, *Household Food Security in the United States, 1998 and 1999*, Food Assistance and Nutrition Research Report no. E-FAN02011 (Washington, D.C.: Economic Research Service, USDA, June 2002); U.S. Department of Agriculture, *WIC Participant and Program Characteristics 2000* (Washington, D.C.: Food and Nutrition Service, USDA, June 2003).

22. U.S. Conference of Mayors, "Sodexho, Inc. Hunger and Homelessness Survey, 1997," news release, December 15, 1997; U.S. Conference of Mayors, "Sodexho, Inc. Hunger and Homelessness Survey, 2004," news release, December 14, 2004; U.S. Conference of Mayors, "Sodexho, Inc. Hun-

ger and Homelessness Survey, 2005," news release, December 19, 2005; Catholic Charities USA, "Seven Million People Knock on Catholic Charities Door for Emergency Relief," news release, December 10, 1997.

23. John B. Cobb Jr., *Sustaining the Common Good: A Christian Perspective on the Global Economy* (Cleveland, Ohio: Pilgrim Press, 1994), 106.

24. Overseas Development Institute, *Global Hunger and Food Security after the World Food Summit,* Briefing Paper no. 1 (London: Overseas Development Institute, February 1997); Majda Bne Saad, *Food Security for the Food Insecure: New Challenges and Renewed Commitments* (position paper for CDS-8, 2000, NGO Women's Caucus, Commission on Sustainable Development, UN, December 1999), 1–3.

25. Worldwatch Institute, *Vital Signs 1999: The Environmental Trends That Are Shaping Our Future* (New York: Norton, 1999), 30–31, 36–39; Saad, *Food Security,* 2.

26. WorldWatch Institute, *Vital Signs 1999,* 18–19.

27. WorldWatch Institute, *State of the World 2002: A Worldwatch Institute Report on Progress Toward a Sustainable Society* (New York: Norton, 2002), 39–58.

28. Worldwatch Institute, *State of the World 2000* (*A Worldwatch Institute Report on Progress Toward a Sustainable Society* (New York: Norton, 2000), 59–78; Worldwatch Institute, *State of the World 2004: A Worldwatch Institute Report on Progress Toward a Sustainable Society* (New York: Norton, 2004), 155–61.

29. Worldwatch Institute, *Vital Signs 1999,* 38, 42.

30. Worldwatch Institute, *Vital Signs 2005: Redefining Global Security* (New York: Norton, 2005), 22–23

31. Charles M. Benbrook, "A Bill of Goods: Agricultural Policy, Trade and Technological Innovation since the Mid-1990's" (paper presented at the Upper Midwest Organic Farming Conference, La Crosse, Wisc., February 26–28, 2004).

CHAPTER 11. FARMING IN DEVELOPING COUNTRIES

The agriculture e-conference described in this chapter was a precursor to the FAO/Netherlands Conference on the Multifunctional Character of Agriculture and Land, Maastricht, the Netherlands, September 12–17, 1999.

CHAPTER 12. THE WTO, NAFTA, CAFTA, AND THE FTAA

1. Deepak Nayyar and Julius Court, *Governing Globalization: Issues and Institutions,* Policy Brief no. 5 (Helsinki, Finland: World Institute for Development Economics Research, UN University, 2002); Eswar Prasad, Kenneth

Rogoff, Shang-Jin Wei, and M. Ayhan Kose, *Effects of Globalization on Developing Countries: Some Empirical Evidence* (Washington, D.C.: International Monetary Fund, March 2003), 9–10.

2. Emilio Lopez Gamez and Jose Luis Alcocer de Leon (speech, Institute for Agriculture and Trade Policy, Minneapolis, July 16, 2003); Laura Theobald, "Mexican Farmers Bring Message to Minnesota," *Agri News*, July 24, 2003.

3. Friends Committee on Unity with Nature, "The Real Costs of 'Free' Trade: An Epistle," *Quaker Eco-Bulletin*, July–August 2003.

4. Robert E. Scott, Carlos Salas, and Bruce Campbell, "NAFTA at Seven: Its Impact on Workers in All Three Nations," EPI Briefing Paper no. 106 (Economic Policy Institute, Washington, D.C., 2001), 6–8, 17–19, 21, 25–27; Mark Weisbrot, Dean Baker, Egor Kraev, and Judy Chen, "The Scorecard on Globalization 1980–2000: Twenty Years of Diminished Progress," briefing paper, Center for Economic and Policy Research, Washington, D.C., July 2001; Public Citizen's Global Trade Watch, *Down on the Farm: NAFTA's Seven-Years War on Farmers and Ranchers in the U.S., Canada and Mexico* (Washington, D.C.: Public Citizen, June 2001); Alejandro Nadal and Timothy A. Wise, "The Environmental Costs of Agricultural Trade Liberalization: Mexico-U.S. Maize Trade under NAFTA," Discussion Paper no. 4 (prepared for the meeting of the Working Group on Development and Environment in the Americas, Brasilia, Brazil, March 29–30, 2004).

5. Sophia Murphy, *Managing the Invisible Hand: Markets, Farmers and International Trade* (Minneapolis: Institute for Agriculture and Trade Policy, April 2002), 30–31.

6. Public Citizen's Global Trade Watch, *Down on the Farm*, 5. IATP monitors trade agreement developments at http://www.tradeobservatory.org/.

7. Adapted from North American Free Trade Agreement, final text, preamble, http://www.nafta-sec-alena.org/DefaultSite/index_e.aspx?DetailID=79.

8. Robert Weissman, "Time to End CEOs' 5-Year NAFTA Party," Focus on the Corporation, *Multinational Monitor*, December 24, 1998.

9. Russell Mokhiber and Robert Weissman, *Corporate Predators: The Hunt for Mega-profits and the Attack on Democracy* (Monroe, Me.: Common Courage Press, 1999), 175; Robert E. Scott, "The High Price of 'Free' Trade: NAFTA's Failure Has Cost the United States Jobs across the Nation," EPI Briefing Paper No. 147 (Economic Policy Institute, Washington, D.C., 2003), 1.

10. Murphy, *Managing the Invisible Hand*, 31–35.

11. Ibid., 34.

12. Public Citizen's Global Trade Watch, *Down on the Farm*, 14, iv, 14–15.

13. Diane J. F. Martz and Wendy Moellenbeck, *Compare the Share Revis-*

ited: The Family Farm in Question (Muenster, Sask.: Centre for Rural Studies and Enrichment, St. Peter's College, 1999); Public Citizen's Global Trade Watch, *Down on the Farm*, 2–3. For a follow-up on Canadian consumer price increases, see Diane J. F. Martz, *The Farmers' Share: Compare the Share 2004* (Muenster, Sask.: Centre for Rural Studies and Enrichment, St. Peter's College, November 2004), 3–15.

14. Glenn Hubbard, "Cafta: A Win-Win Case," *Business Week*, July 4, 2005; Murphy, *Managing the Invisible Hand*, 22–23.

15. Public Citizen's Global Trade Watch, *Down on the Farm*, 2–12. For more on mergers, see William D. Heffernan, Mary K. Hendrickson, and Robert Gronski, "Consolidation in the Food and Agricultural System" (report to the National Farmers Union, University of Missouri–Columbia, February 5, 1999).

16. Alejandro Nadal, *The Environmental and Social Impacts of Economic Liberalization on Corn Production in Mexico* (London: Oxfam Great Britain; Godalming, UK: WWF International, September 2000), 4–5.

17. Ibid., 5–6.

18. Ibid., 6.

19. Ibid., 7. Also see Shapiro, "Sowing Disaster."

20. Nadal, *Environmental and Social Impacts*, 8.

21. Ibid., 99.

22. Gamez and Alcocer de Leon (speech); Theobald, "Mexican Farmers."

23. Joseph E. Stiglitz, *Globalization and Its Discontents* (New York: Norton, 2003), 44.

24. John Madeley, *Trade and Hunger: An Overview of Case Studies on the Impact of Trade Liberalisation on Food Security*, ed. Johanna Sandahl, Globala Studier no. 4 (Church of Sweden Aid, Diakonia, Forum Syd, Swedish Society for Nature Conservation, Programme of Global Studies, October 2000), 17, 66.

25. Greg Palast, *The Best Democracy Money Can Buy: An Investigative Reporter Exposes the Truth about Globalization, Corporate Cons, and High-Finance Fraudsters*, rev. ed. (New York: Plume, 2003), 157; *UN Wire*, April 21, 2003.

26. Kazembe quoted in Madeley, *Trade and Hunger*, 30.

27. Stiglitz, *Globalization and Its Discontents*, 46.

28. Institute for Agriculture and Trade Policy, *The TRIPs Agreement: Who Owns and Controls Knowledge and Resources?* WTO Cancún Series Paper no. 5 (Minneapolis: IATP, 2003), 2–3.

29. Caulder quoted in Neil E. Harl, "The Age of Contract Agriculture: Consequences of Concentration in Input Supply," *Journal of Agribusiness* 18,

no. 1 (March 2000): 126; Northern Plains Sustainable Agriculture Society, "Defining Mission: Calling upon Land Grant Institutions to Serve Rural America in the Twenty-first Century" (position paper), http://www.npsas. org/DefiningMission.html.

30. Murphy, *Managing the Invisible Hand*, 34–35.

31. Ibid., 35, 37, 37 n. 69, 35.

32. Ibid., 36.

33. Ibid., 37–38.

34. Ibid., 38; Public Citizen's Global Trade Watch, *Down on the Farm*, 35.

35. Murphy, *Managing the Invisible Hand*, 38.

36. Fatima V. Mello, *Brazil and the FTAA—The State of the Debate since Lula's Victory* (Santiago, Chile: Estudios sobre el ALCA, November 2002). Also see Rainer Falk, ed., *The Post-Cancún Debate: Options, Views, and Perspectives from South and North with Contributions from 9 Countries*, Global Issue Papers no. 6 (Berlin: Heinrich Böll Foundation, February 2004); J. F. Hornbeck, *A Free Trade Area of the Americas: Status of Negotiations and Major Policy Issues*, Congressional Research Service Report no. RS20864 (Washington, D.C.: Congressional Research Service, Library of Congress, January 2005).

37. *Economist*, "Clouds over Quito: Prospects for the Free-Trade Area of the Americas Do Not Look Good," October 31, 2002; WTO Accountability Review Committee, "WTO Accountability Review: Report to the Seattle City Council," September 14, 2000, http://www.cityofseattle.net/wtocommittee/cover.htm; Lisa Arthur, "Panel Faults Police During FTAA," *Miami Herald*, June 3, 2004.

38. Dominican Republic–Central America–United States Free Trade Agreement, final text, preamble, http://www.ustr.gov/assets/Trade_Agreements/Bilateral/CAFTA/CAFTA-DR_Final_Texts/asset_upload_file308_3917.pdf.

39. Free Trade Agreement of the Americas, second draft agreement (November 1, 2002), general article 4, "Application and Scope of Coverage of Obligations," http://www.ftaa-alca.org/FTAADraft02/preamb_e.asp.

40. Ibid., agriculture article 15, "Domestic Support Measures," http://www.ftaa-alca.org/FTAADraft02/ngag1_e.asp.

41. Ibid.

42. "Ministerial Declaration of Quito" (Free Trade Agreement of the Americas meeting of ministers of trade, Quito, Ecuador, November 1, 2002), http://www.ftaa-alca.org/Ministerials/Quito/Quito_e.asp.

43. Fatima V. Mello, "Cancún Points the Direction of Lula's Foreign Policy," in Rainer, *Post-Cancún Debate*, 49–50.

44. Robert B. Zoellick, "America Will Not Wait for the Won't-Do Countries," *Financial Times*, September 22, 2003.

45. Murphy, *Managing the Invisible Hand*, 23–24.

46. Ibid., 27.

47. Ibid., 13.

48. Mark Muller, "Export Myth Sold to Farmers," *IATP News*, July 2001; Mark Muller, "Renewable Energy Production," *IATP News*, January 2003.

49. Mark Muller, *Mississippi River Navigation: Helping the Midwest Compete with South American Soybeans?* (Minneapolis: Institute for Agriculture and Trade Policy, September 2000); Daryll E. Ray, "Cargill Opens Soybean Terminal at Santorem on the Amazon in Brazil," *MidAmerica Farmer Grower*, April 18, 2003; Daryll E. Ray, "Brazilian Acreage Potential Is Larger Than Previously Thought," *MidAmerica Farmer Grower*, April 4, 2003. For a brief history of the Mississippi River and the dam project, see Dennis Lien, "River on the Edge," *St. Paul (Minn.) Pioneer Press*, June 27, 2004.

50. *Manufacturing News*, "U.S. Food Industry Begins to Feel the Sting of Chinese Imports: China's 800 Million Farmers Are Feasting on the U.S. Market," July 3, 2003.

51. Ibid.

52. Madeley, *Trade and Hunger*, 65.

53. Richard A. Levins, "An Essay on Farm Income," Staff Paper P01-1 (Department of Applied Economics, University of Minnesota, St. Paul, April 2001); Bill McMillin (untitled paper, Kellogg, Minn.; photocopy in author's possession).

54. Barry Krissoff, Fred Kuchler, Kenneth Nelson, Janet Perry, and Agapi Somwaru, *Country of Origin Labeling: Theory and Observation*, Outlook Report no. WRS04-02 (Washington, D.C.: Economic Research Service, USDA, January 2004), 2.

CHAPTER 13. HEALTHY FOOD, HEALTHY ECONOMICS

1. Kamyar Enshayan (speech, annual meeting of the Southeast Minnesota Sustainable Farming Association, Plainview, Minn., February 8, 2003).

2. Ken Roseboro, "Congress Weakens Organic Standards," Savvy Vegetarian, http://www.savvyvegetarian.com/articles/congress-weakens-organic-standards.php; Stephanie Riddick and Claire Klotz, "Congress Repeals Section 771 of the Omnibus Appropriations Act of 2002," *Organic Perspectives*, May–June 2003.

3. Rich Pirog, Timothy Van Pelt, Kamyar Enshayan, and Ellen Cook, *Food, Fuel, and Freeways: An Iowa Perspective on How Far Food Travels, Fuel*

Usage, and Greenhouse Gas Emissions (Ames: Leopold Center for Sustainable Agriculture, Iowa State University, June 2001), 1–2.

4. Brian Halweil, *Home Grown: The Case for Local Food in a Global Market,* Worldwatch Paper no. 163 (Washington, D.C.: Worldwatch Institute, November 2002), 11–12.

5. Ken Meter and Jon Rosales, *Finding Food in Farm Country: The Economics of Food and Farming in Southeast Minnesota* (Lanesboro, Minn.: Hiawatha's Pantry Project, Community Design Center, March 2001), 5, 8–11.

6. David Wallinga, Navis Bermudez, and Edward Hopkins, *Poultry on Antibiotics: Hazards to Human Health,* 2nd ed. (Minneapolis: Institute for Agriculture and Trade Policy; San Francisco: Sierra Club, December 2002), 13–20.

7. Memorandum circulated at a meeting of the Minnesota Institute for Sustainable Agriculture, 1999; Mark Muller, "A Healthier, Smarter Food System," Commentary, Institute for Agriculture and Trade Policy, Minneapolis, May 1, 2006, http://www.iatp.org/iatp/library/admin/uploadedfiles/Healthier_Smarter_Food_System_A.pdf.

8. Kathryn Herzog, "Crop Diversity," Minnesota Public Radio, November 23, 1998; National Organic Program, "Organic Feed for Poultry and Livestock: Availability and Prices" (National Organic Program, Agricultural Marketing Service, USDA, Washington, D.C., June 2003), http://www.ams.usda.gov/nop/ProdHandlers/FeedStudyJune2003.pdf.

9. Barbara Johnson, *Conservation Reserve Program: Status and Current Issues,* Congressional Research Service Report no. RS21613 (Washington, D.C.: Congressional Research Service, Library of Congress, February 2005), 5, 3, 5–6.

10. Gary Snyder, "Indra's Net as Our Own," in *For a Future to Be Possible: Commentaries on the Five Mindfulness Trainings,* by Thich Nhat Hanh (Berkeley, Calif.: Parallax Press, 1993), 132, 127.

CHAPTER 14. ALTERNATIVES FOR AGRICULTURE AND THE WHOLE CULTURE

1. "Alternatives for the Americas" (working paper prepared for the Peoples' Summit of the Americas, Santiago, Chile, April 15–18, 1998; second draft updated April 1, 2005), summary, http://www.globalexchange.org/campaigns/alternatives/americas/Summary.html.

2. Ibid., "Agriculture," http://www.globalexchange.org/campaigns/alternatives/americas/Agriculture.html.

3. Ibid.

4. National Family Farm Coalition, "Farm Aid: A Declaration for a New Direction for American Agriculture and Agricultural Trade" (NFFC, Washington, D.C., September 7, 2003), http://www.nffc.net/resources/statements/farmaid.pdf.

5. Wes Jackson, *Becoming Native to This Place* (Washington, D.C.: Counterpoint, 1996), 8, 13.

6. Land Institute, "Why Natural Systems Agriculture?" http://www.landinstitute.org/vnews/display.v/RT/2000/12/01/37e288b43.

7. Jackson, *Becoming Native*, 83.

8. Land Institute, "Why Natural Systems Agriculture?"; Land Institute, "Research Agenda: Natural Systems Agriculture," http://www.landinstitute.org/vnews/display.v/ART/2000/12/01/37dffb033.

9. Wes Jackson, Right Livelihood Award for 2000 acceptance speech (Stockholm, Sweden, December 8, 2000), published in *Land Report*, Spring 2001.

10. James quoted in Robert Pollin, *The Contours of Descent: U.S. Economic Fractures and the Landscape of Global Austerity*, new ed. (New York: Verso, 2005), 173.

CHAPTER 16. AN ECOLOGY OF HOPE

1. Paul Rogat Loeb, *Soul of a Citizen: Living with Conviction in a Cynical Time* (New York: St. Martin's Griffin, 1999), 7.

2. David W. Orr, "The Urban-Agrarian Mind," in Freyfogle, *New Agrarianism*, 106, 101.

3. Wendell Berry, "The Agrarian Standard," *Orion*, Summer 2002, 55.

4. J. Hector St. John de Crèvecoeur, *Letters from an American Farmer and Sketches of Eighteenth-Century America*, ed. Albert E. Stone (New York: Penguin Books, 1981), 56.

5. Loeb, *Soul of a Citizen*, 118–19.

6. Freyfogle, *New Agrarianism*, xiii, xv, xvii.

7. Orr, "Urban-Agrarian Mind," 107.

8. Ibid.

9. Snyder, "Indra's Net," 132.

10. Curtis White, *The Middle Mind: Why Americans Don't Think for Themselves* (San Francisco: HarperSanFrancisco, 2003), 146, 161–62, 200.

Sources and Resources

BOOKS

Aberley, Doug, ed. *Futures by Design: The Practice of Ecological Planning.* Gabriola Island, B.C.: New Society, 1994.

Achebe, Chinua. *Things Fall Apart.* New York: Anchor Books, 1994.

Amato, Joseph A. *Servants of the Land: God, Family, and Farm; The Trinity of Belgian Economic Folkways in Southwestern Minnesota.* Marshall, Minn.: Crossings Press, 1990.

Amato, Joseph A., and John W. Meyer. *The Decline of Rural Minnesota.* Marshall, Minn.: Crossings Press, 1993.

Ames, Roger T., and Henry Rosemont Jr. *The Analects of Confucius: A Philosophical Translation.* New York: Ballantine, 1998.

Banks, James, as told to Irving Wallace. *Wing of Scarlet.* Washington, D.C.: Pioneer, 1946.

Beck, Roy. *The Case Against Immigration: The Moral, Economic, Social, and Environmental Reasons for Reducing U.S. Immigration Back to Traditional Levels.* New York: Norton, 1996.

Berman, Morris. *The Twilight of American Culture.* New York: Norton, 2000.

Bernard, Ted, and Jora Young. *The Ecology of Hope: Communities Collaborate for Sustainability.* Gabriola Island, B.C.: New Society, 1997.

Berry, Thomas. *The Dream of the Earth.* San Francisco: Sierra Club Books, 1988.

Berry, Wendell. *The Art of the Commonplace: Agrarian Essays of Wendell Berry.* Washington, D.C.: Counterpoint, 2002.

Sources and Resources

————. *A Continuous Harmony: Essays Cultural and Agricultural.* New York: Harcourt Brace Jovanovich, 1972.

————. *Recollected Essays, 1965–1980.* San Francisco: North Point Press, 1981.

————. *Sex, Economy, Freedom and Community: Eight Essays.* New York: Pantheon Books, 1993.

Bowers, C. A. *Educating for an Ecologically Sustainable Culture: Rethinking Moral Education, Creativity, Intelligence, and Other Modern Orthodoxies.* Albany: State University of New York Press, 1995.

————. *Elements of a Post-liberal Theory of Education.* New York: Teachers College Press, Columbia University, 1987.

Brody, Hugh. *The Other Side of Eden: Hunters, Farmers, and the Shaping of the World.* New York: North Point Press, 2001.

Brown, Lester R. *Building a Sustainable Society.* New York: Norton, 1981.

Brown, Lester R., Janet Larsen, and Bernie Fischlowitz-Roberts. *The Earth Policy Reader.* New York: Norton, 2002.

Brown, Norman O. *Apocalypse and/or Metamorphosis.* Berkeley: University of California Press, 1991.

Cajete, Gregory, ed. *A People's Ecology: Explorations in Sustainable Living.* Santa Fe, N.M.: Clear Light, 1999.

Capra, Fritjof. *The Hidden Connections: Integrating the Biological, Cognitive, and Social Dimensions of Life into a Science of Sustainability.* New York: Doubleday, 2002.

The Case Against "Free Trade": GATT, NAFTA, and the Globalization of Corporate Power. San Francisco: Earth Island, 1993.

Cato, Marcus Portius. *On Farming = De Agricultura.* Translated by Andrew Dalby. Blackawton, UK: Prospect Books, 1998.

Cobb, John B., Jr. *Sustaining the Common Good: A Christian Perspective on the Global Economy.* Cleveland, Ohio: Pilgrim Press, 1994.

Colborn, Theo, Dianne Dumanoski, and John Peterson Myers. *Our Stolen Future: Are We Threatening Our Fertility, Intelligence, and Survival? A Scientific Detective Story.* New York: Plume, 1997.

Corselius, Kristen, Suzanne Wisniewski, and Mark Ritchie. *Sustainable Agriculture: Making Money, Making Sense; Twenty Years of Research and Results.* Minneapolis: Institute for Agriculture and Trade Policy, 2001.

Crèvecoeur, J. Hector St. John de. *Letters from an American Farmer and Sketches of Eighteenth-Century America.* Edited by Albert E. Stone. New York: Penguin Books, 1981.

Davidson, Osha Gray. *Broken Heartland: The Rise of America's Rural Ghetto.* New York: Doubleday, 1990.

Dawkins, Kristin. *Gene Wars: The Politics of Biotechnology.* Minneapolis: Institute for Agriculture and Trade Policy, 2003.

Decker, Peter R. *Old Fences, New Neighbors.* Tucson: University of Arizona Press, 1998.

Diamond, Henry L., and Patrick F. Noonan, eds. *Land Use in America: The Report of the Sustainable Use of the Land Project.* Washington, D.C.: Island Press, 1996.

Diamond, Jared. *Guns, Germs, and Steel: The Fates of Human Societies.* New York: Norton, 1998.

Downs, Anthony. *New Visions for Metropolitan America.* Washington, D.C.: Brookings Institution, 1994.

Duncan, David James. *My Story as Told by Water: Confessions, Druidic Rants, Reflections, Bird-Watchings, Fish-Stalkings, Visions, Songs and Prayers Refracting Light, from Living Rivers, in the Age of the Industrial Dark.* San Francisco: Sierra Club Books, 2001.

Elder, John. *Imagining the Earth: Poetry and the Vision of Nature.* 2nd ed. Athens: University of Georgia Press, 1996.

Elgin, Duane. *Promise Ahead: A Vision of Hope and Action for Humanity's Future.* New York: Morrow, 2000.

Forbes, Peter. *The Great Remembering: Further Thoughts on Land, Soul, and Society.* San Francisco: Trust for Public Land, 2001.

Freyfogle, Eric T., ed. *The New Agrarianism: Land, Culture, and the Community of Life.* Washington, D.C.: Island Press, Shearwater Books, 2001.

Gayton, Don. *The Wheatgrass Mechanism: Science and Imagination in the Western Canadian Landscape.* Saskatoon, Sask.: Fifth House, 1990.

Gilkey, Langdon. *Nature, Reality, and the Sacred: The Nexus of Science and Religion.* Minneapolis: Fortress Press, 1993.

Guillaumin, Emile. *The Life of a Simple Man.* Edited by Eugen Weber; revised translation by Margaret Crosland. Hanover, N.H.: University Press of New England, 1983.

Hall, David L., and Roger T. Ames. *Thinking Through Confucius.* Albany: State University of New York Press, 1987.

Halweil, Brian. *Home Grown: The Case for Local Food in a Global Market.* Worldwatch Paper no. 163. Washington, D.C.: Worldwatch Institute, November 2002.

Hanh, Thich Nhat, with Robert Aitken. *For a Future to Be Possible: Commentaries on the Five Mindfulness Trainings.* Rev. ed. Berkeley, Calif.: Parallax Press, 1998.

Hanson, Victor Davis. *The Other Greeks: The Family Farm and the Agrarian Roots of Western Civilization.* New York: Free Press, 1995.

Harman, Willis W., and Elisabet Sahtouris. *Biology Revisioned*. Berkeley, Calif.: North Atlantic Books, 1998.

Hesiod. *Theogony, Works and Days, Shield*. Translated by Apostolos N. Athanassakis. Baltimore: Johns Hopkins University Press, 1983.

Hinrichs, H. C. "As I Remember: A Treatise on Early Rural Life in Goodhue County, Minnesota." Unpublished manuscript, Public Library, Red Wing, Minn., 1982.

Hoffbeck, Steven R. *The Haymakers: A Chronicle of Five Farm Families*. St. Paul: Minnesota Historical Society, 2000.

Holm, Bill. *The Music of Failure*. Marshall, Minn.: Plains Press, 1985.

Hyams, Edward. *Soil and Civilization*. New York: Harper and Row, 1976.

I'll Take My Stand: The South and the Agrarian Tradition, by Twelve Southerners. Baton Rouge: Louisiana State University Press, 1930.

International Forum on Globalization, Alternatives Task Force. *Alternatives to Economic Globalization: A Better World Is Possible*. San Francisco: Berrett-Koehler, 2002.

Jackson, Wes. *Altars of Unhewn Stone: Science and the Earth*. San Francisco: North Point Press, 1987.

————. *Becoming Native to this Place*. Washington, D.C.: Counterpoint, 1996.

Jackson, Wes, Wendell Berry, and Bruce Colman, eds. *Meeting the Expectations of the Land: Essays in Sustainable Agriculture and Stewardship*. San Francisco: North Point Press, 1984.

Jawara, Fatoumata, and Aileen Kwa. *Behind the Scenes at the WTO: The Real World of International Trade Negotiations*. London: Zed Books, 2003.

Kains, M. G. *Five Acres and Independence: A Practical Guide to the Selection and Management of the Small Farm*. Rev. ed. New York: Dover, 1973.

Katakis, Michael, ed. *Sacred Trusts: Essays on Stewardship and Responsibility*. San Francisco: Mercury House, 1993.

Kemmis, Daniel. *Community and the Politics of Place*. Norman: University of Oklahoma Press, 1990.

Kimbrell, Andrew, ed. *The Fatal Harvest Reader: The Tragedy of Industrial Agriculture*. Washington, D.C.: Island Press, 2002.

Kingsolver, Barbara. *Small Wonder*. New York: HarperCollins, 2002.

Kirshner, Orin, ed. *The Bretton Woods-GATT System: Retrospect and Prospect after Fifty Years*. Armonk, N.Y.: Sharpe, 1996.

Kittredge, William. *The Nature of Generosity*. New York: Knopf, 2000.

Kline, David. *Great Possessions: An Amish Farmer's Journal*. Wooster, Ohio: Wooster Book Co., 2001.

————. *Scratching the Woodchuck: Nature on an Amish Farm*. Athens: University of Georgia Press, 1999.

Lao Tsu. *Tao Te Ching.* Translated by Gia-fu Feng and Jane English. New York: Vintage, 1987.

Licht, Daniel S. *Ecology and Economics of the Great Plains.* Lincoln: University of Nebraska Press, 1997.

Little, Charles E., ed. *Louis Bromfield at Malabar: Writings on Farming and Country Life.* Baltimore: Johns Hopkins University Press, 1988.

Loeb, Paul Rogat. *Soul of a Citizen: Living with Conviction in a Cynical Time.* New York: St. Martin's Griffin, 1999.

Logan, Ben. *The Land Remembers: The Story of a Farm and Its People.* 25th anniv. ed. Minnetonka, Minn.: NorthWord Press, 1999.

Logsdon, Gene. *Living at Nature's Pace: Farming and the American Dream.* White River Junction, Vt.: Chelsea Green, 2000.

Lucretius. *On the Nature of the Universe.* Translated by R. E. Latham. Revised by John Godwin. New York: Penguin, 1994.

Margalit, Avishai. *The Decent Society.* Translated by Naomi Goldblum. Cambridge, Mass.: Harvard University Press, 1996.

Marsh, George Perkins. *Man and Nature, or Physical Geography as Modified by Human Action.* Edited by David Lowenthal. Cambridge, Mass.: Belknap Press of Harvard University Press, 1965.

McDonagh, Sean. *To Care for the Earth: A Call to a New Theology.* Santa Fe, N.M.: Bear, 1986.

McGrath, Alister. *The Reenchantment of Nature: The Denial of Religion and the Ecological Crisis.* New York: Doubleday, 2002.

Mencius. *Mencius.* Translated by D. C. Lau. New York: Penguin, 1970.

Meyers, Kent. *The Witness of Combines.* Minneapolis: University of Minnesota Press, 1998.

Mitchell, W. H. *Geographical and Statistical Sketch of the Past and Present of Goodhue County, Together with a General View of the State of Minnesota.* Minneapolis: King's Book and Printing House, 1869.

Mokhiber, Russell, and Robert Weissman. *Corporate Predators: The Hunt for Mega-profits and the Attack on Democracy.* Monroe, Me.: Common Courage Press, 1999.

Nabhan, Gary Paul. *Coming Home to Eat: The Pleasures and Politics of Local Foods.* New York: Norton, 2002.

———. *Cultures of Habitat: On Nature, Culture, and Story.* Washington, D.C.: Counterpoint, 1997.

———. *The Desert Smells Like Rain: A Naturalist in O'Odham Country.* Tucson: University of Arizona Press, 2002.

———. *Songbirds, Truffles, and Wolves: An American Naturalist in Italy.* New York: Penguin, 1994.

Nishitani, Keiji. *Religion and Nothingness.* Translated by Jan Van Bragt. Berkeley: University of California Press, 1982.

Nottingham, Stephen. *Eat Your Genes: How Genetically Modified Food Is Entering Our Diet.* 2nd ed. New York: Zed Books, 2003.

Olson, Steven. *The Prairie in Nineteenth-Century American Poetry.* Norman: University of Oklahoma Press, 1994.

Orr, David W. *Earth in Mind: On Education, Environment, and the Human Prospect.* Washington, D.C.: Island Press, 1994.

Palast, Greg. *The Best Democracy Money Can Buy: An Investigative Reporter Exposes the Truth about Globalization, Corporate Cons, and High-Finance Fraudsters.* Rev. ed. New York: Plume, 2003.

Platt, Rutherford H. *Land Use and Society: Geography, Law, and Public Policy.* Washington, D.C.: Island Press, 1996.

Rasula, Jed. *This Compost: Ecological Imperatives in American Poetry.* Athens: University of Georgia Press, 2002.

Rees, Martin. *Just Six Numbers: The Deep Forces That Shape the Universe.* New York: Basic Books, 2000.

Renner, Michael. *Fighting for Survival: Environmental Decline, Social Conflict, and the New Age of Insecurity.* New York: Norton, 1996.

Sanders, Scott Russell. *Hunting for Hope: A Father's Journeys.* Boston, Mass.: Beacon Press, 1998.

Schlosser, Eric. *Fast Food Nation: The Dark Side of the All-American Meal.* New York: Perennial, 2002.

Smith, Jeffrey M. *Seeds of Deception: Exposing Industry and Government Lies about the Safety of the Genetically Engineered Foods You're Eating.* Fairfield, Iowa: Yes Books, 2003.

Steingraber, Sandra. *Living Downstream: A Scientist's Personal Investigation of Cancer and the Environment.* New York: Vintage Books, 1998.

Stiglitz, Joseph E. *Globalization and Its Discontents.* New York: Norton, 2003.

Stoll, Steven. *Larding the Lean Earth: Soil and Society in Nineteenth-Century America.* New York: Hill and Wang, 2002.

Suzuki, David, with Amanda McConnell. *The Sacred Balance: Rediscovering Our Place in Nature.* Vancouver, B.C.: Greystone Books, 1997.

Theobald, Paul. *Teaching the Commons: Place, Pride, and the Renewal of Community.* Boulder, Colo.: Westview Press, 1997.

Thompson, William Irwin. *Imaginary Landscape: Making Worlds of Myth and Science.* New York: St. Martin's Press, 1989.

Thurman, Howard. *The Search for Common Ground: An Inquiry into the Basis of Man's Experience of Community.* Richmond, Ind.: Friends United Press, 2000.

Tu Wei-Ming. *Centrality and Commonality: An Essay on Confucian Religiousness.* Rev. ed. Albany: State University of New York Press, 1989.

Tucker, Mary Evelyn, and John Berthrong, eds. *Confucianism and Ecology: The Interrelation of Heaven, Earth, and Humans.* Cambridge, Mass.: Harvard University Center for the Study of World Religions, 1998.

U.S. Department of Agriculture. *Grass: The Yearbook of Agriculture, 1948.* Washington, D.C.: U.S. Government Printing Office, 1948.

———. *Soils and Men: Yearbook of Agriculture, 1938.* Washington, D.C.: U.S. Government Printing Office, 1938.

———. *Water: Yearbook of Agriculture, 1955.* Washington, D.C.: U.S. Government Printing Office, 1955.

Vitek, William, and Wes Jackson, eds. *Rooted in the Land: Essays on Community and Place.* New Haven, Conn.: Yale University Press, 1996.

Walters, Charles. *Unforgiven: The American Economic System SOLD for Debt and War.* 2nd ed. Austin, Tex.: Acres USA, 2003.

Welton, Neva, and Linda Wolf. *Global Uprising: Confronting the Tyrannies of the 21st Century; Stories from a New Generation of Activists.* Gabriola Island, B.C.: New Society, 2001.

White, Curtis. *The Middle Mind: Why Americans Don't Think for Themselves.* San Francisco: HarperSanFrancisco, 2003.

White, Gilbert. *The Natural History of Selborne.* London: Dent, 1993.

Wilson, David Sloan. *Darwin's Cathedral: Evolution, Religion, and the Nature of Society.* Chicago: University of Chicago Press, 2002.

Wirzba, Norman, ed. *The Essential Agrarian Reader: The Future of Culture, Community, and the Land.* Lexington: University Press of Kentucky, 2003.

Wolf, Robert. *The Triumph of Technique: The Industrialization of Agriculture and the Destruction of Rural America.* Halls, Tenn.: Ruskin Press, 2003.

Zuckermann, Wolfgang. *End of the Road: From World Car Crisis to Sustainable Transportation.* Post Mills, Vt.: Chelsea Green, 1993.

ARTICLES, REPORTS, AND ADDRESSES

Abramovitz, Janet N. *Imperiled Waters, Impoverished Future: The Decline of Freshwater Ecosystems.* Worldwatch Paper 128. Washington, D.C.: Worldwatch Institute, March 1996.

Adams, Ann, ed. *Holistic Management: A New Environmental Intelligence.* Albuquerque, N.M.: Allan Savory Center for Holistic Management, June 2001.

Akbar, Arifa. "From Lab to Table: How Biotechnology Breeds Hopes, Doubts and Fears Down the Food Chain." *Independent*, October 17, 2003.

Alliance for Responsible Trade. "Sector Analysis of the Free Trade Area of the Americas." 2001. http://www.foodfirst.org/progs/global/trade/Sector_Analysis_FTAA.html.

Allmann, Laurie. *Land Protection Options: A Handbook for Minnesota Landowners*. Minneapolis: Nature Conservancy, 1996.

"Alternatives for the Americas." Working paper prepared for the Peoples' Summit of the Americas, Santiago, Chile, April 15–18, 1998. Second draft updated April 1, 2005. http://www.globalexchange.org/campaigns/alternatives/americas.

Altieri, Miguel A. "Agroecology: Principles and Strategies for Designing Sustainable Farming Systems." Agroecology in Action. http://www.agroeco.org/doc/new_docs/Agroeco_principles.pdf.

———. "Multifunctional Dimensions of Ecologically Based Agriculture in Latin America." Paper prepared for the FAO/Netherlands Conference on the Multifunctional Character of Agriculture and Land, Maastricht, the Netherlands, September 12–17, 1999.

———. "The Myths of Agricultural Biotechnology: Some Ethical Questions." Agroecology in Action. http://www.agroeco.org/doc/the_myths.html.

Altieri, Miguel A., Peter Rosset, and Lori Ann Thrupp. *The Potential of Agroecology to Combat Hunger in the Developing World*. 2020 Brief no. 55. Washington, D.C.: International Food Policy Research Institute, October 1998.

Anderson, Sarah, and John Cavanaugh. *State of the Debate on the Free Trade Area of the Americas*. Washington, D.C.: Institute for Policy Studies, October 14, 2002.

Andrews, David. "Eating Is a Moral Act: Dairy Farmers Deserve Support." Lecture presented at the annual dinner of the California Dairy Campaign, Hanford, Calif., October 26, 2003.

Arner, Audrey, and Brian DeVore. *A New Dawn of Farming: The Sustainable Farming Association of Minnesota's Formation and Growth*. White Bear Lake: Land Stewardship Project and Sustainable Farming Association of Minnesota, 1998.

Associated Press. "Audit Critical of USDA, ConAgra in Meat Recall: Conclusion Is Both Failed to Act Properly." *Agri News*, October 9, 2003.

———. "Equipment Sales Slow across U.S." *Agri News*, March 27, 2003.

Audley, John J., Demetrios G. Papademetriou, Sandra Polaski, and Scott

Vaughan. *NAFTA's Promise and Reality: Lessons from Mexico for the Hemisphere.* Washington, D.C.: Carnegie Endowment for International Peace, 2003.

Balthasar, Christopher W. *Free Trade, Farmers, and the Killer Tomato: An Analysis of the "Hidden" Environmental and Social Costs of Modern Agribusiness, and the Ways in Which These Costs Are Amplified by Free Trade Agreements.* Santa Monica, Calif.: Balthasar, 1995.

Barboza, David. "Monsanto and Pioneer Collaborate." *New York Times,* January 6, 2004.

Barker, Debi, and Jerry Mander. *Invisible Government: The World Trade Organization; Global Government for the New Millennium?* San Francisco: International Forum on Globalization, October 1999.

Barlow, Maude, and Tony Clarke. "Who Owns Water?" *Nation,* September 2, 2002.

Becker, Elizabeth. "Cheney Nips New Farm Subsidies in the Bud." *Rochester Post-Bulletin,* February 13, 2003.

Benbrook, Charles M. *The Bt Premium Price: What Does It Buy?* Special report for the Institute for Agriculture and Trade Policy, Minneapolis, February 2002.

———. *When Does It Pay to Plant Bt Corn? Farm-Level Economics of Bt Corn, 1996–2001.* Special report for the Institute for Agriculture and Trade Policy, Minneapolis, December 2001.

Blumner, Robyn E. "The Best Hope for Working Poor." *St. Petersburg Times,* April 25, 2004.

———. "Miami Crowd Control Would Do Tyrant Proud." *St. Petersburg Times,* November 30, 2003.

Bronfenbrenner, Kate. *Uneasy Terrain: The Impact of Capital Mobility on Workers, Wages, and Union Organizing.* Prepared for the U.S. Trade Deficit Review Commission. Ithaca, N.Y.: Cornell University, September 2000.

Bruckner, Traci. "Agricultural Reforms in Europe." *Center for Rural Affairs Newsletter,* November 2002.

Campany, Chris. "Grassroots Deliver BIG WINS in Senate Farm Bill." *Inspectors' Report,* Winter 2002.

Canadian Broadcasting Corporation. "Farm Income Hits 25-Year Low in 2003." May 27, 2004. http://www.cbc.ca/story/business/national/2004/05/27/farmincome_040527.html.

Carney, Diana. *Approaches to Sustainable Livelihoods for the Rural Poor.* Poverty Briefing 2. London: Overseas Development Institute, January 1999.

Carrell, Severin. "Farmers Can Set Up GM Free Zones." *Independent*, October 12, 2003.

———. "GM Threatens a Superweed Catastrophe." *Independent*, July 1, 2003.

Caspers-Simmet, Jean. "Farmland Owners Are Older: Who Owns Iowa Farmland?" *Agri News*, August 14, 2003.

Cavanagh, John, and Robin Broad. "A Turning Point for World Trade?" *Baltimore Sun*, September 18, 2003.

Center for Rural Affairs Newsletter. "New Data on Rural Poverty." September 2002.

Christison, Bill. "Food and Trade Policy: Are We Trading Away the Future?" *In Motion Magazine*, November 30, 2002.

Cochrane, Willard W. "A Food and Agricultural Policy for the 21st Century." November 16, 1999. http://www.tradeobservatory.org/showFile.php?Ref ID=29697.

Connor, Steve. "Biggest Study into the Effects of GM Crops Gives Critics a Field Day." *Independent*, October 17, 2003.

———. "Crops Giant Retreats from Europe Ahead of GM Report." *Independent*, October 16, 2003.

———. "Separating the Wheat from the Chaff." *Independent*, October 17, 2003.

Cook, Christopher D. "Drilling for Water in the Mojave." *Progressive*, October 2002.

DeNavas-Walt, Carmen, Bernadette D. Proctor, and Cheryl Hill Lee. *Income, Poverty, and Health Insurance Coverage in the United States: 2004.* Current Population Reports, P60-229. Washington, D.C.: U.S. Government Printing Office, 2005.

DeVore, Brian. "Agriculture's Untapped Potential." *Land Stewardship Letter*, September–October 2001.

———. "Selling the Farm Down Contract Creek." *In Motion Magazine*, July 27, 1999.

Diamond, Jared. "Easter's End." *Discover*, August 1995.

DiGangi, Joseph. *U.S. Intervention in EU Chemical Policy*. Jamaica Plain, Mass.: Environmental Health Fund, September 2003.

Dinham, Barbara. "Merger Mania in World Agrochemicals Market." *Pesticide News*, September 2000.

Doak, Richard. "Imagine Iowa Fields If Farmers Weren't Tied to Corn, Soybeans." *Des Moines Register*, May 11, 2002.

Dobbs, Thomas L., and Jules N. Pretty. "The United Kingdom's Experience with Agri-Environmental Stewardship Schemes: Lessons and Issues for the United States and Europe." Staff Paper 2001-1. South Dakota State University, Brookings, 2001.

Economist. "Global Agenda: What Happened to Free Trade?" May 2002.

Epp, Raymond. "'Farming Like the Forest' and 'The Fields Need the Forest'—Complimentary Ideas in the Search for a Sustainable Agriculture." *Land Report,* Spring 2000.

ETC Group. "Fear-Reviewed Science: Contaminated Corn and Tainted Tortillas—Genetic Pollution in Mexico's Centre of Maize Diversity." *ETC Group Communiqué,* January–February 2002.

———. "Globalization, Inc.: Concentration in Corporate Power." *ETC Group Communiqué,* July–August 2001.

———. "New Enclosures: Alternative Mechanisms to Enhance Corporate Monopoly and Bio-Serfdom in the 21st Century." *ETC Group Communiqué,* November 2001.

———. "The New Genomics Agenda: A Political Epilogue to the Book of Life." *ETC Group Communiqué,* September–October 2001.

Etka, Steve. *Contract Agriculture: Serfdom in Our Time.* Pine Bush, N.Y.: National Campaign for Sustainable Agriculture Update, June 2001.

European Commission. "Commission Fines Five Companies in Citric Acid Cartel." News release, December 5, 2001.

———. "Competition: Commission Welcomes Judgment of the Court of Justice in Archer Daniels Midland Case (Amino Acids Cartel)." News release, May 18, 2006.

Everyone's Backyard. "Replacing Risk Assessment: New Approaches to Decision Making." Summer 1999.

Fanjul, Gonzalo, and Arabella Fraser. *Dumping without Borders: How US Agricultural Policies Are Destroying the Livelihoods of Mexican Corn Farmers.* London: Oxfam International, August 2003.

Farm Foundation Issue Report. "Production Contracts." May 2004.

Farrell, Alex. "Sustainability and the Design of Knowledge Tools." *IEEE Technology and Society Magazine,* Winter 1996–1997.

Fennessy, Steve, Scott Harrell, and Kelly Benjamin. "Police State: Last Week in Miami, the Battle Was the Story." *Weekly Planet,* November 26–December 2, 2003.

Finnegan, William. "Letter from Bolivia: Leasing the Rain." *New Yorker,* April 8, 2002.

Fishman, Ted C. "Making a Killing: The Myth of Capitalism's Good Intentions." *Harper's Magazine,* August 2002.

Food First. "Myths of the Free Trade Area of the Americas Agreement." June 28, 2001. http://www.foodfirst.org/progs/global/trade/ftaamyths.html.

Food Research and Action Center. "Hunger in the U.S." October 28, 2005. http://www.frac.org/html/hunger_in_the_us/hunger_index.html.

Friends Committee on Unity with Nature. "The Real Costs of 'Free' Trade: An Epistle." *Quaker Eco-Bulletin*, July–August 2003.

Friends of the Earth International. *Tackling GMO Contamination: Making Segregation and Identity Preservation a Reality*. Amsterdam: FOEI, 2005.

Gittins, Ross. "How Big, Bad Globalisation Can Help Poorer Nations: Those Dreaded Multinationals Often Speed a Country's Growth." *Sydney Morning Herald*, May 31, 2003.

Gold, Mary V. *Sustainable Agriculture: Definitions and Terms*. Alternative Farming Systems Information Center Special Reference Briefs Series no. SRB 99-02. Beltsville, Md.: AFSIC, Agricultural Research Service, U.S. Department of Agriculture, September 1999.

Goodman, James. "World's Farmers Stand in Solidarity Against WTO." *Madison (Wis.) Capital Times*, October 29, 2003.

Groth, Edward, III, Charles M. Benbrook, and Karen Lutz. *Do You Know What You're Eating? An Analysis of U.S. Government Data on Pesticide Residues in Foods*. New York: Consumers Union of United States, February 1999.

Guebert, Alan. "American Farmers Don't Stand to Gain from FTAA Gabfest." *Agri News*, November 21, 2003.

———. "Cargill Says Thank You." *Burlington (Iowa) Hawk Eye*, May 16, 2004.

Haar, Daniel, with Richard Levins and Michael Darger. *Rice and Steele Counties Agriculture Business Retention and Enhancement Program Summary Report*. Department of Applied Economics, University of Minnesota, St. Paul, November 2001.

Haklik, James E. "ISO 14001 and Sustainable Development." Transformation Strategies. August 20, 1997. http://www.trst.com/sustainable.htm.

Hall, Bart, and George Kuepper. *Making the Transition to Sustainable Farming: Fundamentals of Sustainable Agriculture*. Fayetteville, Ark.: Appropriate Technology Transfer for Rural Areas, National Center for Appropriate Technology, December 1997.

Harl, Neil E. "The Structural Transformation of the Agricultural Sector." Paper presented at the Master Farmer Awards Ceremony, West Des Moines, Iowa, March 20, 2003.

Harper, Allen. "Contract Hog Production as an Alternative Farm Enterprise." *Livestock Update*, April 2005.

Hassebrook, Chuck. "The State of Rural America." *Center for Rural Affairs Newsletter*, January 2001.

Hathaway, Dale E. *The Perils and Promises of World Markets: Learning to Live with Globalization.* Washington, D.C.: National Center for Food and Agricultural Policy, n.d.

Heffernan, William D. "Agriculture and Monopoly Capital." *Monthly Review*, July–August 1998.

Heffernan, William D., and Mary K. Hendrickson. "Multi-National Concentrated Food Processing and Marketing Systems and the Farm Crisis." Paper presented at the annual meeting of the American Academy of Arts and Sciences, May 8, 2002.

Heffernan, William D., Mary K. Hendrickson, and Robert Gronski. "Consolidation in the Food and Agricultural System." Report to the National Farmers Union, University of Missouri–Columbia, February 5, 1999.

Hemispheric Social Alliance. "Statement by the Hemispheric Social Alliance on the Declaration by Trade Ministers Meeting in Quito on the FTAA Negotiations." October 24, 2002. http://www.asc-hsa.org/article.php3?id_article=44.

Hightower, Jim. "Taking Our Common Wealth and Selling It: Stop the Corporate Takeover of Our Water." *Hightower Lowdown*, June 2002.

Ho, Mae-Wan, and Lim Li Ching. *The Case for a GM-Free Sustainable World.* London: Independent Science Panel, June 2003.

Hugoson, Gene. "Developing Export Markets Is Critical to Minnesota Farmers." *Agri News*, August 14, 2003.

Ikerd, John. "The Real Costs of Globalization to Farmers, Consumers, and Our Food System." Paper presented at the annual conference of the Sustainable Farming Association of Minnesota, Northfield, Minn., February 23, 2002.

———. "Reclaiming the Spiritual Roots of Farming." Paper presented at the Soul of Agriculture: New Movements in New England Food and Farming, University of New Hampshire, Durham, November 18–20, 2001.

Institute for Agriculture and Trade Policy. *Addressing the Root Causes of the Farm Crisis: The Farmer Summit Platform.* Minneapolis: IATP, February 2000.

Johnson, Barbara. *Conservation Reserve Program: Status and Current Issues.* Congressional Research Service Report no. RS21613. Washington, D.C.: National Council for Science and the Environment, February 2005.

Johnson, Paul D. "Our Agriculture Needs a Public Threshing." Prairie Writers Circle, Land Institute, May 23, 2003. http://www.landinstitute.org/vnews/display.v/ART/2003/05/23/3edd1a7a65539.

Jolliffe, Dean. *Comparisons of Metropolitan-Nonmetropolitan Poverty During the 1990s.* Rural Development Research Report no. 96. Washington, D.C.: USDA, June 2003.

Just-Food. "Rush to Adopt GMO Crops in China Can Save Lives Damaged by Pesticides." October 11, 2000. http://www.just-food.com/article.aspx?art=278&type=2.

Keeler, Barbara. "Buried Data in Monsanto's Study on Roundup Ready Soybeans." *Whole Life Times,* August 2000.

Kielkopf, James J. *Estimating the Economic Impact of the Latino Workforce in South Central Minnesota.* Mankato: Center for Rural Policy and Development, Minnesota State University, September 2000.

King, Robert, and Gigi DiGiacomo. *Collaborative Marketing: A Roadmap and Resource Guide for Farmers.* St. Paul: Minnesota Institute for Sustainable Agriculture, 2000.

Kinsley, Michael J., and L. Hunter Lovins. *Paying for Growth, Prospering from Development.* Snowmass, Colo.: Rocky Mountain Institute, 1996.

Kirschenmann, Frederick. "Agriculture's Uncertain Future: Unfortunate Demise or Timely Opportunity?" Keynote address, Strategic Planning, Institute of Food and Agricultural Sciences, University of Florida, Gainesville, June 27, 2001.

———. "Can Organic Agriculture Feed the World? ... And Is That the Right Question?" Keynote address, Organic Conference, University of Guelph, Guelph, Ont., January 25, 1997.

———. "Feeding the Village First." Position paper. Northern Plains Sustainable Agriculture Society. http://www.npsas.org/feeding.html.

———. "The Future of Agrarianism: Where Are We Now?" Keynote address, The Future of Agrarianism: Considering the Unsettling of America 25 Years Later, Georgetown College, Georgetown, Ky., April 25, 2002.

———. "A New Science for a New Agriculture." Paper presented at Mountain Sky, Bozeman, Mont., October 12, 2004.

Korrow, Christy. "The Basics of Biodynamics from Steiner to Stella Natura." *Acres USA,* February 2002.

Krebs, A. V. "ADM Seeks to Reduce or Void $48.3 Million European Fine for Fixing Citric Acid Prices." *Agribusiness Examiner,* June 11, 2004.

Kummer, Corby. "Back to Grass: The Old Way of Raising Cattle Is Now the New Way—Better for the Animals and Better for Your Table." *Atlantic Monthly,* May 2003.

Land Stewardship Project. *Farming in Goodhue County (Wells Creek): Row Crops Increase and Farms Consolidate.* White Bear Lake, Minn.: Land Stewardship Project, August 2002.

Lappe, Anne. "Last Meals? How Corporate Power Taints Safety Rules." *San Francisco Chronicle,* March 30, 2003.

Le Page, Michael. "Superbug Strain Hits the Healthy." *New Scientist,* March 2003.

Lean, Geoffrey. "Flaw in Crop Trials Destroys Government Case for GM." *Independent,* October 12, 2003.

———. "How GM Crop Trials Were Rigged: Ministers Knew of the Environmental Dangers, but the Tests Were Designed Not to Focus on This." *Independent.* October 12, 2003.

Lederman, Daniel, William Maloney, and Luis Servén. *Lessons from NAFTA for Latin American and the Caribbean Countries: A Summary of Research Findings.* World Bank, December 2003.

Levins, Dick. *Monitoring Sustainable Agriculture with Conventional Financial Data.* White Bear Lake, Minn.: Land Stewardship Project, June 1996.

Levins, Richard A. "An Essay on Farm Income." Staff Paper P01-1. Department of Applied Economics, University of Minnesota, St. Paul, April 2001.

———. "Family Farm Legislation: Who Are We Protecting?" *Sustainable Agriculture,* June 2001.

———. "New Policy Goal Should Be More Farmers, Not Fewer." *Sustainable Agriculture,* December 2001.

Lewandowski, Ann, and Mark Zumwinkle. *Assessing the Soil System: A Soil Quality Literature Review.* Edited by Alison Fish. St. Paul: Energy and Sustainable Agriculture Program, Minnesota Department of Agriculture, June 1999.

Lewis, W. J., J. C. van Lenteren, Sharad C. Phatak, and J. H. Tumlinson III. "A Total System Approach to Sustainable Pest Management." *Proceedings of the National Academy of Sciences USA* 94 (November 1997): 12243–48.

Lien, Dennis. "River on the Edge." *St. Paul (Minn.) Pioneer Press,* June 27, 2004.

Lyman, Francesca. "Twelve Gates to the City: A Dozen Ways to Build Strong, Livable and Sustainable Urban Areas." *Sierra Magazine,* May–June 1997.

Lynn, Barry. "Unmade in America: The True Cost of a Global Assembly Line." *Harper's Magazine,* June 2002.

Madeley, John. *Trade and Hunger: An Overview of Case Studies on the Impact of Trade Liberalisation on Food Security.* Edited by Johanna Sandahl. Globala Studier no. 4. Church of Sweden Aid, Diakonia, Forum Syd, Swedish Society for Nature Conservation, Programme of Global Studies, October 2000.

Manufacturing News. "U.S. Food Industry Begins to Feel the Sting of Chinese Imports: China's 800 Million Farmers Are Feasting on the U.S. Market." July 3, 2003.

Margoluis, Richard, Vance Russell, Mauricia Gonzales, Oscar Rojas, Jaime Magdaleno, Gustavo Madrid, David Kaimowitz. *Maximum Yield? Sustainable Agriculture as a Tool for Conservation.* Biodiversity Support Program Publication no. 117. Washington, D.C.: World Wildlife Fund, 2001.

Martz, Diane, J. F. *The Farmers' Share: Compare the Share 2004.* Muenster, Sask.: Centre for Rural Studies and Enrichment, St. Peter's College, November 2004.

Martz, Diane J. F., and Wendy Moellenbeck. *Compare the Share Revisited: The Family Farm in Question.* Muenster, Sask.: Centre for Rural Studies and Enrichment, St. Peter's College, 1999.

Mayer, Sue, Julie Hill, Robin Grove-White, Sue Mayer, and Brian Wynne. *Uncertainty, Precaution and Decision Making: The Release of Genetically Modified Organisms into the Environment.* Briefing no. 8. Brighton, UK: Global Environmental Change Programme, June 1996.

Mazullo, Jim. *Neighborhood Planning for Community Revitalization: Sources of Funding for the Development of Affordable Housing.* Minneapolis: Central Community Housing Trust, January 2002.

McCarthy, Michael. "Are We Going to Sacrifice a Growing Market for Organic Crops by Risking Contamination?" *Independent,* September 25, 2003.

————. "GM Crops? No Thanks: Britain Delivers Overwhelming Verdict after Unprecedented Public Opinion Exercise." *Independent,* September 25, 2003.

————. "New Blow to Government on GM Food as Public Debate Confirms Skepticism." *Independent,* July 18, 2003.

————. "Proven: The Environmental Dangers That May Halt GM Revolution." *Independent,* October 17, 2003.

————. "US Firms 'Tried to Lie' Over GM Crops Says EU." *Independent,* October 14, 2003.

McGovern, George, and Rudy Boschwitz. "High Yield Farming Can Defeat Hunger." *St. Paul (Minn.) Pioneer Press,* June 5, 2002.

Meacher, Michael. "Cold Comfort for Those Attending the Earth Summit." *Independent,* August 8, 2002.

Meacher, Michael, and Paul Rylott. "GM Crops: The Arguments For and Against." *Independent*, October 17, 2003.

Meadows, Donella H. "Two Mindsets, Two Visions of Sustainable Agriculture." Paper prepared for the FAO/Netherlands Conference on the Multifunctional Character of Agriculture and Land, Maastricht, the Netherlands, September 12–17, 1999.

Mekay, Emad. "North American Deal Dismal after a Decade." News release, Inter Press Service News Agency, December 26, 2003.

Mello, Fatima V. *Brazil and the FTAA—The State of the Debate since Lula's Victory*. Santiago, Chile: Estudios sobre el ALCA, November 2002.

Meter, Ken, and Jon Rosales. *Finding Food in Farm Country: The Economics of Food and Farming in Southeast Minnesota*. Lanesboro, Minn: Hiawatha's Pantry Project, Community Design Center, March 2001.

Minnesota Department of Natural Resources. *Oak Savannah: The Power of the Place*. Owatonna, Minn.: Oak Savannah Landscape Project, Minnesota DNR, 1999.

Minnesota Sustainable Development Initiative. "A Briefing Paper on Sustainable Activities in Minnesota." Paper presented to the Environmental Quality Board, Minnesota Planning, St. Paul, November 1996.

Mintz, Sidney W. *Crops and Human Culture*. Marshall, Minn.: Center for Rural and Regional Studies, Southwest State University, 1994.

Moeller, David. *Livestock Production Contracts: Risks for Family Farmers*. St. Paul, Minn.: Farmers' Legal Action Group, March 2003.

Moffatt, Ian, Nick Hanley, Mike Wilson, and Robin Faichney. *Measuring and Modelling Sustainability*. Briefing no. 26. Brighton, UK: Global Environmental Change Programme, June 1999.

Muller, Mark. "Export Myth Sold to Farmers." *IATP News*, July 2001.

Muller, Mark, and Richard Levins. *Feeding the World? The Upper Mississippi River Navigation Project*. Minneapolis: Institute for Agriculture and Trade Policy, December 1999.

Mulvaney, Patrick. *Cultivating Our Futures: The FAO/Netherlands Conference on the Multifunctional Character of Agriculture and Land: A Report*. Ottawa, Ont.: Rural Advancement Foundation International, September 28, 1999.

Murphy, Sophia. *Managing the Invisible Hand: Markets, Farmers and International Trade*. Minneapolis: Institute for Agriculture and Trade Policy, April 2002.

———. *Market Power in Agricultural Markets: Some Issues for Developing Countries*. Geneva, Switzerland: South Centre, 2002.

Sources and Resources

Nadal, Alejandro. *The Environmental and Social Impacts of Economic Liberalization on Corn Production in Mexico.* London: Oxfam Great Britain; Godalming, UK: WWF International, September 2000.

Nadal, Alejandro, and Timothy A. Wise. "The Environmental Costs of Agricultural Trade Liberalization: Mexico-U.S. Maize Trade under NAFTA." Discussion Paper no. 4. Prepared for the meeting of the Working Group on Development and Environment in the Americas, Brasilia, Brazil, March 29–30, 2004.

National Commission on the Environment. *Choosing a Sustainable Future: The Report of the National Commission on the Environment.* Washington, D.C.: Island Press, 1993.

National Family Farm Coalition. "Farm Aid: A Declaration for a New Direction for American Agriculture and Agricultural Trade." September 2003. http://www.nffc.net/resources/statements/farmaid.pdf.

National Farmers Union. "The Effects of Export-Oriented Agriculture on Canadian Farm Families, Canadian Consumers, and Farmers around the World." Submission to the Canadian House of Commons Standing Committee on Foreign Affairs and International Trade, April 26, 1999.

Nayyar, Deepak, and Julius Court. *Governing Globalization: Issues and Institutions.* Policy Brief no. 5. Helsinki, Finland: World Institute for Development Economics Research, UN University, 2002.

Nerbonne, Julia Frost, and Ralph Lentz. "Rooted in Grass: Challenging Patterns of Knowledge Exchange as a Means of Fostering Social Change in a Southeast Minnesota Farm Community." *Agriculture and Human Values* 20 (2003): 65–78.

Nestle, Marion. "The Ironic Politics of Obesity." *Science*, February 7, 2003.

New York Times. "Biotechnology = Hunger." November 8, 1999.

Nickel, Raylene F. "Helping Hands: Community Banks Offer Tutorials on Farming for Beginners." *Independent Banker*, March 2003.

Nixon, Will. "The Big Fix: Mirrors in Orbit, Geritol for the Oceans, and Other Environmental Magic Bullets." *Amicus Journal*, Winter 1995.

Northern Plains Sustainable Agriculture Society. "NDSU Research Extension Centers Say NO to Transgenic Wheat Trials." News release, May 16, 2002.

———. "New Directions for Ag Research." Position paper. http://www.npsas.org/Research.html.

———. "Foundation Seedstocks Not Exempt from Contamination." News release, November 11, 2002.

Olson, Dennis R. "ATSDF Agricultural Tent Recommendations." Recommendations presented at the Americas Trade and Sustainable Development Forum, Free Trade Area of the Americas eighth ministerial meeting, Miami, Fla., November 19, 2003.

Olson, Dennis R., Giel Ton, Jacques Berthelot, Marita Wiggerthale, Arze Glipo-Carasco, R. Jessica Cantos Concepcion, Anne Mutisya, Indrani Thuraisingham, and Catherine Gaudard. "Towards Food Sovereignty: Constructing an Alternative to the World Trade Organization's Agreement on Agriculture." Paper developed from the Farmers, Food and Trade International Workshop on the Review of the AoA, Geneva, Switzerland, February 19–21, 2003.

O'Reilly, Brian. "Biotech Corn: Reaping a Biotech Blunder," *Fortune*, February 8, 2001.

Oxfam America. *Like Machines in the Fields: Workers without Rights in American Agriculture.* Boston: Oxfam America, March 2004.

Patchett, James M., and Gerould S. Wilhelm. "The Ecology and Culture of Water." Conservation Design Forum, July 1999. http://www.cdfinc.com/CDF_Resources/Ecology_and_Culture_of_Water.pdf.

Patel, Raj. "Bad Farm Policies Starve Millions: The 'Export Model of Agriculture' Is a Globally Destructive Force." *Los Angeles Times*, December 26, 2003.

People's Recovery, Empowerment and Development Assistance Foundation. "Contributions of Fair Trade to Poverty Reduction: A Case Study from the Philippines." 1997. http://www.preda.net/article/contents.html.

Peters, Scott J. "Rousing the People on the Land: The Roots of the Educational Organizing Tradition in Extension Work." *Journal of Extension* 40, no. 3 (June 2002).

Pirog, Rich, Timothy Van Pelt, Kamyar Enshayan, and Ellen Cook. *Food, Fuel, and Freeways: An Iowa Perspective on How Far Food Travels, Fuel Usage, and Greenhouse Gas Emissions.* Ames: Leopold Center for Sustainable Agriculture, Iowa State University, June 2001.

Policy Commission on the Future of Farming and Food. *Farming and Food: A Sustainable Future.* London: Crown, January 2002.

Pollan, Michael. "The (Agri)Cultural Contradictions of Obesity." *New York Times Magazine*, October 12, 2003.

Porter, Adam. "Is the World's Oil Running Out Fast?" BBC News, June 7, 2004.

Posner, Richard A. "The University as Business." *Atlantic Monthly*, June 2002.

Poverty & Race Research Action Council. "Shattered Promises: Immigrants

Sources and Resources

and Refugees in the Meatpacking Industry." *Poverty and Race,* November 1992. http://www.prrac.org/full_text.php?text_id=644&item_id=6514&newsletter_id=5&header=Search%20Results.

Prasad, Eswar, Kenneth Rogoff, Shang-Jin Wei, and M. Ayhan Kose. *Effects of Globalization on Developing Countries: Some Empirical Evidence.* Washington, D.C.: International Monetary Fund, March 2003.

Pretty, Jules. *Farmer-Based Agroecological Technology.* 2020 Focus 07: Appropriate Technology for Sustainable Food, Brief no. 02. Washington, D.C.: International Food Policy Research Institute, August 2001.

Pretty, Jules, and Rachel Hine. *Reducing Food Poverty with Sustainable Agriculture: A Summary of New Evidence.* SAFE-World Research Project. Wivenhoe Park, UK.: Centre for Environment and Society, University of Essex, August 2001.

Public Citizen's Global Trade Watch. *Down on the Farm: NAFTA's Seven-Years War on Farmers and Ranchers in the U.S., Canada and Mexico.* Washington, D.C.: Public Citizen, June 2001.

Randall, Gyles. "Our Corn-Soybean System Fails the Sustainability Test on All Fronts." *Land Stewardship Letter,* September–October 2001.

Raskin, Paul, Tariq Banuri, Gilberto Gallopín, Pablo Gutman, Al Hammond, Robert Kates, and Rob Swart. *Great Transition: The Promise and Lure of the Times Ahead.* A report by the Global Scenario Group. SEI PoleStar Series Report no. 10. Boston: Stockholm Environment Institute, August 2002.

Rasmussen, Donna. "Early Studies Key to Understanding Ground Water." *Fillmore County (Minn.) Journal,* January 27, 2003.

———. "Living on the Edge: The Decorah Shale." *Fillmore County (Minn.) Journal,* March 3, 2003.

Ray, Daryll E. "Brazilian Acreage Potential Is Larger Than Previously Thought." *MidAmerica Farmer Grower,* April 4, 2003.

———. "Cargill Opens Soybean Terminal at Santorem on the Amazon in Brazil." *MidAmerica Farmer Grower,* April 18, 2003.

———. "Export Values Up, Net Income Down: Which Is More Important?" *MidAmerica Farmer Grower,* September 13, 2002.

Reinke, Darrell. "Farming for Place and Community." *Fremont (Idaho) Current,* August 20, 1998.

Ritchie, Mark. *The Loss of Our Family Farms: Inevitable Results or Conscious Policy?* Minneapolis: League of Rural Voters, 1979.

Rockwell, R. F., K. F. Abraham, and R. L. Jefferies. "Tundra under Siege." *Natural History Magazine,* November 1996.

Rosset, Peter M. "Benefits of Small Farm Agriculture." *Sustainability Review,* January 24, 2000.

———. "The End of the Styrofoam Strawberry." *New Internationalist,* January–February 2002.

———. *The Multiple Functions and Benefits of Small Farm Agriculture in the Context of Global Trade Negotiations.* Food First Policy Brief no. 4. Prepared for the FAO/Netherlands Conference on the Multifunctional Character of Agriculture and Land, Maastricht, the Netherlands, September 12–17, 1999.

Ruminator Review. "Cultivation: Rural Lives, Global Issues." Summer 2003.

Rural Advancement Foundation International. "Captain Hook, the Cattle Rustlers, and the Plant Privateers: Biopiracy of Marine, Plant, and Livestock Continues." *RAFI Communiqué,* May–June 2000.

———. "Golden Rice and Trojan Trade Reps: A Case Study in the Public Sector's Mismanagement of Intellectual Property." *RAFI Communiqué,* September–October 2000.

———. "2001: A Seed Odyssey." *RAFI Communiqué,* January–February 2001.

Rylott, Paul. "British Consumers Should Have the Option of Buying Cheaper, More Convenient Food." *Independent,* September 25, 2003.

Saad, Majda Bne. *Food Security for the Food Insecure: New Challenges and Renewed Commitments.* Dublin, Ireland: Centre for Development Studies, University College Dublin, December 1999.

Saunders, Peter, and Mae-Wan Ho. "The Precautionary Principle Is Coherent." Institute of Science and Society. October 31, 2000. http://www.i-sis.org.uk/precautionary-pr.php.

Schecter, Arnold, Hoang Trong Quynh, Marian Pavuk, Olaf Päpke, Rainer Malisch, and John D. Constable. "Food as a Source of Dioxin Exposure in the Residents of Bien Hoa City, Vietnam." *Journal of Occupational and Environmental Medicine* 45, no. 8 (2003): 781–88.

Schlosser, Eric. "The Cow Jumped Over the U.S.D.A." *New York Times,* January 2, 2004.

Shapiro, Mark. "Sowing Disaster: GMO's and Mexican Corn." *Nation,* October 28, 2002.

Shiva, Vandana. "Tradeoffs." *OnEarth Magazine,* Summer 2002.

Snyder, Gary. "Indra's Net as Our Own." In *For a Future to Be Possible: Commentaries on the Five Mindfulness Trainings,* by Thich Nhat Hanh, 127–35. Berkeley, Calif.: Parallax Press, 1993.

Sperbeck, Jack. "Farmer Bargaining Unit Would Increase Economic Power, Economist Says." *Sustainable Agriculture,* January 2000.

———. "Hog Farms Can Be Hazardous to the Health." *Sustainable Agriculture,* May 2000.

———. "Many Farm Women Pessimistic and Fatalistic, Study Shows." *Sustainable Agriculture*, May 2001.

———. "New Markets from Alternative Swine Systems Offer Hope for Farmers." *Sustainable Agriculture*, July 2000.

———. "New Research Shows Green Manure Crops Enhance Natural Disease Control in Soils." *Sustainable Agriculture*, December 2000.

———. New Study Shows Sustainable Farms Profitable, Help the Environment." *Sustainable Agriculture*, November 2001.

———. "World Food Supply Will Suffer from a Global, Industrialized Agriculture." *Sustainable Agriculture*, May 2000.

Steelman, Toddi A., Brian Page, and Lloyd Burton. "Change on the Range: The Challenge of Regulating Large-Scale Hog Farming in Colorado." Working paper, Graduate School of Public Affairs, University of Colorado, Denver, October 1998.

Stender, Carol. "Consumers Prefer Locally Grown Food in Tough Economic Times." *Agri News*, July 10, 2003.

Stewart, Julie, Kevin O'Connell, Marian Ciborski, and Mathew Pacenza. *A People Damned: The Impact of the World Bank Chixoy Hydroelectric Project in Guatemala*. Washington, D.C.: Witness for Peace, May 1996.

Stith, Pat, Joby Warrick, and Melanie Sill. "Boss Hog: North Carolina's Pork Revolution." *Raleigh (N.C.) News and Observer*, February 19, 21, 22, 24, 26, 1995.

Strange, Marty, and Cheryl Miller. *A Better Row to Hoe: The Economic, Environmental, and Social Impact of Sustainable Agriculture*. Saint Paul, Minn.: Northwest Area Foundation, December 1994.

Theobald, Laura. "Mexican Farmers Bring Message to Minnesota: They Say NAFTA Is Hurting Both Sides of Border." *Agri News*, July 24, 2003.

Thorson, Gregory R., and Jacqueline Edmondson. "Making Difficult Times Worse: The Impact of Per Pupil Funding Formulas on Rural Minnesota Schools." Center for Rural Policy and Development, St. Peter, Minn., 2000. http://www.morris.umn.edu/~gthorson/crp.pdf.

Torres, Filemon, Martin Piñeiro, Eduardo Trigo, and Roberto Martinez Nogueira. "Agriculture in the Early Twenty-first Century: Agrodiversity and Pluralism as a Contribution to Address Issues on Food Security, Poverty and Natural Resource Conservation—Reflections on Its Nature and Implications for Global Research." Draft of issues paper commissioned by the Global Forum on Agricultural Research, Dresden, Germany, May 21–23, 2000.

Tsybine, Alex. *Water Privatization: A Broken Promise*. Special report by Public Citizen's Critical Mass Energy and Environment Program. Washington, D.C.: Public Citizen, October 2001.

UN Economic Commission for Latin America and the Caribbean. "ECLAC Forecasts a Modest Recovery in Latin American and Caribbean Economies during 2003." News release, August 7, 2003.

———. *Economic Survey of Latin America and the Caribbean, 2002–2003.* Prepared by the Economic Development Division. Santiago, Chile: UN, August 2003.

———. *Economic Survey of Latin America and the Caribbean, 2003–2004.* Prepared by the Economic Development Division. Santiago, Chile: UN, November 2004.

UN Foundation. "ECLAC Calls for New Economic Approach." *UN Wire,* December 20, 2000.

———. "EU Approves UN Protocol Allowing Regulation of GM Food." *UN Wire,* June 5, 2003.

———. "Water: UNEP Study Finds Groundwater at Risk." *UN Wire,* June 5, 2003.

United Methodist Church General Board of Church and Society. *U.S. Agriculture and Rural Communities in Crisis.* Washington, D.C.: UMC, 1996.

U.S. Department of Agriculture. *Farm Production Expenditures Summary.* Washington, D.C.: National Agricultural Statistics Service, Agricultural Statistics Board, USDA, July 1999.

———. "Rural Earnings." Rural Labor and Education Briefing Room. Economic Research Service. September 15, 2003. http://www.ers.usda.gov/Briefing/LaborAndEducation/earnings/.

———. "U.S. and Regional Cost and Return Data." Commodity Cost and Returns. Economic Research Service. http://www.ers.usda.gov/Data/CostsAndReturns/testpick.htm.

U.S. House Committee on Governmental Reform. *Human Pesticides Experiments.* 109th Cong., 1st sess., June 16, 2005. http://www.democrats.reform.house.gov/Documents/20050617123506-17998.pdf.

VanMeveren, Paul. *In Search of the 1920s Farm Depression: A Study of Rock Lake Township, Lyon County, Minnesota.* Marshall: Center for Rural and Regional Studies, Southwest Minnesota State University, 1994.

Vatovec, Christine. *The Sustainable Farming Association: Making a Difference for Farmers in Minnesota.* Starbuck: Sustainable Farming Association of Minnesota, September 2001.

Vetsch, Jeff, and Gyles Randall. "Conservation Tillage Systems for a Corn-

Soybean Rotation." 2001 Soil Fertility Research Summaries. Southern Research and Outreach Center, Waseca, Minn., 2001.

Vorley, Bill, and Julio Berdegué. "The Chains of Agriculture." Opinion paper developed in preparation for the World Summit on Sustainable Development, Johannesburg, South Africa, August 26–September 4, 2002. London: International Institute for Environment and Development, May 2001.

Warrick, Joby, and Pat Stith. "Boss Hog 2: Corporate Takeovers." *Raleigh (N.C.) News and Observer,* February 21, 1995.

Weeks, Priscilla, and Jane M. Packard. "Acceptance of Scientific Management by Natural Resource Dependent Communities." *Conservation Biology* 11, no. 1 (February 1997): 236–45.

Weisbrot, Mark. "NAFTA Has Been a Big Flop." *Philadelphia Inquirer,* December 24, 2003.

Weisbrot, Mark, Dean Baker, Egor Kraev, and Judy Chen. "The Scorecard on Globalization 1980–2000: Twenty Years of Diminished Progress." Briefing paper. Center for Economic and Policy Research, Washington, D.C., July 2001.

Western Producer. "Farm Exports, Farm Income—the Gap Widens." September 14, 2000.

Wilkinson, Fenton, and David Van Seters. *Adding Values to Our Food System: An Economic Analysis of Sustainable Community Food Systems.* Everson, Wash.: Integrity Systems Cooperative, February 1997.

Willette, Janet Kubat. "Harl: Federal Policy Top Challenge." *Agri News,* February 13, 2003.

Wilmes, Mychal. "Time for an Alternative Crop to Soybeans?" *Agri News,* October 9, 2003.

Wilson, Charles. "Asphalt Eden." *Preservation Magazine,* May–June, 2002.

Wirzba, Norman. "Soil Cultivating Citizens: An Agrarian Contribution to Civic Life." *Land Report,* Spring 2000.

Wolfensohn, James. "The Growing Threat of Global Poverty." *International Herald Tribune,* April 23, 2004.

WorldWatch Institute. "Agribusiness Concentration, Not Low Prices, Is Behind Global Farm Crisis." News release, September 1, 2000.

———. "All-You-Can-Eat Economy Is Making World Sick." News release, May 24, 2001.

———. "Emerging Water Shortages Threaten Food Supplies, Regional Peace." News release, July 17, 1999.

———. "Global Environment Reaches Dangerous Crossroads." News release, January 13, 2001.

Yes! A Journal of Positive Futures. "Whose Water?" Winter 2004.

Young, Trevor. *Adoption of Sustainable Agricultural Technologies: Economic and Non-Economic Determinants.* Briefing no. 21. Brighton, UK: Global Environmental Change Programme, July 1998.

Zaro-Moore, Kyla. *Sustaining Green Space in the Rural-Urban Fringe: A Landowner's Guide.* St. Paul: Green Sprawl Working Group, University of Minnesota, May 2001.

Ziebarth, Ann, and Jaehyun Byun. *Migrant Worker Housing: Survey Results from South Central Minnesota.* Centro Campesino, University of Minnesota Center for Urban and Regional Affairs, and Hispanic Advocacy and Community Empowerment through Research, 2000.

Index

Index